# Social Ecology
# State of the Art and Future Prospects

## Special Issue Editors

Johanna Kramm
Melanie Pichler
Anke Schaffartzik
Martin Zimmermann

MDPI • Basel • Beijing • Wuhan • Barcelona • Belgrade

**MDPI**

*Special Issue Editors*

Johanna Kramm
ISOE—Institute for Social-
Ecological Research
Germany

Melanie Pichler
Institute of Social Ecology
Austria

Anke Schaffartzik
Institute of Social Ecology
Austria

Martin Zimmermann
ISOE—Institute for Social-
Ecological Research
Germany

*Editorial Office*
MDPI AG
St. Alban-Anlage 66
Basel, Switzerland

This edition is a reprint of the Special Issue published online in the open access journal *Sustainability* (ISSN 2071-1050) in 2017 (available at: http://www.mdpi.com/journal/sustainability/special_issues/social_ecology).

For citation purposes, cite each article independently as indicated on the article page online and as indicated below:

Author 1; Author 2. Article title. *Journal Name* **Year**, *Article number*, page range.

**First Edition 2017**

**ISBN 978-3-03842-546-5 (Pbk)**
**ISBN 978-3-03842-547-2 (PDF)**

Photo courtesy of iStock.com/VladSokolovsky

# Table of Contents

# About the Special Issue Editors

**Johanna Kramm** is leading an inter- and transdisciplinary research group dealing with risks of plastics in the environment (PlastX) at ISOE—Institute for Social-Ecological Research and is a member of the research unit Water Resources and Land Use. She holds a PhD in Geography from the University of Bonn and has a regional interest in East- and Southern Africa as well as Southeast Asia. Her research is conceptually inspired by social ecology as well as social theory. Her research focus is on environmental governance, social aspects of water resource management and social-ecological risk research.

**Melanie Pichler** is a post-doc researcher at the Institute of Social Ecology (SEC) in Vienna, Austria, and a member of the research group on international political ecology at the University of Vienna. She holds a PhD in Political Science from the University of Vienna and an MA in International Development Studies from the same university. Her research focuses on socio-ecological transformations and conflicts, applying political ecology and critical state, hegemony, and democracy theories with a regional focus on Southeast Asia and Europe. She is editor-in-chief of the Austrian Journal of South-East Asian Studies (ASEAS) and interested in the interdisciplinary integration of social and natural science.

**Anke Schaffartzik** enjoys finding meaning through number crunching, focusing on the interplay between societal organization and resource use. She holds a PhD in Social Ecology from Alpen-Adria University for research on the socio-metabolic patterns in international trade. As a senior scientist at the Institute of Social Ecology Vienna (SEC), she is immersed in an interdisciplinary environment in which she collaboratively investigates linkages between resource use and socio-economic development across levels of spatial scale. Anke Schaffartzik teaches a number of postgraduate classes on social ecology's tools for quantitative analysis of resource use as well as on the socio-ecological aspects of trade relations.

**Martin Zimmermann** is a senior researcher at the ISOE—Institute for Social-Ecological Research since 2014 and is working in the research unit Water Infrastructure and Risk Analyses. He studied civil engineering and economics at Technische Universität Darmstadt with a focus on environmental, spatial, and infrastructural planning. He obtained his PhD at the TU Darmstadt with a thesis on sustainable transformations of central northern Namibia's water supply system. After this, he has been working as a postdoc at the Department of Geography, University of Munich. Further research interests are sustainability assessment, novel water infrastructures, water reuse and the vulnerability of critical infrastructures

# Preface to "Social Ecology. State of the Art and Future Prospects"

Over the last decades, social ecology has made important contributions to interdisciplinary sustainability studies. Established in the late 1980s, social ecology was developed as a deliberate provocation to the more 'disciplined' natural and social science environmental research. With its focus on the specific interrelations between societies and their natural environment (consisting of social and biophysical processes), it has challenged disciplinary assumptions about environmental problems. The particular conceptualization of societal relations to nature in social ecology yields strong arguments for the necessity of inter- and transdisciplinary analyses of and responses to the ecological crisis which integrate different knowledge types and stakeholder perspectives.

While the hybrid subject matter as much as the inter- and transdisciplinary approach were highly contested at the beginning, social ecology is now widely accepted within sustainability research and beyond. The contributions to this special issue take stock of these developments and evaluate major conceptual and empirical achievements and current frontiers of social ecology. The authors come from a variety of academic backgrounds. Most of them are affiliated with ISOE — Institute for Social-Ecological Research in Frankfurt/Main, Germany or the Institute of Social Ecology (SEC) in Vienna, Austria. The theoretical concepts, methods, and empirical results reflected in the contributions to this special issue, however, are in no way limited to those developed at ISOE and SEC.

Inter- and transdisciplinarity are fundamentally integrated into the social ecology research framework. This integration rests on the development of concepts and methods for the specific purpose of this type of research. Through the transdisciplinary participation of societal actors, social-ecological research is faced with both the advantage and challenge of working with heterogeneous knowledge. ISOE is one of the leading institutes engaged in method- and theory-based transdisciplinary sustainability research. The focus of ISOE's empirical research is on the crisis-driven changes of patterns of society-nature interrelations within a systemic framework. The concept of regulation and transformation of societal relations to nature and the model of social-ecological provisioning systems (SEPS) specify these interrelations and are combined with analytical tools such as an ideal model of a transdisciplinary research process and social-ecological lifestyle analysis.

A strong focus on the systemic framework within which society-nature relations can be researched has guided the work at SEC, making the institute one of the powerhouses of development of social-economic metabolism research. The concepts of metabolism and of colonization help to characterize society-nature relations and are complemented by analytical tools such a material flow accounting and the human appropriation of net primary production.

The approaches to social ecology at ISOE and SEC are highly complementary, in part due to their productive differences. The aim of this special issue is not to present one monolithic approach to social ecology but to present the variety in the existing research, to discuss how mutual irritation can be productive and to reflect on how the different conceptual achievements must be understood in light of the current social-ecological challenges to which they respond. Social ecology has identified evidence for biophysical and societal prerequisites for social-ecological transformations, thereby linking academic endeavour to public interest and concern.

**Johanna Kramm, Melanie Pichler, Anke Schaffartzik and Martin Zimmermann**

*Special Issue Editors*

*sustainability*

MDPI

*Editorial*

# Societal Relations to Nature in Times of Crisis—Social Ecology's Contributions to Interdisciplinary Sustainability Studies

Johanna Kramm [1,*], Melanie Pichler [2], Anke Schaffartzik [2] and Martin Zimmermann [1]

1    ISOE—Institute for Social-Ecological Research, 60486 Frankfurt am Main, Germany; zimmermann@isoe.de
2    Institute of Social Ecology, Alpen-Adria University Klagenfurt-Wien-Graz, A-1070 Vienna, Austria;
     melanie.pichler@aau.at (M.P.); anke.schaffartzik@aau.at (A.S.)
*    Correspondance: kramm@isoe.de; Tel.: +49-697076919-16

Received: 8 February 2017; Accepted: 15 February 2017; Published: 26 June 2017

**Abstract:** During the second half of the 20th century, the crisis of societal relations to nature emerged as the subject of an international scientific, political, and popular debate. Anthropogenic climate change, loss of biodiversity, resource peaks, or local air and water pollution are symptoms of this crisis. Social ecology provides an inter- and transdisciplinary take on sustainability research and is well-equipped to respond to the research challenges associated with this crisis. Social ecology comprises different schools of thought, of which two initiated this special issue on "State of the Art and Future Prospects" for the research field. The approaches to social ecology of the ISOE—Institute for Social-Ecological Research in Frankfurt, Germany, and the Institute of Social Ecology (SEC) in Vienna, Austria are based on a common understanding of the challenges posed by social-ecological crises. In how these social ecologies tackle their research questions, conceptual differences become evident. In this article, we provide an overview of social ecology research as it is conducted in Frankfurt and in Vienna. We discuss how this research responds to the ongoing crisis and conclude by identifying important future prospects for social ecology.

**Keywords:** social ecology; societal relations to nature; colonization; metabolism; regulation; transformation; social-ecological crisis

---

## 1. Introduction

During the second half of the 20th century, a deepening *crisis* of the societal relations to nature emerged as the subject of an international scientific, political, and popular debate [1]. In the early 1970s, the Club of Rome commissioned "The Limits to Growth" study [2]. During that same decade, large-scale social and environmental disasters occurred, for example, at the nuclear reactors in Harrisburg and Chernobyl and the chemical plants in Bhopal and Seveso. In the 1980s, the Brundtland Report [3] popularized the term "sustainable development" and initiated an increasingly mainstream discourse on the interaction between socio-economic development and environmental change. The UN Sustainable Development Goals (SDGs) are the most recent and currently highly visible attempt to address the social, economic, and environmental pillars of sustainability.

The development of policy interest in sustainability was accompanied by an intensifying social-ecological crisis: anthropogenic climate change, widespread and extremely rapid loss of biodiversity, resource peaks and fluctuating supply, and local air and water pollution are symptoms of this crisis. In all these symptoms, humans are both 'perpetrators' and 'victims'. It is human activity that has caused far-reaching and hazardous change to the environment, so much so that a new geological epoch—the Anthropocene—has been defined [4]. The challenge for sustainability research is immense. To identify the underlying drivers of detrimental (global) environmental change is a prerequisite for any contribution to much-needed transformations of societal relations to nature [5]. Such research is conducted in the face of controversy, as the role of human activity in causing epochal environmental change has repeatedly been disputed; the contestation of *anthropogenic* climate change by certain societal groups is among the most well-known examples. Research on the crisis of the societal relations to nature must therefore not only address questions of societal organization and drivers of environmental change, it must also be able to consider research and academia, disciplines and schools of thought as its object to contribute to transformative knowledge creation.

Social ecology provides an inter- and transdisciplinary take on sustainability research and is well-equipped to respond to these challenges. Social ecology is, however, not a conceptual monolith but comprises different schools of thought [6]. This special issue predominantly brings together contributions from two such schools: the ISOE—Institute for Social-Ecological Research in Frankfurt, Germany, and the Institute of Social Ecology (SEC) in Vienna, Austria. In this editorial, we introduce the institutes' past and recent advances in social ecology, focusing in particular on their respective conceptual research programs, approaches, and methodologies. Based on this review, we identify how social ecology addresses the crisis of the societal relations to nature and discuss some of the research frontiers that have recently emerged. We conclude by reflecting on important future perspectives for social ecology.

## 2. Frankfurt Social Ecology

Frankfurt Social Ecology defines itself as a critical, transdisciplinary science of *societal relations to nature* (SRN) (for a detailed account on Frankfurt Social Ecology, see Hummel et al., this issue). The concept of SRN emerged almost 30 years ago within the political context of the 1980s and 1990s. This period was marked by the debates of the environmental movement, the women's movement and other social movements. Corresponding debates centered on various versions of Marxist, feminist and ecological critiques of ecological destruction, dehumanizing modes of production, patriarchal rule, of the naive belief in progress, and of an objectivist concept of knowledge. Furthermore, challenging theoretical questions raised by the "new social movements" were discussed. Referring to Critical Theory, the ecological crisis was understood in several dimensions as a crisis of politics, gender relations, and science. Moreover, it became clear that disciplinary research was often not able to tackle social-ecological problems. Frankfurt Social Ecology therefore pursues a mode of transdisciplinary research that aims to bring together knowledge from the natural and social sciences with non-scientific knowledge.

The SRN concept refers to patterns that emerge from culturally specific and historically variable forms and practices in which individuals, groups, and cultures design and regulate their relations to nature. SRN are formed either directly through the interaction of individual actors or are mediated by institutions and functional systems. The spectrum of forms and practices ranges from the appropriation of natural resources to the aesthetic contemplation of nature, from physical measurement to environmental education. In the SRN concept, physical-material and cultural-symbolic attributes of the patterns are differentiated. This distinction emphasizes the materiality of all natural relations under consideration and, at the same time, takes the relations' embeddedness in symbolic orders, interpretive contexts and social constructions into account. Material aspects do not simply exist as part of a reality independent of interpretation. Rather, they are the result of social and cognitive processes of construction. This is of particular relevance since the material regulation of modern societies' relations to nature is increasingly dependent on scientific models and technical principles. Similar distinctions are also made in other research approaches, for instance, in Vienna Social Ecology [7], in the theory of reflexive modernization [8], or in actor-network theory [9,10].

At its most fundamental level, the SRN concept evolves around the idea of basic needs. SRN should be regulated in a way guaranteeing that all human beings can satisfy their basic needs. This dimension is closely related to the ideas of justice, equity, and sustainable development. Such "basal societal relations to nature" are essential for individual and societal reproduction, as well as for the capacity of individuals and societies to thrive. Basal SRN are, for instance, work and production, land use and nutrition, sexuality and reproduction, hygiene and health care, or movement and mobility. If the regulation of basal SRN fails, spatially, temporally, and socially extensive crises or societal collapse might occur. In the case of global water or food crises, for instance, large parts of the world population cannot satisfy their vital needs.

In Frankfurt Social Ecology, the idea of the regulation of SRN is incorporated in the concept of provisioning systems: as a reinterpretation of the social-ecological systems approach, this concept allows for the empirical analysis of SRN (Hummel et al., this issue).

The notions of regulation and transformation play an important role in the context of the crisis of societal relations to nature. In this context, regulation has a normative dimension since there has to be a vision of successful regulation. Different disciplinary approaches to regulation can be pursued, for instance, through a technological cybernetic understanding or political-economic theories of regulation. Frankfurt Social Ecology distinguishes patterns and modes of regulation: "Whereas patterns of regulation stand for the material and symbolic aspects of the organization of the individual and social satisfaction of needs, modes of regulation represent a second order regulation, which mirrors the norms and power structures of a society" (Hummel et al., this issue). Changes in the modes and patterns of regulation are conceptualized as social-ecological transformations (see Görg et al., this issue).

The focus on real-world problems (in contrast to purely scientific problems) and the development of possible solutions are constitutive for Frankfurt Social Ecology. In transdisciplinary research, real-world problems are the starting point of the research, and contributing to their solution is the research objective [11,12]. Research is conducted to gain a better understanding of social-ecological problems and to study alternative options for action that can show the way out of problematic states or processes. Transdisciplinarity is a mode of research in which the main assumption is that societal and scientific problems are interlinked. The model of an ideal transdisciplinary research process comprises three phases [13]: In the first phase of problem transformation, societal and scientific problems are set in relation to each other to form a common research object. In the second phase of interdisciplinary integration, new knowledge is produced by the integration of scientific and non-scientific knowledge. In the third phase of transdisciplinary integration, the results of the second phase are assessed in terms of their contribution to societal and scientific progress.

Frankfurt Social Ecology's research results provide options for sustainable solutions, such as integrated management strategies for water and land resources (see Liehr et al. in this issue), or innovative concepts for adapting aged water infrastructures to changing conditions and strategies, in order to minimize critical substances, such as pharmaceuticals, in ground- and drinking water [14]. Designs for transport and mobility, or urban developments, can result from research projects by integrating lifestyle concepts and the quality of life in urban areas [15]. Studies of changes in consumption patterns, lifestyles, and everyday practices are conducted to identify recommendations for reducing carbon emission in households [16,17]. In the field of biodiversity, the interactions between biodiversity, ecosystem services and population dynamics are in the focus of current research [18,19]. Finally, scientific and practice-related foundations for transdisciplinary research are developed, and the impacts of its application on research and cognitive processes are examined [20,21]. Concepts such as ecosystem services (see Mehring et al. and Schleyer et al. in this issue), (social-ecological) risks (see Völker et al. in this issue, [22]), vulnerability [21], resilience or lifestyles [23] are fruitfully integrated. These different approaches stem from the humanities, social, natural and engineering sciences. Their joint application allows for a broader picture of SRN, in addition to the non-scientific view contributed by stakeholders participating in the transdisciplinary research process. Public debates can be encouraged by bringing together different kinds of knowledge.

Frankfurt Social Ecology's contribution to coping with crises of societal relations to nature evolves in a twofold way: First, it aims at developing appropriate solutions to ecological crises by tackling their social, political, and economic causes. Second, it offers solutions to the epistemological and methodological challenges of sustainability science by advancing a critical, transdisciplinary mode of research.

### 3. Social Ecology Vienna

Social ecology as studied at the Institute of Social Ecology in Vienna shares with the ISOE in Frankfurt the identification of real-world problems as its point of departure. In the current social-ecological crisis, the underlying causes are systemic. The crisis is not the aggregate effect of individual, personal resource use but of the societal organization of resource use, i.e., of societal relations to nature. Social ecology in Vienna deliberately has a name that distinguishes it from human ecology. Social ecology's object of study is not humans per se—distinct from plants and other animals—but the societies within which these humans organize their reproduction [6]. From high rates of deforestation, water extraction, and fossil fuel combustion to soil degradation, and the pollution of aquifers and air, the "Great Acceleration" since the 1950s [24] has led to unprecedented patterns of global resource use. From a social-ecological perspective, this new quality and magnitude of environmental impact is associated with a comparable shift in societal organization and in the socio-economic drivers of this environmental impact.

Social ecology in Vienna conceptualizes human societies as "hybrid", as simultaneously subject to biophysical and socio-cultural spheres of causation [25] (Figure 1). Society has a "biophysical compartment", containing its human population, livestock, its infrastructure and artefacts, and must entertain biophysical relations with its natural environment in order to maintain and reproduce this compartment. This conceptualization of society is a prerequisite to the systematic study of societies across time and space in terms of their material and energetic inputs and outputs [26], i.e., of their social metabolism, and of their colonizing interventions in the environment. In this context, social metabolism is used to refer to a society's energetic and material inputs, their transformation, and either integration into societal stocks or output in the shape of exports to other socio-economic systems, or discharge to the environment as wastes and emissions [27]. Colonization describes societal interventions in the environment that aim to render that environment more societally useful than it was prior to the intervention, or than it would be if the intervention were to cease. Deforestation and tilling are examples of colonizing activities that enable the agricultural use of land for society's biomass metabolism. Social-ecological research is currently advancing to specify the decisive role that society's biophysical stocks play for resource flows [28], constituting a legacy of past decisions for future development options (Haberl et al., this issue). The material exchange with the environment is not only the result of biophysical needs but is shaped by societal perceptions of nature (and what constitutes a resource), by patterns of consumption, and by socially mediated access to resources. Social metabolism must therefore also be understood as the result of a type of communication within society's cultural sphere. This communication, in turn, is enabled and shaped by the use of material and energy. The two spheres do not exist independently of one another but are mutually dependent and influential.

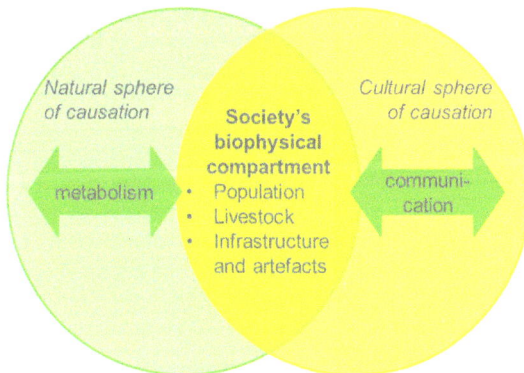

**Figure 1.** Schematic conceptualization of society as simultaneously shaped by a natural and a cultural sphere of causation in Vienna's social ecology.

Social-ecological research scrutinizes patterns of resource use and societal mechanisms that, intentionally or unintentionally, shape these patterns. In turn, changes to these patterns constitute a change to the societal relations to nature, including an altered form of communication about the environment and the resources it provides. To understand and—as has become especially pressing in the current social-ecological crisis—address patterns and drivers of societal resource use, both the natural and the cultural spheres of causation must be taken into account. In doing so, it becomes apparent that specific socio-cultural (political, economic, etc.) mechanisms are in place through which society organizes its metabolism. These mechanisms are simultaneously the product of and the prerequisite for specific forms of resource use [5].

Social-ecological research has uncovered important dynamics of this societal organization of metabolism through the study of past transitions between so-called socio-metabolic modes. Socio-metabolic modes are a meta-distinction between societies according to their resource basis and were introduced by the environmental historian Rolf-Peter Sieferle [29,30]. Hunters and gatherers, agrarian societies, and industrial societies have distinct socio-metabolic modes. Hunters and gatherers passively use solar energy (via biomass hunted or gathered), agrarian societies engage in a controlled solar energy use (biomass cultivated in agricultural systems), and industrial societies mainly obtain their energy via the use of fossil fuels, i.e., by harnessing the products of solar energy influx of the past. The average human requires three billion Joules (Gigajoules GJ) of energy per year in order to survive, as direct nutritional energy and with some use of firewood in order to reduce the amount of energy expended in the digestion of that biomass. Humans living together in a society of hunters and gatherers, with minimal material possessions, use almost four times that amount of energy per person (11 GJ/cap) per year. Agrarian societies use almost 17 times as much (50 GJ/cap), and industrial societies use an average of 200 GJ/cap, i.e., more than 66 times the basic metabolic rate [31]. The particular form of societal organization, and not the sum of basic needs of its inhabitants, is decisive for societal relations to nature. Social-ecological research has demonstrated that sustainability (or lack thereof) is shaped socio-culturally and through resource use patterns (Gizicki-Neundlinger and Güldner, Haas and Andarge, Schaffartzik and Pichler, Haberl et al., all this issue), which must be understood as being interconnected.

Social ecology research in Vienna has focused strongly, but not exclusively, on the quantification of society's metabolism. This has included studies of global resource use [32,33] as well as of resource use by world regions [34,35], countries [36], and local communities [37,38], (Haas and Andarge, this issue). Data availability differs strongly along these levels of scale, resulting in vast differences in the type of research conducted. Across spatial levels of scale, social ecology is also characterized by a broad range of applications to different time scales. These range from long-term socio-ecological research (LTSER) [39] covering millennia of societal resource use [40], (Gizicki-Neundlinger and Güldner in this issue) to studies focusing on current levels of resource use and their development in the more immediate past (Haas and Andarge, Schaffartzik and Pichler, in this issue).

While at and above the national level and for shorter, more recent periods of time, research can rely on national statistics (with some modelling of unreported flows), work at the subnational level and for a historically distant past always requires either estimation procedures or primary data collection. Especially at the community level, social-ecological research has been characterized by a much stronger (and indispensable) link between quantitative and qualitative approaches, where the quantitative data could only be gathered or verified through historical archive material (see Gizicki-Neundlinger and Güldner in this issue) or through interviews and participant observation (see Haas and Andarge in this issue). The latter practices in particular have allowed this branch of social-ecological research to contribute to questioning and reflecting on the role of the researcher [41]. There has never been any strong idea of an "objective" understanding of sustainability, even in the data-driven analyses within social ecology.

Research across levels of scale continues to provide a challenge in the field of social ecology. In rising to this challenge, much has already been discovered about links across levels of scale. Past legacies shape current resource use (Haberl et al., this issue) and drivers of resource use may originate at a spatial distance from where they take effect [42]. The fast pace and strong environmental implications of globalization for sustainability that characterize the current crisis require this type of multi-scalar approach. Future social ecology research is likely to be concerned to a greater degree with the links across different levels of scale [5].

The ability to quantify society's metabolism has developed into one of the strengths of the research conducted at the Institute of Social Ecology in Vienna and has been decisive in generating insights into the biophysical dimensions of particular patterns of economic growth, of land use or of trade relations. This research was essential in fostering the notion of strong sustainability in which natural resources are not considered to be substitutable by financial capital. Indicators derived from material flow accounting, the tool with which social metabolism is measured [43], have been implemented in national and, for example, European statistics [44], and now provide a measure of the biophysical size of an economy which is statistically independent from the leading monetary indicators (including GDP).

## 4. Where to in Times of Crisis?

The two approaches to social ecology presented in this special issue provide a productive framework for conceptualizing different dimensions, dynamics, and scales of the social-ecological crisis. At the same time, persisting and new dynamics of non-sustainable societal relations to nature constitute challenges for social-ecological research. Such challenges comprise, for instance, inequality and power asymmetries (e.g., due to climate change, or in terms of access to resources or ecosystem services), dysfunctional technological systems (e.g., due to path dependencies, inertia or lock-in situations in the energy, water or traffic sector), unjust transformations (e.g., when the introduction of innovations leads to discrimination between user groups or a disparate distribution of benefits, costs and risks), or interdependencies of different scales (e.g., local and regional actions and practices versus transregional and global impacts of pollutants such as pharmaceuticals or microplastics). In the following, we highlight some emerging clusters of challenges to which social ecology is making important contributions. These are (1) the role of power relations in enabling and maintaining unsustainable resource use patterns; (2) the role of social-ecological innovation within transformation processes; and (3) transregional interdependencies crucial for the analysis of societal relations to nature and their transformation towards sustainability.

In considering and contributing to research on power relations that shape access to, control of, and distribution of resources and environmental benefits, social ecology has developed important, fruitful links to the field of political ecology. Political ecology highlights the political and economic dimensions of crisis phenomena by asking whose crisis it is [45]. In highlighting "that unequal relations between actors are a key factor in understanding patterns of human-environment interaction and the associated environmental problems" [45], detailed studies analyze how powerful groups of actors control access to nature, natural resources and ecosystems through specific mechanisms (e.g., capital accumulation, commodification, gender relations, institutional settings) and how marginalized groups of actors (e.g., peasants, indigenous people, fisherfolk, urban dwellers) react to these inequalities [46–48]. The focus of social ecology on the environmental impact of a society's resource use (see Section 3) does not imply that every member of a society contributes equally to that resource use. In fact, the dominant capitalist mode of production and consumption limits the power of individuals to directly affect change so that political and economic drivers and interwoven power relations are decisive for understanding the social-ecological crisis. Focusing on these political and economic dynamics of crisis phenomena implies taking related conflicts into account [49]. Conflicts serve "as a prime form and expression of politics" [50], where underlying relations of power and domination and (contradictory) interests are revealed. Research has shown that societal elites (e.g., state representatives, companies, local elites), for example, actively produce resource scarcities through enclosure processes and the commodification of resources, which exclude other actors from their lands and resources [51,52]. Hence, social-ecological conflicts and crisis phenomena do not necessarily emerge from universal and abstract "limits to growth" or planetary boundaries [2,53] but through socio-economic and political processes that shape societal relations to nature.

*4.1. Transformation through Social-Ecological Innovation*

Sustainability transformations play a crucial role in coping with social-ecological crises (see Görg et al., this issue). Since societal relations to nature are often facilitated through technological regimes (e.g., supply and disposal infrastructures, production and consumption patterns), it is important to understand the factors and processes that foster or hamper regime transitions and/or transformations which are in turn largely influenced by innovation processes. Technological regimes are a sum of institutions and infrastructures such as technologies, engineering practices, skills and procedures, and problem definitions [54]. The concept of technological regimes can describe and explain the predominance of a (dominant) technique and the rules that enable or constrain directions of development [55]. The concept can be applied to better understand innovation or to develop options for steering or managing technology. Especially if resource inefficiency is identified as the cause of environmental problems, technological innovations are often considered as a potential solution. Technology by itself, however, is not able to facilitate sustainable transformation. In the sustainable transformation of water infrastructures [56,57], for example, the invention of water-saving reuse technologies will only be achieved by "reorganizing the world around these technological inventions" [58]. A technological invention and the pursuant innovation must be accompanied by social, economic, and organizational innovations. These comprise, for instance, innovations in routine economic cycles of production and consumption, organizational structures, and acceptance by the users. In addition, the dimension of socio-technical systems necessarily needs to be complemented by an ecological dimension in order to fully grasp the societal relations to nature, i.e., the social-ecological conditions and impacts of technological innovations. The activities, perceptions, and interests of actors such as engineers, researchers, political decision-makers, users, and customers have to be integrated productively. Technological innovation can cause conflict among the involved actors. Research and practice (societal actors) must collaborate closely and negotiate the innovation process to achieve sustainable results. Research and planning become reflexive and iterative processes, which embrace openness and unanticipated contingencies (see Liehr et al. in this issue). Obstacles to innovation must be identified in advance, e.g., institutional barriers, which require other innovations, such as new institutional arrangements. In particular, measures leading to cooperative management have the potential to support the necessary restructuring of institutional arrangements and pave the way for transformation [59,60].

*4.2. Global and Transregional Interdependencies*

Global resource use regimes, transregional economic and commodity chains, and global information flows increasingly connect people, states, economies, and regions. The realms of everyday life in countries of the Global North depend on imported resources, which often stem from the Global South ([61], see Schaffartzik and Pichler in this issue). Flows and connections include, for example, exchange of resources through international trade [35,62], such as coal shipped from South America to be burned in German power plants. The social [63,64] and the environmental dimension [53,65] of these interconnections are objects of intensive research on the social-ecological impacts of scale interdependencies. Ecological and biophysical processes like wind systems, ocean currents, or the conditions of resources like soil and water, are impacted by and have impacts on these international resource flows. For example, microplastics and other chemicals are distributed around the world, posing a global risk [22], and invasive species carried by ships or long-distance traffic on land upset local ecosystemic balances [66]. In order to study such interdependencies, social and ecological analyses, including spatial and temporal dimensions, must be combined, considering natural processes (such as ocean currents, wind), technical processes (technologies), and social processes (communication, practices, trade, politics). The debate on the Anthropocene has triggered questions of scale regarding the tension between the diagnosis of a global social-ecological crisis with responsibility and potential for action on a local and regional scale [67]. Social ecology has begun to integrate concepts which have been used to fruitfully study multi-scalar interdependencies and which include social-ecological systems [68,69], tele-coupling [70–72], virtual water [73], and the flat ontology approach developed by ANT researchers [74].

## 5. Future Prospects for Social-Ecological Research

Each of the approaches to social ecology presented in this special issue contribute to our understanding of the crisis of the societal relations to nature. Each approach has also developed its specific foci. Frankfurt Social Ecology has developed social ecology as a critical, transdisciplinary science which addresses real-world problems in order to shape societal relations to nature in a sustainable way. The work at the Institute of Social Ecology in Vienna has been instrumental in the consideration of society as biophysical and as therefore subject to the impact of a changed environment. Both approaches to social ecology share an understanding of the problems and challenges posed by social-ecological crises and the need to develop responses, solutions, and coping strategies that must include shaping the scientific and public discourse on societal relations to nature. Social-ecological research has identified that and how the sustainability of societal resource use patterns (or more appropriately: the lack of sustainability therein) is systemic and—partially—unintended. These research insights raise serious concern as to the ability of many of the mainstream political and policy measures to adequately address the current crisis. Achieving sustainability is not a question of a few 'tweaks' to the system, it is a question of transformations of our very fundamental societal relations to nature. Social-ecological research can generate knowledge and instate academic practices that greatly contribute to such transformations by rising to the challenge of the current crisis. This is undoubtedly a tall order, and the research in social ecology is far from completed. Nor will or should the research on sustainability transformations be understood as the domain of social ecology alone: New cooperation across disciplinary boundaries and in the public domain must play a catalytic role in the future of this research field. At the same time, the intensified exchange among social ecologists, in which this special issue constitutes but a small step forward, provides a positive challenge for the further development of the field.

**Acknowledgments:** The authors wish to thank Thomas Jahn and Florian Keil for their helpful suggestions and comments.

**Author Contributions:** J.K., M.P., A.S., and M.Z. developed the concept and wrote the article. Section 2 was written mainly by J.K. and M.Z. Section 3 was written mainly by A.S. and M.P.

**Conflicts of Interest:** The authors declare no conflict of interest.

## References

1. Becker, E. Soziale Ökologie—Konstitution und Kontext. In *Soziale Ökologie. Grundzüge einer Wissenschaft von den Gesellschaftlichen Naturverhältnissen*; Becker, E., Jahn, T., Eds.; Campus Verlag: Frankfurt, Germany; New York, NY, USA, 2006; pp. 29–53.
2. Meadows, D.H.; Meadows, D.H.; Randers, J.; Behrens, W.W., III. *The Limits to Growth: A Report to the Club of Rome (1972)*; Universe Books: New York, NY, USA, 1972.
3. World Commission on Environment and Development. Our Common Future. In *Report of the World Commission on Environment and Development*; United Nations: New York, NY, USA, 1987.
4. Crutzen, P.J. Geology of mankind. *Nature* **2002**, *415*, 23. [CrossRef] [PubMed]
5. Pichler, M.; Schaffartzik, A.; Haberl, H.; Görg, C. Drivers of society-nature relations in the Anthropocene and their implications for sustainability transformations. *Curr. Opin. Environ. Sustain.* **2017**, *26–27*, 32–36. [CrossRef]
6. Fischer-Kowalski, M.; Weisz, H. The archipelago of social ecology and the island of the Vienna school. In *Social Ecology. Society-Nature Relations across Time and Space*; Haberl, H., Fischer-Kowalski, M., Krausmann, F., Winiwarter, V., Eds.; Springer International Publishing: Cham, Germany, 2016; Volume 5, pp. 3–28.

7.  Fischer-Kowalski, M.; Weisz, H. Society as Hybrid between Material and Symbolic Realms. Towards a Theoretical Framework of Society-nature Interaction. *Adv. Hum. Ecol.* **1999**, *8*, 215–251.
8.  Beck, U. *Risk Society: Towards a New Modernity*; Theory, culture & society; Sage Publications: London, UK; Newbury Park, CA, USA, 1992.
9.  Latour, B. *Reassembling the Social: An Introduction to Actor-Network-Theory*; Clarendon lectures in management studies; Oxford Univ. Press: Oxford, UK, 2007.
10. Voss, M.; Peuker, B. (Eds.) *Verschwindet Die Natur? Die Akteur-Netzwerk-Theorie in der Umweltsoziologischen Diskussion*; Transcript: Bielefeld, Germany, 2006.
11. Jahn, T. Transdisziplinarität in der Forschungspraxis. In *Transdisziplinäre Forschung. Integrative Forschungsprozesse Verstehen und Bewerten*; Bergmann, M., Schramm, E., Eds.; Campus Verlag: Frankfurt, Germany; New York, NY, USA, 2008; pp. 21–37.
12. Mittelstrass, J. Methodische Transdisziplinarität. *Tech. Theor. Prax.* **2005**, *14*, 18–23.
13. Jahn, T.; Bergmann, M.; Keil, F. Transdisciplinarity: Between mainstreaming and marginalization. *Ecol. Econ.* **2012**, *79*, 1–10. [CrossRef]
14. Winker, M.; Schramm, E.; Schulz, O.; Zimmermann, M.; Liehr, S. Integrated water research and how it can help address the challenges faced by Germany's water sector. *Environ. Earth Sci.* **2016**, *75*, 1226. [CrossRef]
15. Deffner, J. Sustainable mobility cultures and the role of cycling planning professionals. *ISOE Policy Brief.* **2015**, *3*, 1–5.
16. Stieß, I.; Dunkelberg, E. Objectives, barriers and occasions for energy efficient refurbishment by private homeowners. *J. Clean. Prod.* **2013**, *48*, 250–259. [CrossRef]
17. Stieß, I.; Rubik, F. Alltagsroutinen klimafreundlicher gestalten. *Ökol. Wirtsch.* **2015**, *30*, 39–45. [CrossRef]
18. Hummel, D.; Lux, A. Population decline and infrastructure. The case of the German water supply system. *Vienna Yearb. Popul. Res.* **2007**, *5*, 167–191. [CrossRef]
19. Mehring, M. How to frame social-ecological biodiversity research. A methodological comparison between two approaches of social-ecological systems. In *Biodiversität und Gesellschaft. Gesellschaftliche Dimensionen von Schutz und Nutzung Biologischer Vielfalt*; Friedrich, J., Halsband, A., Minkmar, L., Eds.; Universitätsverlag: Göttingen, Germany, 2013; pp. 91–98.
20. Jahn, T.; Keil, F. An actor-specific guideline for quality assurance in transdisciplinary research. *Futures* **2015**, *65*, 195–208. [CrossRef]
21. Lütkemeier, R.; Liehr, S. *Drought: Research and Science-Policy Interfacing*; Andreu, J., Solera, A., Paredes-Arquiola, J., Haro-Monteagudo, D., van Lanen, H., Eds.; CRC Press: London, UK, 2015; pp. 41–48.
22. Kramm, J.; Völker, C. Understanding risks of microplastics. A social-ecological risk perspective. In *Freshwater Microplastics. Emerging Contaminants?* Handbook of Environmental Chemistry; Wagner, M., Lambert, S., Eds.; Springer: Berlin, Germany, 2017, in press.
23. Götz, K.; Ohnmacht, T. Research on Mobility and Lifestyle—What are the Results? In *Mobilities: New Perspectives on Transport and Society*; Grieco, M., Urry, J., Eds.; Ashgate Publishing, Ltd.: Farnham, UK, 2011; pp. 91–108.
24. Steffen, W.; Broadgate, W.; Deutsch, L.; Gaffney, O.; Ludwig, C. The trajectory of the Anthropocene: The Great Acceleration. *Anthr. Rev.* **2015**, *2*, 81–98. [CrossRef]
25. Fischer-Kowalski, M.; Weisz, H. Society as hybrid between material and symbolic realms: Toward a theoretical framework of society-nature interaction. *Adv. Hum. Ecol.* **1999**, *8*, 215–252.
26. Ayres, R.U.; Kneese, A.V. Production, consumption, and externalities. *Am. Econ. Rev.* **1969**, *59*, 282–297.
27. Fischer-Kowalski, M.; Haberl, H. Social metabolism: A metric for biophysical growth and degrowth. In *Handbook of Ecological Economics*; Martinez-Alier, J., Ed.; Edward Elgar Publishing: Cheltenham, UK, 2015; pp. 100–138.
28. Krausmann, F.; Wiedenhofer, D.; Lauk, C.; Haas, W.; Tanikawa, H.; Fishman, T.; Miatto, A. Global socioeconomic material stocks rise 23-fold over the 20th century and require half of annual resource use. *Proc. Natl. Acad. Sci. USA* **2017**, *114*, 1880–1885. [CrossRef] [PubMed]
29. Sieferle, R.P. Sustainability in a world history perspective. In *Exploitation and Overexploitation in Societies Past and Present*; Benzig, B., Ed.; LIT Publishing House: Münster, Germany, 2003; pp. 123–142.
30. Sieferle, R.P. *Rückblick auf Die Natur. Eine Geschichte des Menschen und Seiner Umwelt*; Luchterhand: Munich, Germany, 1997.

31. Fischer-Kowalski, M.; Schaffartzik, A. Energy availability and energy sources as determinants of societal development in a long-term perspective. *MRS Energy Sustain. A Rev. J.* **2015**, *2*. [CrossRef]
32. Krausmann, F.; Erb, K.-H.; Gingrich, S.; Haberl, H.; Bondeau, A.; Gaube, V.; Lauk, C.; Plutzar, C.; Searchinger, T.D. Global human appropriation of net primary production doubled in the 20th century. *Proc. Natl. Acad. Sci. USA* **2013**, *110*, 10324–10329. [CrossRef] [PubMed]
33. Krausmann, F.; Gingrich, S.; Eisenmenger, N.; Erb, K.-H.; Haberl, H.; Fischer-Kowalski, M. Growth in global materials use, GDP and population during the 20th century. *Ecol. Econ.* **2009**, *68*, 2696–2705. [CrossRef]
34. Erb, K.-H.; Krausmann, F.; Lucht, W.; Haberl, H. Embodied HANPP: Mapping the spatial disconnect between global biomass production and consumption. *Ecol. Econ.* **2009**, *69*, 328–334. [CrossRef]
35. Schaffartzik, A.; Mayer, A.; Gingrich, S.; Eisenmenger, N.; Loy, C.; Krausmann, F. The global metabolic transition: Regional patterns and trends of global material flows, 1950–2010. *Glob. Environ. Chang.* **2014**, *26*, 87–97. [CrossRef] [PubMed]
36. Niedertscheider, M.; Kuemmerle, T.; Müller, D.; Erb, K.H. Exploring the effects of drastic institutional and socio-economic changes on land system dynamics in Germany between 1883 and 2007. *Glob. Environ. Chang.* **2014**, *28*, 98–108. [CrossRef] [PubMed]
37. Ringhofer, L. Time, Labour and the Household: Measuring "Time Poverty" through a Gender Lens. *Dev. Pract.* **2015**, *25*, 321–332. [CrossRef]
38. Singh, S.J.; Grünbühel, C.M.; Schandl, H.; Schulz, N. Social Metabolism and Labour in a Local Context: Changing Environmental Relations on Trinket Island. *Popul. Environ.* **2001**, *23*, 71–104. [CrossRef]
39. Haberl, H.; Winiwarter, V.; Andersson, K.; Ayres, R.U.; Boone, C.G.; Castillo, A.; Cunfer, G.; Fischer-Kowalski, M.; Freudenburg, W.R.; Furman, E.; et al. From LTER to LTSER: Conceptualizing the Socioeconomic Dimension of Long-term Socioecological Research. *Ecol. Soc.* **2006**, *11*, 13. [CrossRef]
40. Gingrich, S.; Kuskova, P.; Steinberger, J.K. Long-term changes in $CO_2$-emissions in Austria and Czechoslovakia—Identifying the drivers of environmental pressures. *Energy Policy* **2011**, *39*, 535–543. [CrossRef] [PubMed]
41. Singh, S.J.; Haas, W. Complex Disasters on the Nicobar Islands. In *Social Ecology. Society-Nature Relations across Time and Space*; Haberl, H., Fischer-Kowalski, M., Krausmann, F., Winiwarter, V., Eds.; Springer International Publishing: Cham, Germany, 2016; Volume 5, pp. 523–538.
42. Schaffartzik, A.; Brad, A.; Pichler, M.; Plank, C. At a Distance from the territory: Distal drivers in the (Re)territorialization of oil palm plantations in Indonesia. In *Land Use Competition. Ecological, Economic and Social Perspectives*; Niewöhner, J., Bruns, A., Hostert, P., Krüger, T., Nielsen, J.Ø., Haberl, H., Lauk, C., Lutz, J., Müller, D., Eds.; Springer International Publishing: Cham, Germany, 2016.
43. Fischer-Kowalski, M.; Krausmann, F.; Giljum, S.; Lutter, S.; Mayer, A.; Bringezu, S.; Moriguchi, Y.; Schütz, H.; Schandl, H.; Weisz, H. Methodology and Indicators of Economy-wide Material Flow Accounting. *J. Ind. Ecol.* **2011**, *15*, 855–876. [CrossRef]
44. Eurostat Economy-wide Material Flow Accounts (EW-MFA). *Compilation Guide*; Eurostat: Luxembourg, 2013.
45. Bryant, R.L.; Bailey, S. *Third World Political Ecology*; Routledge: London, UK; New York, NY, USA, 1997.
46. Pichler, M. Legal Dispossession: State Strategies and Selectivities in the Expansion of Indonesian Palm Oil and Agrofuel Production. *Dev. Chang.* **2015**, *64*, 508–533. [CrossRef]
47. Schäfer, M.; Schultz, I.; Wendorf, G. (Eds.) *Gender-Perspektiven in der Sozial-ökologischen Forschung. Herausforderungen und Erfahrungen aus Inter- und Transdisziplinären Projekten*; Oekom Verlag: München, Germany, 2006.
48. Dietz, K.; Engels, B.; Pye, O.; Brunnengräber, A. (Eds.) *The Political Ecology of Agrofuels*; Routledge ISS studies in rural livelihoods; Routledge: Abingdon, Oxon, UK; New York, NY, USA, 2015.
49. Pichler, M.; Brad, A. Political Ecology and Socio-Ecological Conflicts in Southeast Asia. *Austrian J. South-East Asian Stud.* **2016**, *9*, 1–10.
50. Le Billon, P. Environmental conflict. In *The Routledge Handbook of Political Ecology*; Perreault, T., Bridge, G., McCarthy, J., Eds.; Routledge: New York, NY, USA, 2015; pp. 598–608.
51. Brad, A.; Schaffartzik, A.; Pichler, M.; Plank, C. Contested territorialization and biophysical expansion of oil palm plantations in Indonesia. *Geoforum* **2015**, *64*, 100–111. [CrossRef]
52. Ley, L. "Dry Feet for All": Flood Management and Chronic Time in Semarang, Indonesia. *Austrian J. South-East Asian Stud.* **2016**, *9*, 107.

53. Rockström, J.; Steffen, W.; Noone, K.; Persson, Å.; Chapin, F.S.; Lambin, E.F.; Lenton, T.M.; Scheffer, M.; Folke, C.; Schellnhuber, H.J.; et al. A safe operating space for humanity. *Nature* **2009**, *461*, 472–475. [CrossRef] [PubMed]

54. Rip, A.; Kemp, R. Technological Change. In *Human Choice and Climate Change*; Rayner, S., Malone, E.L., Eds.; Battelle Press: Columbus, OH, USA, 1998; pp. 327–399.

55. Van de Poel, I.R.; Franssen, M.P.M. Understanding technical development. The concept of "Technological Regime". *IJTPM* **2002**, *2*, 355. [CrossRef]

56. Kluge, T.; Libbe, J. (Eds.) *Transformationsmanagement für eine Nachhaltige Wasserwirtschaft. Handreichung zur Realisierung Neuartiger Infrastrukturlösungen im Bereich Wasser und Abwasser*; [Ergebnisse des Forschungsverbunds netWORKS]; SÖF—Sozial-Ökologische Forschung; Difu: Berlin, Germany, 2010.

57. Kluge, T.; Schramm, E. *Wasser 2050. Mehr Nachhaltigkeit Durch Systemlösungen*; oekom: München, Germany, 2016.

58. Kaghan, W.; Bowker, G. Out of machine age? Complexity, sociotechnical systems and actor network theory. *J. Eng. Technol. Manag.* **2001**, *18*, 253–269. [CrossRef]

59. Kerber, H.; Schramm, E.; Winker, M. *Transformationsrisiken Bearbeiten: Umsetzung Differenzierter Wasserinfrastruktursysteme Durch Kooperation*; Forschungsverbund netWORKS, Ed.; netWORKS-Papers; Deutsches Institut für Urbanistik Difu: Berlin, Germany, 2016; Volume 28.

60. Schramm, E.; Kerber, H.; Trapp, J.H.; Zimmermann, M.; Winker, M. Novel urban water systems in Germany: Governance structures to encourage transformation. *Urban Water J.* **2017**, in press. [CrossRef]

61. Brand, U.; Wissen, M. Crisis and continuity of capitalist society-nature relationships: The imperial mode of living and the limits to environmental governance. *Rev. Int. Polit. Econ.* **2013**, *20*, 687–711. [CrossRef]

62. Giljum, S.; Eisenmenger, N. North-South Trade and the Distribution of Environmental Goods and Burdens: A Biophysical Perspective. *J. Environ. Dev.* **2004**, *13*, 73–100. [CrossRef]

63. Appadurai, A. *Modernity at Large. Cultural Dimensions of Globalization*; University of Minnesota Press: Minneapolis, MN, USA, 1996.

64. Castells, M. The rise of the network society. In *The Information Age: Economy, Society, and Culture*; Wiley-Blackwell: Oxford, UK, 2011.

65. Schellnhuber, H.J.; Crutzen, P.J.; Clark, W.C.; Hunt, J. Earth System Analysis for Sustainability. *Environ. Sci. Policy Sustain. Dev.* **2005**, *47*, 10–25. [CrossRef]

66. Hulme, P.E. Trade, transport and trouble: Managing invasive species pathways in an era of globalization. *J. Appl. Ecol.* **2009**, *46*, 10–18. [CrossRef]

67. Jahn, T.; Hummel, D.; Schramm, E. Sustainable science in the anthropocene. In *ISOE Discussion Paper*; ISOE—Institute for Social-Ecological Research: Frankfurt am Main, Germany, 2016.

68. Becker, E. Social-Ecological systems as epistemic objects. In *Human Nature Interactions in the Anthropocene: Potentials of Social-Ecological Systems Analysis*; Glaser, M., Krause, G., Ratter, B., Welp, M., Eds.; Routledge: London, UK, 2012; pp. 37–59.

69. Berkes, F.; Colding, J.; Folke, C. *Navigating Social-Ecological Systems: Building Resilience for Complexity and Change*; Cambridge University Press: Cambridge, UK, 2008.

70. Bruckner, M.; Fischer, G.; Tramberend, S.; Giljum, S. Measuring telecouplings in the global land system. A review and comparative evaluation of land footprint accounting methods. *Ecol. Econ.* **2015**, *114*, 11–21.

71. Haberl, H.; Erb, K.-H.; Krausmann, F.; Berecz, S.; Ludwiczek, N.; Martinez-Alier, J.; Musel, A.; Schaffartzik, A. Using embodied HANPP to analyze teleconnections in the global land system: Conceptual considerations. *Geogr. Tidsskr. Dan. J. Geogr.* **2009**, *109*, 119–130. [CrossRef]

72. Friis, C.; Østergaard Nielsen, J.; Otero, I.; Haberl, H.; Niewöhner, J.; Hostert, P. From teleconnection to telecoupling: Taking stock of an emerging framework in land system science. *J. Land Use Sci.* **2016**, *11*, 131–153. [CrossRef]

73. Chapagain, A.K.; Hoekstra, A.Y. The global component of freshwater demand and supply: An assessment of virtual water flows between nations as a result of trade in agricultural and industrial products. *Water Int.* **2008**, *33*, 19–32. [CrossRef]

74. Latour, B. On actor-network theory: A few clarifications. *Soz. Welt* **1996**, *47*, 369–381.

*sustainability*

MDPI

*Article*

# Social Ecology as Critical, Transdisciplinary Science—Conceptualizing, Analyzing and Shaping Societal Relations to Nature

**Diana Hummel [1,2], Thomas Jahn [1,2,*], Florian Keil [3], Stefan Liehr [1,2] and Immanuel Stieß [1]**

[1]  ISOE—Institute for Social-Ecological Research, 60486 Frankfurt/Main, Germany; hummel@isoe.de (D.H.); liehr@isoe.de (S.L.); stiess@isoe.de (I.S.)
[2]  Senckenberg Biodiversity and Climate Research Centre, 60325 Frankfurt/Main, Germany
[3]  keep it balanced, 10999 Berlin, Germany; keil@kib-research.org
*   Correspondence: jahn@isoe.de; Tel.: +49-69-7076-9190

Received: 31 January 2017; Accepted: 8 April 2017; Published: 26 June 2017

**Abstract:** The sustainability discourse is, essentially, centered on the question of how complex relations between nature and society can be conceptualized, analyzed and shaped. In this paper, we present a specific interpretation of social ecology as an attempt to address this question. For this purpose, we establish Frankfurt Social Ecology (FSE) as a formal research program, which is based on the concept of societal relations to nature (SRN). The basic idea of the SRN concept is to put the modern distinction between nature and society at the start of a critical analysis. Such an analysis, we argue, has to focus on the interplay between what we call patterns and modes of regulation. Whereas patterns of regulation stand for the material and symbolic aspects of the organization of the individual and societal satisfaction of needs, modes of regulation mirror the norms and power structures of a society. Using an approach that is based on reformulating social-ecological systems as provisioning systems, we show how this interplay can be analyzed empirically. Finally, we propose critical transdisciplinarity as the research mode of choice of FSE. To conclude, we discuss how FSE can contribute to the development of a research program for a sustainable Anthropocene.

**Keywords:** Anthropocene; critical theory; social-ecological systems; social ecology; societal relations to nature; sustainable development; transdisciplinarity

---

## 1. Introduction

The implementation of the Sustainable Development Goals (SDG) by the United Nations as of 1 January 2016 is paradigmatic for two rather recent developments in science and in its relation to society. The SDG themselves have, on the one hand, re-emphasized the importance of research approaches, which enable an integrated perspective on global change and sustainability [1]. Such programs, like sustainability science, earth system analysis, human ecology or social ecology, have been around for quite some time now. They all try to answer the question of how we can conceptualize and analyze the complex relations between nature and society. On the other hand, the process of defining the SDG shows that the relations between science and society have fundamentally changed in the past three decades. Science today is expected to do more than pursue excellent basic research: It is supposed to align its research interests more closely with the knowledge demands of society [2]. In other words, science should contribute to both understanding and shaping sustainable relations between nature and society. Yet science is, in its current constitution, not only part of the solution; it is also part of the problem. As a functional system of society, science provides the means for the domination of nature and thus, ultimately, is one of the drivers that has turned mankind into a geological force, which now threatens to push the planet to its boundaries [3,4]. Moreover, the disciplinarity that the science system still maintains and incentivizes prevents an understanding of the ways in which the very technology it helped create impacts the complex, interconnected systems we depend on [5].

The challenge for science that this brief description entails calls for nothing less than a research program for a sustainable Anthropocene (cf. [6,7]). Such a program is, presumably, still a long way off. Based on a cursory literature review as well as on our own experience in the field, we can, however, already name a few requirements such a program would have to fulfill. First, it would have to work on those theoretical problems that emerge in places where the relations between nature and society are perturbed in such a way that a sustainable development is blocked [8]. Second, it would have to be based on an integrative mode of research, which allows for the combination or synthesis of different theoretical or methodological approaches and various forms of knowledge [9]. Third, it would have to combine scientific research with societal practice, so that it could offer solutions for real-world problems and, at the same time, produce generalizable knowledge [10]. Finally, a research program for a sustainable Anthropocene would have to be (self)reflexive and (self)critical with respect to the double role of science as a provider of evidence and a driver of problems [11,12]. Against this background this paper introduces a particular interpretation of social ecology as developed by the Institute for Social-Ecological Research (ISOE) in Frankfurt (henceforth referred to as "Frankfurt Social Ecology" or FSE). Its aim is to show how these four requirements, in principle, can be integrated into the formal structure of a research program for a critical, transdisciplinary science.

## 1.1. Frankfurt Social Ecology in Historical Context

Social ecology as a field of scientific inquiry has a long and multifarious history (for an overview see, for example, [13]), which we cannot do justice to within the limited scope of this article. In order to put Frankfurt Social Ecology into context, however, we briefly refer to a few important antecedents and relationships here.

Social ecology today is, in the most basic terms, conceived as a scholarly approach, which starts off with the idea of separated material and constructed systems or dimensions and certain forms of transactions that mediate between the two [14]. On the epistemological level, this approach comes with a strong emphasis on the need to integrate scientific disciplines, theories and methods. In fact, the term "social ecology" was suggested by Milla A. Alihan [15] as the name of an analytical framework for studies of the relations between humans and their environments, which was intended to be more integrative than the field of human ecology as put forward by the Chicago School in the 1920s. This focus on integration is equally important for FSE.

In ways too widely ramified to be traced comprehensively here, the Chicago School is a key reference for contemporary human as well as social ecology. Founded mainly by the works of Robert E. Park, Ernest W. Burgess, Roderick D. McKenzie and Louis Wirth [15–17], the Chicago School set out to understand the principles of organization of the urban metropolis. Drawing on insights from plant ecology, the relations of human beings with their environments were described as the products of competition and selection [18]. The accepted development model was that of a cyclic change of states of balance and imbalance whereby community problems like social segregation, increasing crime rates, ethnic conflicts, or public health issues were interpreted as manifestations of the latter. The city was characterized as a "psychophysical mechanism", which possesses "a moral as well as a physical organization, and these two mutually interact in characteristic ways to mould and to modify one" [19] (p. 578). Although here it sounds like Park conceptualized the material environment as a distinct counterpart to the social aggregate, he (and most of his contemporaries) ultimately saw the city—drawing on the works of Charles Darwin—as a "web of life" [20] in which a distinction between what is natural and what is social is no longer possible.

There are two aspects of the Chicago School that are important for our discussion here: First, strongly influenced by American pragmatism, its scientific works were mainly aimed at providing the theoretical and methodological means to solve the real-world problems of contemporary societies [14]. Consequently, the proponents of the Chicago School were less concerned with laying out grand theories of the likes of Auguste Comte, Herbert Spencer or Charles Darwin, but with advancing interdisciplinary, empirical, and contextual research on community problems [21,22]. This pragmatic, problem-oriented approach has lived on in contemporary human and social ecology as it is represented by the College of Human Ecology at Cornell University, the School of Social Ecology at the University of California, Irvine and the Institute for Social Ecology at the University of Vermont, for example. As we will discuss in detail in Section 4, the problem-oriented approach is also a mainstay of FSE.

The second important aspect in the context of this paper is the use of the term "environment" in the tradition of the Chicago School. Even contemporary scholars [23] criticized that the term remained ambiguous in most papers written by Chicago School sociologists, in that it could both refer to the social and the extra-social environments of human beings. This ambiguity is still present in today's versions of social and human ecology, although more recent works routinely use the term so that it explicitly encompasses social and built environments, as well as natural ones [14]. Obviously, "environment" is a relative and definable concept. In order to avoid ambiguity, therefore, "environment" is not used in theoretical reflections in Frankfurt Social Ecology.

It is interesting to note that some important predecessors of the Chicago School, like Albion W. Small and George E. Vincent, already held this broader understanding of environment when they ascertained that "[s]ociety, in order to maintain its coherence and continue its development, must constantly readjust itself to natural and artificial conditions" and that, forestalling a systemic perspective, "[n]atural circumstances make an impression upon society, which in turn effects modification in nature" [24] (p. 336). For these forefathers of North American sociology, nature was a counterpart to society whereas in the organicistic approach of the Chicago School, nature was essentially conceived of as part of society [25].

The onward loss of nature as an analytical counterpart to society in North American sociology culminated in the structural functionalism of Talcott Parsons and was only redeemed with the rise to prominence of the New Ecological Paradigm (NEP) of Riley Dunlap and William R. Catton [26]. Although the programmatic stance of the NEP was to focus on the degradation of nature caused by the human appropriation of resources, it led to a "sociology of ecological problems" [27] in which the idea of interrelationships between nature and society took a backseat. In the following chapter, we will show that the distinction between nature and society, and a critical analysis of the relations between the two, is constitutive for our approach to social ecology.

In the German-speaking world, the development of social ecology is closely related to the discourse about the ecological crisis [28] and the formation of the corresponding political movements of the 1970s [29]. Part of this discourse was the critique that the then-pressing environmental problems were not adequately addressed by the established institutions and systems of knowledge production. In Germany, this has led to a differentiation of the science system. Until the end of 1970s, this system consisted of two sectors: the universities and the state-funded non-university research facilities. The foundation of institutes such as the ISOE by private, civil society initiatives lead to the formation of a third sector. These independent institutes pioneered the allocation of research capacities dedicated to dealing with concrete problems in the relations between nature and society. Unlike, for example, the Institute for Social Ecology at the University of Vermont, however, they conceived of themselves as being part of a scientific rather than a political movement. In Austria, the critique on traditional modes of knowledge production led to the foundation of dedicated departments within universities, most notably the Institute of Social Ecology at the Klagenfurt University (SEC). SEC's and FSE's interpretation of social ecology share the same overall approach to exploring the relations between nature and society but differ in important theoretical and conceptual details (for a comparison see Kramm et al., this issue).

*1.2. Outline of the Article*

The article is organized as follows: In Section 2, we introduce Frankfurt Social Ecology and her fundamental concept of societal relations to nature (in German "Gesellschaftliche Naturverhältnisse").

In Section 3, we show how a systems theory approach can help to make this concept productive for empirical research. At the center of this approach lies our model of social ecological systems as provisioning systems. In Section 4, we outline how FSE can be established as a critical, transdisciplinary science of societal relations to nature. Using the example of micro contaminants in municipal water cycles, we also demonstrate here how her main concepts can be applied in transdisciplinary research practice. To conclude the article, we briefly discuss a few open questions we consider to be important for the further development of a research program for a sustainable Anthropocene.

## 2. Frankfurt Social Ecology as a Research Program

Following Imre Lakatos (1970), we conceptualize Frankfurt Social Ecology as a research program. In his examination of Karl Popper's falsificationism, Lakatos characterized a scientific research program by its "hard core" [30] (p. 133). In terms of a methodological decision by the program's proponents, the hard core is conceived of as irrefutable. A "negative heuristic" is supposed to ensure that theoretical and empirical research avoids paths that would question the core's rationale. Instead, by means of articulating or inventing "auxiliary hypotheses", the program's proponents form a "protective belt" around the hard core and direct their research efforts at these ("positive heuristic"). Finally, a research program is successful if it leads to a "progressive problemshift", that is, if the program has "excess empirical content over its predecessor" and if "some of this excess empirical content is also corroborated" [30] (p. 113).

Lakatos developed his concept of research programs by reconstructing the history of the natural sciences in certain domains (mechanics and gravitation). Social ecology, in all its past and present diversity, however, cannot look back at a long sequence of attempts to advance the understanding of a shared and precisely defined object of research. Lakatos' concept thus cannot be transferred straightforwardly to social ecology. Yet, in an ex-ante manner, its basic structural ideas can serve as formative principles for a science in statu nascendi. This is how Frankfurt Social Ecology came about as an open, continuously evolving research program over the past 30 years at ISOE. Its core ideas became the basis of a dedicated, still ongoing research funding initiative by the German Federal Ministry for Education and Research [31].

An important prerequisite for this development was, and still is, the existence of a specified hard core. Generally, the hard core can effectively serve as a guarantor for stability and identity over time, so that a productive scientific community can develop around a research program in the first place. In the case of FSE, the concept of societal relations to nature serves that purpose. Obviously, SRN, as much as any other hard core, did not come into existence fully fledged. It emerges rather in the course of an ongoing "process of trial and error" [30]—a process which is driven by, for example, changes in the scientific or societal contexts of the program. Reflexivity and critical faculty are key to shaping this process so that the program can develop progressively. What Lakatos' analysis also teaches us is that a successful research program has to provide a clear, workable link between the theoretical reflection on and the practice of empirical inquiry into its subject matter. (In Section 4 we will show how this link is realized in Frankfurt Social Ecology.) Moreover, we consider the notion of a positive heuristic to be prolific: It should consist of procedures for identifying constructive research questions and hypotheses, help structure research processes, and provide guidance for the treatment of research problems.

Before we define the SRN concept and show how it helps to form FSE as a research program, we note that Frankfurt Social Ecology differs in another crucial aspect from the sciences Lakatos discussed: FSE positions herself in an explicitly normative context, which consists of two related components: First, FSE assumes that the current global relations between nature and society are fundamentally perturbed—a diagnosis of the time that is of course well known under various terms such as "ecologic crisis" [28], "non-sustainable development" [32] or, in some interpretations, the Anthropocene [4]. We refer to this normative component as the crisis of societal relations to nature. Second, social-ecology is a problem-oriented research program. That is, it aims at empowering societies in dealing with historically and geographically specific manifestations of this crisis. For a research program that adheres to the scientific method, it is crucial to separate such normative components from the purely descriptive ones [33]. In Frankfurt Social Ecology this operation is explicitly defined as a self-critical and self-reflexive task [34].

## 2.1. Societal Relations to Nature

The hard core of Frankfurt Social Ecology as a research program is the concept of "societal relations to nature". We define the term to represent the relations between nature and society in their historical development [35]. Our definition includes three postulates: First, nature and society are different things. Second, although distinct, nature and society cannot be treated as being independent from one another. Third, relations between nature and society are given as characteristic, observable patterns, which, in any given location, might transform over time. We refer to such patterns as "patterns of regulation". They have both material and symbolic attributes. While the latter, in the most general terms, refer to flows of information and meanings, the former account for flows of matter and energy. In the following, we will discuss these postulates and their implications for establishing FSE as a research program in more detail. We note that our definition of SRN does not contain any implicit normativity.

With the first postulate, Frankfurt Social Ecology takes a Western, essentially Cartesian, cultural tradition as its point of departure. At its core, this tradition is about the question of what can be associated with human action and decision-making and what cannot. In most general terms, the answer is that there is a universal nature and a contingent culture [36]. This tradition is still highly influential, as the current discourse about the Anthropocene demonstrates. Here, it is about nothing less than a redefinition of what it means to be human in relation to an outside natural world, which is thought to be manageable [37]. The achievements of this tradition are well known. They culminate in what has been termed modernity. Equally well known are, however, its ramifications on the social, ecological, political and scientific levels. Frankfurt Social Ecology is set up as a critique of this dualistic tradition. This manifests itself in a scientific procedure we call "double-sided critique". Double-sided critique can be understood as a functional equivalent of dialectics and aims at scrutinizing binary oppositions such as realism/constructivism, male/female, subject/object etc. as powerful social constructs on the levels of both societal and scientific practice. As far as the latter is concerned, such practices materialize in the "Great Divide" [38] between the natural sciences, on one side, and the social sciences and humanities, on the other. (This divide is synonymous with a methodological rather than ontological distinction between nature and society [35].) Double-sided critique plays the role of the negative heuristic in the research program: It should help to avoid lines of research that reproduce analytical dualisms [9]. Conversely, this implies that FSE has to establish herself as an integrative research program. In Section 4, we will, therefore, introduce transdisciplinarity as the research mode of choice of Frankfurt Social Ecology.

The distinction between nature and society is historically, geographically and culturally variable [37]. What counts as "natural" or "social" in any given context depends, among other things, on power and gender relations, institutional arrangements, political-economic regimes as well as on the type and availability of resources for the satisfaction of individual and social needs. The first postulate thus also implies that FSE has to examine cultural practices of distinction empirically and, in so doing, make the social orders that underlie such practices accessible for critical analysis. This line of research is part of the positive heuristic of Frankfurt Social Ecology. We note that the resilience approach as put forward by the Resilience Alliance sees the delineation between natural and social systems as arbitrary and conceives humans as parts of the ecosystem [33,39]. From the perspective of FSE, this risks ignoring the real-world effects of prevailing practices of distinction.

The second postulate refers to something that has strongly influenced the development of the sciences during the 20th century and that scholars like Ernst Cassirer, Alfred N. Whitehead or Gregory Bateson reflected upon philosophically in their works: the concentration of the epistemological interest on relations instead of things with specific intrinsic features and thus on functions instead of substances and on processes instead of structures [40]. What it means epistemologically and ontologically, however, to analyze relations is a non-trivial and still open question [40]. In the following chapter, we will, therefore, in a pragmatic approach, use the concept of social-ecological systems to make SRN accessible for empirical research.

Moreover, it is possible that directing research towards practices of distinction generates or reproduces dualisms and, therefore, thwarts the identification of relations or, as Brandom notes, "[a] distinction becomes a dualism when its components are distinguished in terms that makes their characteristic relations to one another ultimately unintelligible." [41] (p. 615). It is important to keep in mind here, however, that a conceptual distinction and the social appraisal of that which has been distinguished are not the same things. In other words, a distinction itself does not necessarily imply a revaluation or devaluation of either side; this only happens in the societal context in which the distinction is made. FSE, therefore, needs explicit and precise procedures for making distinctions. Moreover, FSE does not stop at analyzing existing binary distinctions and their effects when dealing with, for example, community problems. She also tries to identify new relations within the established dualisms. Consequently, FSE has to conceptualize the operations of making distinctions and identifying relations as two phases of a formal cognitive process. In Section 4 we will show how such a process can be organized in research practice.

The third postulate introduces the term "patterns of regulation". This establishes, on the one hand, the basic hypothesis that the relations between nature and society are always given in (self)organized forms. On the other hand, in using the term "regulation", we refer to corresponding theoretical approaches for understanding the principles that govern such forms (like, for example, second order cybernetics [42]). In the following section, we will take a closer look at these two aspects. Yet no matter how regulation is treated theoretically, the third postulate implies that FSE has to provide a procedure for distinguishing between the patterns' symbolic and material attributes. Although most approaches for understanding and analyzing nature-society-interactions make this fundamental distinction [8,14,39,43,44], to our knowledge, such a procedure is currently not available. At this point, therefore, we can only indicate a rather coarse approach. It is based on characterizing material relations by their causal effects and symbolic ones by their meanings for an observer [40].

So far, with the SRN concept, Frankfurt Social Ecology appears to only operate on the macro level of society—an objection, which is also brought forward against systems theory approaches in the field [45]. In fact, by putting the distinction between nature and society first, and not that between the human and the non-human, she initially ignores many aspects, which, for example, anthropology, economics, sociology or psychology examine extensively. In the following section, we describe how FSE conceptualizes the relations between the individual, society and nature in more detail. The basic approach is to relate SRN, at the most fundamental level, to the processes of satisfying basic human needs such as food, shelter and reproduction (in this context, we also use the term "basal societal relations to nature" or, in German, "basale gesellschaftliche Naturverhältnisse"; [46]); yet these processes always exist, for the socialized individual, only in culturally and thus socially interpreted forms. Besides the use of the concept of (basic) human needs [47], FSE also applies concepts such as "everyday routines" [48] and "social practices" [49,50] in order to make the individual actor perspective accessible for theoretical and empirical research. Moreover, FSE reflects gender/sexual reproduction as a primary pattern of social order. A basic assumption is that SRN are strongly structured by gender relations and vice versa [51].

In the previous section we introduced Lakatos' idea of a protective belt, which shields the hard core of a research program. In Frankfurt Social Ecology the protective belt is realized as a network of higher order concepts, which all refer either directly or indirectly to SRN. Regulation, as introduced above, is an example of such a higher order concept. Others are the concept of social-ecological systems as a means of analyzing SRN empirically (see the following chapter) as well as the concepts of human needs, everyday routines and gender/sexual reproduction. The higher order concepts and their links among each other or with SRN are the results of generalizations in the outcomes of contextual or problem-oriented empirical research. They form, together with any specific theory they might be associated with, the theoretical knowledge of FSE. Reformulating existing higher order concepts or adding new ones to the protective belt is part of the ongoing development of Frankfurt Social Ecology.

## 2.2. Regulation and Transformation of Societal Relations to Nature

Societies have to shape their relations to nature in order to sustain the process of life intra- and inter-generationally. If a society fails in this effort, it collapses [52]. In this sense, the patterns of regulation introduced above can be interpreted as preconditions for viable social orders. When societies' interactions with nature become dysfunctional (for example as a result of the overexploitation of natural resources or the failure of a mechanism for their cost-efficient provision), they react by regulating the corresponding patterns of regulation. In any society, however, the ways in which such second order regulations can be realized are constrained by what we call "modes of regulation". They mirror the power relations, cultural norms and conflicts of interest in a society and thus represent "social relations" [53,54]. With respect to the terminology introduced above, this means: Whereas patterns of regulation always have both material and symbolic attributes, modes of regulation only have symbolic ones. A key scientific problem of Frankfurt Social Ecology is to understand the interplay of these two levels of regulation in order to identify or anticipate looming crises of SRN. We note that with this approach it becomes possible to analyze local crises of SRN in an overarching societal context such as capitalist modes of production.

What this approach does not specify is, however, on which level of social aggregation patterns and modes of regulation are located. They emerge from both the interactions between individuals and the interplay of institutions or fully differentiated functional systems. Consequently, understanding crises of SRN requires analyzing not only complex interactions on various spatial and temporal scales, but also on social ones (cf. [46,55]):

- On the micro level of individual actions, patterns of regulation are tightly knit to the corporeality of humans and psycho-physical processes—for example to feelings of deprivation, ways of perception and ideas of identity. This level thus primarily addresses the forms of satisfaction of individual needs. These forms are expressed by, and dependent on, practices and routines of everyday life as well as on the norms and power structures that are associated with various modes of regulation on higher social levels.
- On the meso level of organizations and institutions, patterns of regulation essentially address the collective needs of society. They materialize as provisioning systems, for example, for water, food and energy, or as techno-structures, like those for mobility and communication. The forms of needs satisfaction on this level depend on the availability of vital goods and services as well as access to, and the usability of, the techno-structures. They are shaped by certain modes of regulation such as property relations.
- On the macro level of society powerful modes of regulation like, for example, relations of production, property and gender provide the contexts and dispositifs for the processes of needs satisfaction on the lower levels of social aggregation. They take the form of political-economic regimes and thus define the limits within which SRN can be regulated on the meso and micro levels.

Ideally, the patterns and modes of regulation interact in ways so that social integration succeeds and societal reproduction is continuable in the long run. Such a condition is, on a descriptive level, analogous to resilience in a social-ecological system [39]. Problems with this interplay occur when the available options for second order regulations are insufficient to counter an unfolding crisis of SRN. In a historical perspective, such problems lead to fundamental changes of the patterns and modes of regulation. We refer to such changes commonly as "transformations". Conceptually, "transformation" denotes a process, which leads from an initial pattern or mode of regulation to a new one with respect to a given context of regulation. In such a process the relations that define a pattern or mode can become rearranged, broken up or replaced entirely (the implementation of centralized water supply systems in Europe at the end of the 19th century and the rather recent debate about the privatization of water services can serve as examples of transformations of patterns and modes of regulations, respectively). Transformation, in Frankfurt Social Ecology, is a higher order concept and thus part of the protective belt of the research program.

Subjects of transformations are natural phenomena and all actors, individual or collective, whose actions or behaviors influence the form or stability of a given pattern of regulation. Transformations can, therefore, have both intentional and unintentional drivers. Frankfurt Social Ecology uses the concept of transformation in a descriptive manner. The historiographic description of moving from one pattern of regulation to another is the basis for analyzing those factors that have, or will have, determined starting point, course and endpoint of a transformation. For such an analysis, various theoretical avenues can be followed. FSE focuses on those that address power relations, like hegemonic or regulation school theories [56–58], gender relations [59,60], or the diffusion of technological or social innovations [61,62]. The aim of the analysis of transformation processes is to discern or anticipate critical changes in the relations between nature and society as well as to identify those levers for interventions that help steer the course of a transformation in a desired direction—an analytical perspective, which is also emphasized in resilience-based studies of social-ecological systems [63,64].

Whereas there has been progress in creating sustainable patterns and modes of regulation on the local and regional levels in the past few decades, those on the global level are undoubtedly non-sustainable. The familiar keywords are here: climate change, social inequality, loss of biodiversity, poverty or land degradation. Frankfurt Social Ecology starts from the assumption that sustainable development becomes possible only when both the corresponding patterns as well as the modes of regulation are changed. We refer to such intentional processes of change towards sustainability as "social-ecological transformations". In FSE, this term, therefore, is used in a normative manner. The definition, passage and implementation of the SDG can be interpreted as the attempt to initiate such social-ecological transformations. The prevailing critique of the SDG that they lack an integrative perspective and that, therefore, goal conflicts are likely [1,65,66], means, in our terminology: The interplay between the patterns and modes of regulation that the various SDG address were not sufficiently taken into account.

## 3. A Systems Approach to Social Ecology

In the previous chapter, we suggested that SRN cannot be analyzed empirically as such. One way to make them proper objects of empirical research is to represent SRN as social-ecological systems (SES). In this chapter, we examine this idea more closely. Drawing on general systems theory, we define a system as "a set of objects together with relationships between the objects and their attributes" [67] (p. 18). An SES-based approach is common in most elaborated frameworks for the (problem-oriented) analysis of the interactions between nature and society (for a comparative overview see, for example, [45]). Although these frameworks differ in general layout, theoretical underpinnings and conceptual details, there appears to be a consensus on how to define and analyze SES. It can be characterized by three core ideas [8,39,43,68–71]: First, SES are a conjunction of anthropocentric (social aspects in an ecological context) and ecocentric (natural aspects in a social context) perspectives; second SES are complex, adaptive systems, which operate on interacting and hierarchically structured spatial, temporal and social scales; third, analyzing SES requires an interdisciplinary, theoretical framework.

Frankfurt Social Ecology adopts this understanding of SES [72] but adds two more key characteristics: First, SES are not real objects, but models of knowledge about real-world phenomena— they are, in other words, "abstract objects in an ideal world" [73]. In the sense of the rationale introduced in the previous chapter, SES are being constructed by relating the results of scientific or societal practices of distinction between nature and society. The historical contingency of such distinctions is, therefore, as much part of the systems as is the interpretation of the modelled reality by those who construct the model or by those who are represented as actors within it. This leads us to the second characteristic of SES that Frankfurt Social Ecology adds to the consensus outlined above: Science itself is an actor of the SES it sets out to analyze, that is, science is a participant observer. SES are, therefore, self-describing and self-referential systems—an aspect that, so far, has received little attention in the pertinent literature [74].

Generating generalized knowledge about its object of interest is a core aspiration of any research program. Because of their formal structure, systems theory approaches seem to be an ideal means towards that end. From what has been outlined up to this point, however, it becomes clear that SES in Frankfurt Social Ecology cannot claim the status of universally valid descriptions of the dynamic relations between nature and society. Rather, SES are always historically and geographically specific representations of such relations, that is, SES can be conceived of as problem-specific social-ecological case studies. Producing generalizable knowledge in FSE, therefore, either requires observing a given case in its temporal development or working out factual analogies to related cases [10]. We return to this in the following chapter.

*Social-Ecological Systems as Provisioning Systems*

Significant progress in modelling techniques like, for example, agent-based modelling [75] notwithstanding, SES have, as yet, had mostly a heuristic function in research approaches for the analysis of human-nature-interactions [33]. Realizing SES as analytical models requires considerable, strongly structuring simplifications. As a first step toward that end, Frankfurt Social Ecology conceptualizes SES as provisioning systems [76,77]. The term "provisioning", in our definition, includes any benefit societies draw from natural resources. More specifically, and terminological ambiguity notwithstanding, it spans all four ecosystem service categories of the Millennium Ecosystem Assessment, that is: supporting, provisioning, regulating and cultural services [78]. In addition, our understanding of "provisioning" involves forms of care work and, therefore, connects to the feminist discourses about, for example, the crisis of reproductive work [76,79].

Conceptualizing a concrete social-ecological system as a provisioning system means, essentially, to realize it as a specific arrangement of patterns and modes of regulation. Figure 1 shows the basic components of SES as provisioning systems. Via the utilization of resources, provisioning systems connect natural objects, like rivers, forests and oceans, to the societal realms of action and decision-making. The process of resource utilization in provisioning systems is, however, not represented by direct links between users and resources, but rather by four contextual factors: practices, knowledge, technologies and institutions. These four factors are, in a nutshell, the building blocks of the patterns and modes of regulation the system is supposed to model. Such a provisioning system can thus be interpreted as representing characteristic SRN. We note that our model of social-ecological systems as provisioning systems has considerable similarities to Elinor Ostrom's social-ecological systems framework [71]. Her framework particularly emphasizes the role of governance and institutions for resource utilization and management.

Basically, we designed our model of SES as provisioning systems to be scale independent. This means that the analysis is neither limited to certain spatial and temporal scales nor to the macro level of society. The depth in which the social and ecological components of the system are described rather depends on the given research problem and, therefore, on the specific constellation of the transdisciplinary research process (see next chapter). In more concrete terms: How and with which variables the components "resources" and "users" as well as the four factors are described, is the result of a learning process among the scientific and, where applicable, the non-scientific actors participating in the research process. Various research projects at ISOE have demonstrated that the translation of the relatively few elements and premises of the model into variables for empirical research is rather straightforward [76,80–82]. As a formal method for realizing SES as provisioning systems, Bayesian belief networks have proven successful [83]. Cross-scale analysis is the crux of the matter of any SES study. As already indicated above, hierarchical approaches to scale are particularly appropriate for FSE because they emphasize the fact that "the observer is critical to defining scale" [84] (p. 782). As Allen and colleagues also note with respect to the panarchy concept [85], however, applying such approaches in empirical research is still in its infancy in FSE.

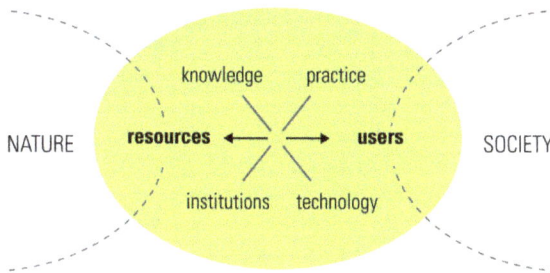

**Figure 1.** Social-ecological systems as provisioning systems (modified according to [76,80]). Note that the green ellipse represents already the social-ecological system. The two dashed ellipses simply indicate the environment of the constructed system. This environment consists of natural and societal aspects that are not part of the system, but can nevertheless have a (significant) impact on it.

## 4. Frankfurt Social Ecology as Critical, Transdisciplinary Science

On the scientific level, Frankfurt Social Ecology aims at a critical analysis of the complex relations between nature and society. In reference to the normative context introduced above, FSE also aims at contributing to shaping these relations in a sustainable way. Inferring from what we discussed so far, the combination of both goals requires a mode of research, which fulfills three requirements: First, it has to be open to theoretical and epistemological pluralism; second, it has to provide a strong link between theoretical reflection and the empirical analysis of real-world problems; third, it has to think practice within theory, that is, it has to reflect upon the implications of science as a functional system within society. A research mode that fulfills all three requirements is critical transdisciplinarity [34,86–89].

The discourse about transdisciplinarity has a long and eclectic history that we cannot reproduce within the limited scope of this article (for a synopsis see [86]). For our purpose here, we thus define transdisciplinarity as being "a critical and self-reflexive research approach that relates societal with scientific problems; it produces new knowledge by integrating different scientific and extra-scientific insights; its aim is to contribute to both societal and scientific progress; integration is the cognitive operation of establishing a novel, hitherto non-existent connection between the distinct epistemic, social–organizational and communicative entities that make up the given problem context" [86] (p. 8f). (Note that this definition of transdisciplinarity implies that, usually, non scientific actors participate in the research process.) The first requirement is implicit in this definition. The second can be met with a formal procedure we will introduce in the following section.

For the third requirement, we draw on elements of the critical theory of the Frankfurt School [90]. An essential feature of critical transdisciplinarity is thus to analyze the functions and status of science and scientific knowledge production within society as well as their role in preserving the current (non-sustainable) state of affairs. From this point of view, science and, for that matter, scientific theories cannot be conceived of as independent from the other branches of production in capitalist societies. A critical theory of society lays bare the ways in which theories that encapsulate knowledge for the domination of nature reflect the social context in which they were developed. For an in-depth discussion of the relation between critical theory, transdisciplinarity and social ecology see, for example, [35].

*Linking Theory and Empirical Research Practice in Frankfurt Social Ecology*

Linking theory and practice in Frankfurt Social Ecology requires relating the SRN concept to transdisciplinary research. The basic approach here is to provide a procedure, which translates historically and geographically unique realms of the relations between nature and society into "epistemic objects" [91,92]. In the following, we present the four basic steps of such a procedure.

We call the starting point of transdisciplinary research a "case". A case consists of an issue and a problem. An issue is a proposition, which conveys the approved state of scientific knowledge with respect to the chosen realm of reality. An example of an issue is the phrase "Active pharmaceutical ingredients are found in water bodies with concentrations in the range of X to Y µg/L". As "problem", we define a statement, which refers to an issue in an evaluative manner. Here, the reference can be implicit or explicit, conscious or unconscious. "Active pharmaceutical ingredients pose a risk for aquatic ecosystems and public health" is an example of such a statement. It is crucial to note here that, generally, such an assessment is made under scientific uncertainty. It is not (entirely) objective but guided by norms, values, or interests. It is precisely for this reason that transdisciplinarity strives for a shared problem description among all actors involved in the research process [86].

Turning a case into an epistemic object now means applying certain formal operations [46]. Figure 2 shows four such operations: distinguish, relate, assess and arrange. Being part of social ecology's protective belt, they are the basic operations with which patterns of relations between nature and society can be identified and made scientifically accessible. The links between the four operations, as shown in Figure 2, describe a four-step process, which takes place at the beginning of the transdisciplinary research process. The first step is to distinguish between the natural and the social attributes of the given case. The guiding question here is: "Which distinctions can be stated that do not contain implicit or hidden assumptions about relations between the distinguished attributes?" Attributes, for which this turns out to be impossible, are called "hybrids" [74]. In such hybrids, a relation between natural and societal entities is, in a manner of speaking, historically embedded. A rectified section of a river is, in this understanding, a hybrid. Hybrids demand particular attention during the ongoing research process as they might need to be disentangled in order to move towards a solution to the problem associated with the given case.

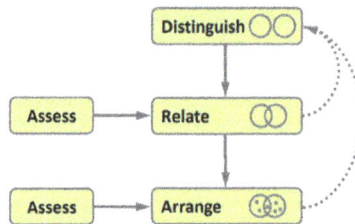

**Figure 2.** The general procedure for translating SRN into epistemic objects for empirical research by means of four formal operations. The dotted arrows indicate that the respective operations have to be applied iteratively. The circles symbolize the domains of nature and society.

In the second step researchers identify or establish relations between the differentiated attributes. The guiding question for this task is: "Which social structures and processes connect those attributes of the given case that were previously classified as 'natural' or 'social'?" Or, in other words: "Which relations can be regarded as characteristic in the sense that they represent a culturally embedded relation between nature and society?" Within the pattern of relations emerging from this second step, researchers then distinguish between material and symbolic relations (the repeated application of the formal operation "distinguish" is represented in Figure 2 by the upper dashed arrow). As we demonstrate in the example (see Table 1), the differentiation between "material" and "symbolic" is neither trivial nor always unambiguously possible.

**Table 1.** Application of the four basic formal operations of FSE to the case of active pharmaceutical ingredients (API) in water bodies (see text for explanations).

| A four-step process for translating SRN into epistemic objects of research |
|---|
| The case can be described as follows: Active ingredients of pharmaceuticals for human use (API) are being increasingly observed in municipal water cycles. Little is known, however, about the adverse effects on wildlife and humans at measured concentrations. Correspondingly, effective measures for reducing APIs in water are lacking. Note that this and the following descriptions are not intended to be exhaustive or to represent the current state of knowledge of the case. They have been strongly simplified to provide a concise example (for more detailed information on this case see [93]). |

**Step 1—Distinguish**

Applying the first formal operation can lead to the following description: Natural attributes of the case are aquatic ecosystems and groundwater reservoirs; social attributes are public health, individual needs for physical and mental health, and providing the population with clean drinking water; hybrids are humans, here in their roles as patients and consumers of drinking water.

**Step 2—Relate**

The attributes distinguished above are connected by a web of relations, which we cannot reproduce in full in this example. The basic relation is established by the functionality of pharmaceuticals: APIs are not fully metabolized by the human body so that a certain fraction of an ingested dose is excreted and ends up in domestic sewage; when these APIs survive sewage treatment, they might reach rivers where they are taken up by aquatic organisms, get deposited in sediments and infiltrate aquifers; if they eventually break through the systems of water purification, they return to humans in the form of contaminated drinking water.

The material side of this specific SRN is characterized by flows of matter (substances). The symbolic side includes technical and institutional arrangements in the healthcare and water supply systems, culturally imparted ideas of physical and mental health with respect to both the purposeful application of drugs as well as the unintended uptake of APIs via contaminated drinking water, personal hygiene, water as the most fundamental life-sustaining natural resource and corresponding notions of purity, and healthy aquatic ecosystems as sources of food and recreation.

Note that the distinction between material and symbolic attributes of the identified relation does not necessarily follow the one between natural and social. If a society, for example, values healthy aquatic ecosystems (symbolic relation), then the definition of what counts as "healthy" is largely independent from the actual concentrations of APIs in water bodies (material relation). Rather, it is the result of societal negotiation processes about safety limits.

**Step 3—Assess**

Drawing on the state of scientific knowledge, the identified relation can be assessed as follows: water cannot be free from contaminants; water purity is always related to the level of pollution and the available measurement accuracy; for a number of reasons, the validity of current risk assessments for APIs in water is principally limited; the chemical properties of most drugs make their partial excretion and slow degradation in the environment unavoidable; environmental engineering today provides no single technology which completely eliminates all APIs from domestic sewage or sources of drinking water; prevailing practices of the prescription, use and disposal of drugs are deeply rooted in ideals and norms of physical or mental fitness and certain lifestyles.

From this description of the case's issue it follows that the corresponding societal problem can neither be reduced to fit the approach of the natural or engineering sciences (for example, in terms of the toxicological assessment of environment and public health hazards by the occurrence of APIs in water, the definition of corresponding safety limits, and the removal of drug residues from municipal sewage by advanced treatment technologies) nor to that of the social sciences (for example, the deliberation of environment and public health risks or the promotion of behavioral change in the handling of pharmaceuticals to keep APIs out of municipal water cycles). The argument of this second assessment (double-sided critique, here greatly abbreviated) is that, as it turns out, the normal mode of operation of the health care system inadvertently causes the occurrence of APIs in municipal water cycles.

The societal problem of the case that can be derived from this description can thus be defined as follows: What are efficient strategies for reducing the occurrence of APIs in water that do not impair the quality of health care? Note that this description can already serve as the epistemic object of research.

**Step 4—Arrange**

It is now possible to represent the identified relations as a social-ecological system by applying the fourth formal operation. For this step it is first necessary to single out the patterns and modes of regulation that are deemed most relevant in addressing the identified problem. Important patterns of regulation are the development and production of pharmaceuticals, their description and (gender-specific) use for the prevention and cure of disease as organized by the healthcare system, and the systems of municipal sewage disposal and water supply. Key modes of regulation are prevalent norms regarding physical and mental fitness as well as policies with respect to drug authorization and environmental protection.

An analysis of the interplay between these patterns and modes of regulation shows that, for example, under current European law, it is not possible to deny authorization of a new drug even when it demonstrably poses a risk for the environment. Moreover, it turns out that, as a rule, safeguarding the quality of healthcare outweighs environmental protection as far as the public opinion is concerned. From the assessment in the step 3 it clearly follows that these critical interplays need to be addressed in order to find sustainable solutions for the identified problem (see Section 3 for further explanations). Once this analysis for relevance is completed, the remaining set of patterns and modes of regulation can be arranged to represent a social-ecological provisioning system (see Section 3 for further explanations).

23

The third formal operation requires assessing the established relations. It consists of two parts: For the first part, the guiding question is: "How do the identified relations reflect the current state of scientific knowledge and to what extent do they represent the case or (justifiably) go beyond it?" (In the latter case the description of the case might have to be revised or extended). Note that from this part the specification of the issue of the given case results. For the second part, researchers apply what we above called "double-sided critique". The guiding question here is: "To what extent are both the analytical distinctions between 'natural' and 'social' as well as those between 'material' and 'symbolic' dependent on the disciplinary or epistemological backgrounds against which they were made?" Those attributes that were characterized as hybrids in the first step deserve special attention in this assessment. They have to be assessed as to which social practices of distinction are inscribed in them as well as to whether, as a unit, they represent, or rather cover, the given problem. If applicable, it might become necessary to disentangle the hybrids so that their embedded relations between natural and social attributes can be critically analyzed. As a consequence of the assessment step it might become necessary to repeat steps one to three (indicated in the figure by the lower dashed arrow). An important result of step three is the description of the societal problem that is associated with the case at hand. In transdisciplinarity, it is determined jointly by the researchers and the participating societal actors [86].

In the final step, researchers arrange the identified attributes and their relations. According to the systemic approach of FSE, the guiding question here is: "Is it possible to describe the identified composition as a provisioning system?" In case this turns out to be possible and expedient, this formal operation amounts to arranging the identified relations between the natural and social attributes in a way that singles out those which are considered important for understanding and solving the given problem. The system, thus defined, is being assessed according to the procedure introduced above (as a result, it might become necessary to rerun the entire process). The social-ecological system that finally emerges from this last step is the epistemic object of research.

It is important to keep in mind here that such an epistemic object only exists temporarily, that is, it might become modified in the course of the transdisciplinary research process. The reason is that identifying a pattern is observer-dependent [84]. Since the perspectives of observers, that is scientists and non-scientific actors, change during the process, something that appeared to be a pattern in the first place, might change its form or even dissolve entirely. It is an essential aspect of good research practice in Frankfurt Social Ecology to deal with the possible instability of the epistemic object and thus of the reference to the initial problem.

Note that the system knowledge produced in transdisciplinary research on social-ecological cases is context-specific initially; it forms the basis for developing orientation and transformation knowledge for dealing with critical developments in the examined SRN [86]. The ultimate scientific goal of the research is to decontextualize or generalize the case knowledge produced. As we already mentioned in the previous chapter, the approach here is to observe a particular case in its historical development (each transdisciplinary research project on a particular case can be viewed as a data point on a timeline) or to carve out factual relations to other cases (Figure 3). Connecting case studies this way leads to new higher order concepts (see Section 2.1) and augments or even entirely discards, existing ones. This changes and strengthens the protective belt of the research program and thus broadens the theoretical knowledge of social ecology. Formalizing and operationalizing the process of moving from case study results to generalized knowledge is an important and still unsolved issue in this context. Experiences and data, such as those gathered by the Long-Term Socio-Ecological Research (LTSER) platforms (cf. [94]), can particularly help to advance it.

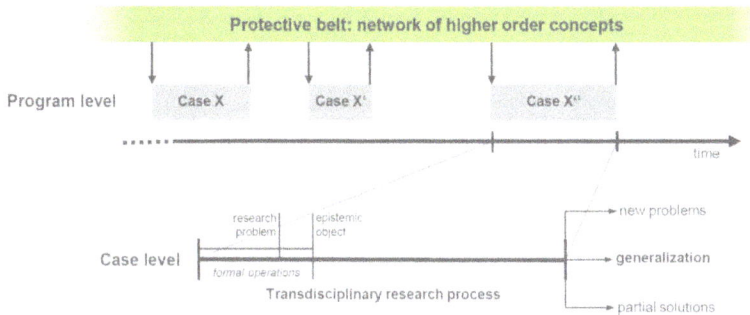

**Figure 3.** The connection between the production of case knowledge and theoretical knowledge in Frankfurt Social Ecology. The light gray rectangles each represent a complete transdisciplinary research project on a given case. The down-arrows indicate that, for defining the problem at the beginning of the research process, the theoretical knowledge of social ecology is used. The up-arrows symbolize the decontextualization and generalization of the case knowledge at the end of the project.

## 5. Conclusions

In this paper, we established Frankfurt Social Ecology as a critical, transdisciplinary scientific research program for studies of societal relations to nature. The aim of the program is to understand these relations and to contribute towards shaping them in a sustainable way. We have argued that a key scientific problem of FSE is to analyze the interplay of patterns and modes of regulation in order to identify or anticipate looming crises of SRN. Whereas patterns of regulation stand for the material and symbolic aspects of the organization of the individual and social satisfaction of needs, modes of regulation mirror the norms and power structures of a society. We introduced the conceptual model of social-ecological systems of provision and showed how it helps to make SRN accessible for empirical research. Finally, we argued that critical transdisciplinarity has to be the research mode of choice of social ecology. This mode, on the one hand, provides the framework that links the production of new scientific knowledge with the development of (partial) solutions for concrete societal problems. In so doing, it forms the basis for the continuous development of FSE as a scientific research program. On the other hand, drawing on critical theory, critical transdisciplinarity allows for a reflection of the social contexts of scientific knowledge production.

We note that, although critical theory was, and is, a key reference for the development of Frankfurt Social Ecology, its precise interpretation for the program has, so far, not been clarified conclusively Yet such a clarification is necessary in order to avoid upholding old mistakes and misconceptions, which have been identified as being intrinsic to critical theory [95]. An approach we currently pursue here is to re-examine the seminal texts on critical theory in order to identify those elements of the theory that are prevailing and adaptable to the basic rationale of social ecology. An important outcome of this endeavor would have to be a theory, which sheds light on how the relations between science and society have changed in the Anthropocene and on what this change implies for the established institutions and procedures of democratically legitimized decision-making—issues that are also discussed in the context of transdisciplinarity, albeit, so far, inconclusively [96,97].

In the introduction we argued for the need of a research program for a sustainable Anthropocene. Our aim here was to show that Frankfurt Social Ecology, along with her founding principles and concepts, could serve as a first step towards such a program. There are, however, problems, which need to be addressed in order to continue on this path. We highlight two of them here: First, FSE and, for that matter, a research program for a sustainable Anthropocene would benefit from a thorough justification for the need of a systems theory approach. An examination of the neo-cybernetic concept of "second order observations" [98] or "observing systems" [99], in our view, has the potential to provide a sound methodological foundation for such a justification. Second, there is currently no compelling social theory, which could explain the development of the epochal crisis of SRN as it manifests itself in the Anthropocene concept. Given the essence of our subject, such a theory cannot spring from a purely sociological endeavor but only from an interdisciplinary one. It would, on the one hand, have to provide the means to describe how SRN form as a result of the satisfaction of (basic) human needs. On the other hand, it would have to posit society as already existing and then help to explain how its structures and processes frame the formation of SRN in the first place. The concepts of patterns and modes of regulation presented here can be interpreted as a pragmatic makeshift for such a social theory.

We established Frankfurt Social Ecology as a critique on the dualistic mindset of modernity. Unlike more recent approaches in, for example, anthropology, however, social ecology does not aim at "understanding the world without dissociating the symbolic from the material" [100]. Instead, FSE tries to make practices of distinction between such binary oppositions and their real-world consequences accessible for critical analysis so that new, more sustainable relations can be established. Nevertheless, we believe a mature research program for a sustainable Anthropocene would eventually have to discuss coexistent, non-Western ontologies [36] and their corresponding epistemologies more thoroughly in order to map the road towards a sustainable future for humanity.

**Acknowledgments:** The authors are grateful to Egon Becker whose pioneering works have defined the development of Frankfurt Social Ecology. The authors also wish to thank the two independent reviewers of this article for their valuable comments.

**Author Contributions:** Diana Hummel contributed to the concepts of regulation and transformation as used in FSE (Section 2.2) and developed the model of social-ecological provisioning systems (Section 3). Thomas Jahn contributed to the concept of societal relations to nature (Section 2.1) and to the link between social ecology, transdisciplinarity, and critical theory (Section 4). Florian Keil compiled the historical context of FSE (Section 1.1) and devised the example for the application of FSE in research practice (Table 1). Stefan Liehr adapted the concept of social-ecological systems for FSE (Section 3). Immanuel Stieß contributed to the concepts of regulation and transformation (Section 2.2). The authors wish to emphasize that FSE is the result of a collective research process by all present and former scientists at the Institute for Social-Ecological Research (ISOE).

**Conflicts of Interest:** The authors declare no conflict of interest.

# References

1. Griggs, D.; Stafford Smith, M.; Rockström, J.; Öhman, M.C.; Gaffney, O.; Glaser, G.; Kanie, N.; Noble, I.; Steffen, W.; Shyamsundar, P. An integrated framework for sustainable development goals. *Ecol. Soc.* **2014**, *19*. [CrossRef]
2. Jahn, T.; Keil, F. An actor-specific guideline for quality assurance in transdisciplinary research. *Futures* **2015**, *65*, 195–208. [CrossRef]
3. Rockström, J.; Steffen, W.L.; Noone, K.; Persson, Å.; Chapin, F.S., III; Lambin, E.; Lenton, T.M.; Scheffer, M.; Folke, C.; Schellnhuber, H.J. Planetary boundaries: Exploring the safe operating space for humanity. *Ecol. Soc.* **2009**, *14*, 32. [CrossRef]
4. Steffen, W.; Crutzen, P.J.; McNeill, J.R. The Anthropocene: Are humans now overwhelming the great forces of nature. *AMBIO J. Hum. Environ.* **2007**, *36*, 614–621. [CrossRef]
5. Vasbinder, J.W.; Nanyang, B.A.; Arthur, W.B. Transdisciplinary EU science institute needs funds urgently. *Nature* **2010**, *463*, 876. [CrossRef] [PubMed]
6. Bai, X.; van der Leeuw, S.; O'Brien, K.; Berkhout, F.; Biermann, F.; Brondizio, E.S.; Cudennec, C.; Dearing, J.; Duraiappah, A.; Glaser, M. Plausible and desirable futures in the Anthropocene: A new research agenda. *Glob. Environ. Chang.* **2016**, *39*, 351–362. [CrossRef]
7. Jahn, T.; Hummel, D.; Schramm, E. Nachhaltige Wissenschaft im Anthropozän. *GAIA* **2015**, *24*, 92–95. English Translation available online http://www.isoe.de/fileadmin/redaktion/ISOE-Reihen/dp/dp-40-isoe-2016.pdf (accessed on 3 April 2017). [CrossRef]
8. Bruckmeier, K. *Natural Resource Use and Global Change. New Interdisciplinary Perspectives in Social Ecology*; Palgrave Macmillan: New York, NY, USA, 2013.

9.    Brondizio, E.S.; O'Brien, K.; Bai, X.; Biermann, F.; Steffen, W.; Berkhout, F.; Cudennec, C.; Lemos, M.C.; Wolfe, A.; Palma-Oliveira, J.; et al. Re-conceptualizing the Anthropocene: A call for collaboration. *Glob. Environ. Chang.* **2016**, *39*, 318–327. [CrossRef]

10.   Van der Leeuw, S.; Costanza, R.; Aulenbach, S.; Brewer, S.; Burek, M.; Cornell, S.; Crumley, C.; Dearing, J.A.; Downy, C.; Graumlich, L.J.; et al. Toward an integrated history to guide the future. *Ecol. Soc.* **2011**, *16*. [CrossRef]

11.   Barad, K. *Meeting the Universe Halfway. Quantum Physics and the Entanglement of Matter and Meaning*; Duke University Press: Durham, UK, 2007.

12.   Jerneck, A.; Olsson, L.; Ness, B.; Anderberg, S.; Baier, M.; Clark, E.; Hickler, T.; Hornborg, A.; Kronsell, A.; Lövbrand, E.; et al. Structuring sustainability science. *Sustain. Sci.* **2011**, *6*, 69–82. [CrossRef]

13.   Fischer-Kowalski, M.; Weisz, H. The archipelago of social ecology and the island of the Vienna School. In *Social Ecology: Society-Nature Relations across Time and Space*; Haberl, H., Fischer-Kowalski, M., Krausmann, F., Winiwarter, V., Eds.; Springer: Cham, Vietnam, 2016; pp. 3–28.

14.   Lejano, R.P.; Stokols, D. Social ecology, sustainability, and economics. *Ecol. Econ.* **2013**, *89*, 1–6. [CrossRef]

15.   Alihan, M.A. *Social Ecology: A Critical Analysis*; Columbia University Press: New York, NY, USA, 1938.

16.   Hawley, A.H. Ecology and human ecology. *Soc. Forces* **1944**, *22*, 398–405. [CrossRef]

17.   Young, G.L. Human ecology as an interdisciplinary concept: A critical inquiry. In *Advances in Ecological Research Volume 8*; Elsevier: Amsterdam, The Netherlands, 1974; pp. 1–105.

18.   McKenzie, R.D. The ecological approach to the study of human community. *Am. J. Sociol.* **1924**, *30*, 287–301. [CrossRef]

19.   Park, R.E. The City: Suggestions for the investigation of human behavior in the city environment. *Am. J. Sociol.* **1915**, *20*, 577–612. [CrossRef]

20.   Park, R.E. Human ecology. *Am. J. Sociol.* **1936**, *43*, 1–5. [CrossRef]

21.   Stokols, D. Toward a science of transdisciplinary action research. *Am. J. Community Psychol.* **2006**, *38*, 63–77. [CrossRef] [PubMed]

22.   Fleury, J.; Lee, S.M. The social ecological model and physical activity in African American women. *Am. J. Community Psychol.* **2006**, *37*, 129–140. [CrossRef] [PubMed]

23.   Bernard, L.L. A classification of environments. *Am. J. Sociol.* **1925**, *29*, 318–332. [CrossRef]

24.   Small, A.W.; Vincent, G.E. *An Introduction to the Science of Society*; American Book Company: New York, NY, USA, 1894.

25.   Cittadino, E. The failed promise of human ecology. In *Science and Nature, Essays in the History of the Environmental Sciences Monographs 8)*. *British Society for the History of Science, London*; Shortland, M., Ed.; British Society for the History of Science: Oxford, UK, 1993; pp. 251–283.

26.   Dunlap, R.E.; Catton, W.R. Environmental sociology. *Annu. Rev. Sociol.* **1979**, *5*, 243–273. [CrossRef]

27.   Groß, M.; Heinrichs, H. Introduction: New trends and interdisciplinary challenges in environmental sociology. In *Environmental Sociology: European Perspectives and Interdisciplinary Challenges*; Groß, M., Heinrichs, H., Eds.; Springer: Dordrecht, The Netherlands, 2010; pp. 1–16.

28.   White, L. The historical roots of our ecologic crisis. *Science* **1967**, *155*, 1203–1207. [CrossRef] [PubMed]

29.   Jahn, T. Krise als gesellschaftliche Erfahrungsform. Versuch der Aktualisierung eines wissenschaftlich-politischen Konzeptes. Ph.D. Dissertation, Frankfurt am Main, Germany, 1989.

30.   Lakatos, I. Falsification and the Methodology of Scientific Research Programmes. In *Criticism and the Growth of Knowledge*; Lakatos, I., Musgrave, A., Eds.; Cambridge University Press: London, UK, 1970; pp. 91–196.

31.   Forschungsgruppe Soziale Ökologie. *Soziale Ökologie. Gutachten zur Förderung der Sozial-Ökologischen Forschung in Hessen. Erstellt im Auftrag der Hessischen Landesregierung (Commissioned by the Government of the State of Hesse)*; ISOE (Institut für sozial-ökologische Forschung): Frankfurt am Main, Germany, 1987.

32.   Jahn, T. Theory of sustainability? Considerations on a basic understanding of "sustainability science". In *Theories of Sustainable Development*; Enders, J.C., Remig, M., Eds.; Routledge: London, UK; New York, NY, USA, 2015; pp. 30–42.

33.   Brand, F.S.; Jax, K. Focusing the meaning(s) of resilience: Resilience as a descriptive concept and a boundary object. *Ecol. Soc.* **2007**, *12*, 23. [CrossRef]

34.   Jahn, T. Wissenschaft für eine Nachhaltige Entwicklung braucht eine kritische Orientierung. *GAIA* **2013**, *1*, 29–33. English translation available online http://www.isoe.de/fileadmin/redaktion/ISOE-Reihen/dp/dp-39-isoe-2016.pdf (accessed on 3 April 2017). [CrossRef]

35. Becker, E.; Jahn, T. (Eds.) *Soziale Ökologie. Grundzüge Einer Wissenschaft von den Gesellschaftlichen Naturverhältnissen*; Campus Verlag: Frankfurt, Germany; New York, NY, USA, 2006.
36. Descola, P.; Lloyd, J. *Beyond Nature and Culture*; University of Chicago Press: Chicago, IL, USA, 2013.
37. Kohn, E. Anthropology of Ontologies. *Annu. Rev. Anthropol.* **2015**, *44*, 311–327. [CrossRef]
38. Latour, B. *We Have Never Been Modern*; Harvard University Press: Cambridge, MA, USA, 1993.
39. Berkes, F.; Colding, J.; Folke, C. (Eds.) *Navigating Social-Ecological Systems. Building Resilience for Complexity and Change*; Cambridge University Press: Cambridge, MA, USA, 2003.
40. Becker, E. Soziale Ökologie: Konturen und Konzepte einer neuen Wissenschaft. In *Wissenschaftstheoretische Perspektiven für die Umweltwissenschaften*; Matschonat, G., Gerber, A., Eds.; Margraf Publishers: Weikersheim, Germany, 2003; pp. 165–195.
41. Brandom, R.B. *Making it Explicit: Reasoning, Representing and Discursive Commitment*, revised ed.; Harvard University Press: Cambridge, MA, USA, 1998.
42. Hummel, D.; Kluge, T. Regulationen. In *Soziale Ökologie. Grundzüge einer Wissenschaft von den Gesellschaftlichen Naturverhältnissen*; Becker, E., Jahn, T., Eds.; Campus Verlag: Frankfurt, Germany; New York, NY, USA, 2006; pp. 248–258.
43. Costanza, R.; Wainger, L.; Folke, C. Modeling complex ecological economic systems. *BioScience* **1993**, *43*, 545–555. [CrossRef]
44. Fischer-Kowalski, M. Society's metabolism: On the childhood and adolescence of a rising conceptual star. In *The International Handbook of Environmental Sociology*; Redclift, M., Woodgate, M., Eds.; Edward Elgar Publishing: Cheltenham, UK; Northhampton, UK, 1997; pp. 119–137.
45. Binder, C.R.; Hinkel, J.; Bots, P.W.G.; Pahl-Wostl, C. Comparison of Framworks for Analyzing Socia-ecological Systems. *Ecol. Soc.* **2013**, *18*, 26. [CrossRef]
46. Becker, E.; Hummel, D.; Jahn, T. Gesellschaftliche Naturverhältnisse als Rahmenkonzept. In *Handbuch Umweltsoziologie*; Groß, M., Ed.; VS Verlag für Sozialwissenschaften: Wiesbaden, Germany, 2011; pp. 75–96. English translation available online http://www.isoe.de/uploads/media/becker-hummel-jahn-soc-rel-nat-en-2012.pdf (accessed on 3 April 2017).
47. Hummel, D.; Becker, E. Bedürfnisse. In *Soziale Ökologie. Grundzüge einer Wissenschaft von den gesellschaftlichen Naturverhältnissen*; Becker, E., Jahn, T., Eds.; Campus Verlag: Frankfurt, Germany; New York, NY, USA, 2006; pp. 198–210.
48. Schultz, I.; Stieß, I. Linking sustainable consumption to everyday life. A social-ecological approach to consumption research. In *Perspectives on Radical Changes to Sustainable Consumption and Production*; Tukker, A., Charter, M., Vezzoli, C., Eds.; Greenleaf Publishing Ltd.: Sheffield, UK, 2008; pp. 288–300.
49. Shove, E. *The Dynamics of Social Practice. Everyday Life and How it Changes*; Sage: Los Angeles, CA, USA, 2012.
50. Schatzki, T. Materiality and Social Life. *Nat. Cult.* **2010**, *5*. [CrossRef]
51. Schultz, I.; Hummel, D.; Hayn, D. Geschlechterverhältnisse. In *Soziale Ökologie. Grundzüge einer Wissenschaft von den Gesellschaftlichen Naturverhältnissen*; Becker, E., Jahn, T., Eds.; Campus Verlag: Frankfurt, Germany; New York, NY, USA, 2006; pp. 224–235.
52. Diamond, J.M. *Collapse. How Societies Choose to Fail or Succeed*; Penguin Books: New York, NY, USA, 2006.
53. Brand, U. Transition und Transformation: Sozialökologische Perspektiven. In *Futuring: Perspektiven der Transformaton im Kapitalismus über ihn Hinaus*; Brie, M., Ed.; Westfälisches Dampfboot: Münster, Germany; Westf, UK, 2014; pp. 242–280.
54. Brand, U. "Transformation" as a New Critical Orthodoxy: The Strategic Use of the Term "Transformation" Does Not Prevent Multiple Crises. *GAIA Ecol. Perspect. Sci. Soc.* **2016**, *25*, 23–27. [CrossRef]
55. Bronfenbrenner, U. Ecological systems theory. In *Six Theories of Child Development: Revised Formulations and Current Issues*; Vasta, R., Ed.; Kingsley: London, UK, 1992; pp. 187–249.
56. Brand, U.; Wissen, M. Crisis and continuity of capitalist society-nature relationships: The imperial mode of living and the limits to environmental governance. *Rev. Int. Political Econ.* **2013**, *20*, 687–711. [CrossRef]
57. Walby, S.; Armstrong, J.; Strid, S. Intersectionality: Multiple inequalities in social theory. *Sociology* **2012**, *46*, 224–240. [CrossRef]
58. Zuindeau, B. Régulation School and environment: Theoretical proposals and avenues of research. *Ecol. Econ.* **2007**, *62*, 281–290. [CrossRef]
59. Schultz, I. The Natural World and the Nature of Gender. In *Handbook of Gender and Women's Studies*; Davis, K., Evans, M., Lorber, J., Eds.; SAGE Publications Ltd.: London, UK, 2006; pp. 376–396.

60. MacGregor, S. (Ed.) *Routledge International Handbook on Gender and Environment*; Routledge: London, UK; New York, NY, USA, 2017.
61. Biggs, R.; Westley, F.R.; Carpenter, S.R. Navigating the Back Loop: Fostering Social Innovation and Transformation in Ecosystem Management. *Ecol. Soc.* **2010**, *15*, 9. [CrossRef]
62. Young, H.P. The dynamics of social innovation. *Proc. Natl. Acad. Sci. USA* **2011**, *108* (Suppl. 4), 21285–21291. [CrossRef] [PubMed]
63. Walker, B.; Holling, C.S.; Carpenter, S.R.; Kinzig, A.P. Resilience, Adaptability and Transformability in Social-ecological Systems. *Ecol. Soc.* **2004**, *9*, 5. [CrossRef]
64. Folke, C.; Carpenter, S.R.; Walker, B.; Scheffer, M.; Chapin, T.; Rockstrom, J. Resilience thinking: Integrating resilience, adaptability and transformability. *Ecol. Soc.* **2010**, *15*, 20. [CrossRef]
65. Griggs, D.; Stafford-Smith, M.; Gaffney, O.; Rockstrom, J.; Ohman, M.C.; Shyamsundar, P.; Steffen, W.; Glaser, G.; Kanie, N.; Noble, I. Policy: Sustainable development goals for people and planet. *Nature* **2013**, *495*, 305–307. [CrossRef] [PubMed]
66. Nilsson, M.; Griggs, D.; Visbeck, M. Policy: Map the interactions between Sustainable Development Goals. *Nature* **2016**, *534*, 320–322. [CrossRef] [PubMed]
67. Hall, A.D.; Fagen, R.E. Definition of system. *Gen. Syst.* **1956**, *1*, 18–28.
68. Glaser, M.; Krause, G.; Ratter, B.; Welp, M. New Approaches to the Analysis of Human-Nature Relations. In *Human Nature Interactions in the Anthropocene: Potentials of Social-Ecological Systems Analysis*; Glaser, M., Krause, G., Ratter, B., Welp, M., Eds.; Routledge: London, UK, 2012; pp. 3–12.
69. Haberl, H.; Fischer-Kowalski, M.; Krausmann, F.; Weisz, H.; Winiwarter, V. Progress towards sustainability?: What the conceptual framework of material and energy flow accounting (MEFA) can offer. *Land Use Policy* **2004**, *21*, 199–213. [CrossRef]
70. Levin, S.; Xepapadeas, T.; Crépin, A.-S.; Norberg, J.; Zeeuw, A.; de Folke, C.; Hughes, T.; Arrow, K.; Barrett, S.; Daily, G.; et al. Social-ecological systems as complex adaptive systems: Modeling and policy implications. *Environ. Dev. Econ.* **2013**, *18*, 111–132. [CrossRef]
71. Ostrom, E. A General Framework for Analyzing Sustainability of Social-Ecological Systems. *Science* **2009**, *325*, 419–422. [CrossRef] [PubMed]
72. Liehr, S.; Becker, E.; Keil, F. Systemdynamiken. In *Soziale Ökologie. Grundzüge einer Wissenschaft von den Gesellschaftlichen Naturverhältnissen*; Becker, E., Jahn, T., Eds.; Campus Verlag: Frankfurt, Germany; New York, NY, USA, 2006; pp. 267–283.
73. Becker, E.; Breckling, B. Border Zones of Ecology and Systems Theory. In *Ecology Revisited. Reflecting on Concepts, Advancing Science*; Schwarz, A., Jax, K., Eds.; Springer: Dordrecht, The Netherlands, 2011; pp. 385–403.
74. Becker, E. Social-Ecological Systems as Epistemic Objects. In *Human Nature Interactions in the Anthropocene: Potentials of Social-Ecological Systems Analysis*; Glaser, M., Krause, G., Ratter, B., Welp, M., Eds.; Routledge: London, UK, 2012; pp. 37–59.
75. Schlüter, M.; Pahl-Wostl, C. Mechanisms of resilience in common-pool resource management systems: An agent-based model of water use in a river basin. *Ecol. Soc.* **2007**, *12*, 4. [CrossRef]
76. Hummel, D. (Ed.) *Population Dynamics and Supply Systems. A Transdisciplinary Approach*; Campus Verlag: Frankfurt, Germany; New York, NY, USA, 2008.
77. Lux, A.; Janowicz, C.; Hummel, D. Versorgungssysteme. In *Soziale Ökologie. Grundzüge Einer Wissenschaft von den Gesellschaftlichen Naturverhältnissen*; Becker, E. Jahn, T. Eds.; Campus Verlag. Frankfurt, Germany; New York, NY, USA, 2006; pp. 423–433.
78. Millennium Ecosystem Assessment. *Ecosystem and Human Well-Being. Synthesis*; Island Press: Washington, DC, USA, 2015.
79. Biesecker, A.; Hofmeister, S. Focus: (Re)productivity: Sustainable relations both between society and nature and between the genders. *Ecol. Econ.* **2010**, *69*, 1703–1711. [CrossRef]
80. Hummel, D.; Jahn, T.; Schramm, E. *Social-Ecological Analysis of Climate Induced Changes in Biodiversity—Outline of a Research Concept*; Bik-F Knowledge Flow Paper No. 11; Frankfurt am Main, Germany, 2011. Available online: http://www.bik-f.de/files/publications/kfp_nr-11_neu__71c3b9.pdf (accessed on 18 April 2017).
81. Hummel, D. Climate change, land degradation and migration in Mali and Senegal—Some policy implications. *Migr. Dev.* **2016**, *5*, 211–233. [CrossRef]

82. Mehring, M. How to Frame Social-Ecological Biodiversity Research—A Methodological Comparison between two Approaches of Social-Ecological Systems. In *Biodiversität und Gesellschaft.Gesellschaftliche Dimensionen von Schutz und Nutzung biologischer Vielfalt*; Friedrich, J., Halsband, A., Minkmar, L., Eds.; Universitätsverlag: Göttingen, Germany, 2013; pp. 91–98.

83. Drees, L.; Liehr, S. Using Bayesian belief networks to analyse social-ecological conditions for migration in the Sahel. *Glob. Environ. Chang.* **2015**, *35*, 323–339. [CrossRef]

84. Manson, S.M. Does scale exist?: An epistemological scale continuum for complex human–environment systems. *Geoforum* **2008**, *39*, 776–788. [CrossRef]

85. Allen, C.R.; Angeler, D.G.; Garmestani, A.S.; Gunderson, L.H.; Holling, C.S. Panarchy: Theory and Application. *Ecosystems* **2014**, *17*, 578–589. [CrossRef]

86. Jahn, T.; Bergmann, M.; Keil, F. Transdisciplinarity: Between mainstreaming and marginalization. *Ecol. Econ.* **2012**, *79*, 1–10. [CrossRef]

87. Frodeman, R. *The Oxford Handbook of Interdisciplinarity*; Oxford University Press: Oxford, UK, 2010.

88. Hadorn, G.H.; Biber-Klemm, S.; Grossenbacher-Mansuy, W.; Hoffmann-Riem, H.; Joye, D.; Pohl, C.; Wiesmann, U.; Zemp, E. (Eds.) *Handbook of Transdisciplinary Research*; Springer Science: Zurich, Switzerland, 2008.

89. Klein, J.T. Evaluation of interdisciplinary and transdisciplinary research: A literature review. *Am. J. Prev. Med.* **2008**, *35*, 116–123. [CrossRef] [PubMed]

90. Horkheimer, M. Traditional and Critical Theory. In *Critical Theory: Selected Essays*; Seabury Press: New York, NY, USA, 1972; pp. 188–243.

91. Knorr-Cetina, K. *Epistemic Cultures. How the Sciences Make Knowledge*; Harvard University Press: Cambridge, MA, USA, 1999.

92. Nicolini, D.; Mengis, J.; Swan, J. Understanding the Role of Objects in Cross-Disciplinary Collaboration. *Organ. Sci.* **2012**, *23*, 612–629. [CrossRef]

93. Keil, F.; Bechmann, G.; Kümmerer, K.; Schramm, E. Systemic risk governance for pharmaceutical residues in drinking water. *GAIA* **2008**, *17*, 355–361. [CrossRef]

94. Haberl, H.; Winiwarter, V.; Andersson, K.; Ayres, R.; Boone, C.; Castillo, A.; Cunfer, G.; Fischer-Kowalski, M.; Freudenburg, W.; Furman, E. From LTER to LTSER: Conceptualizing the socioeconomic dimension of long-term socioecological research. *Ecol. Soc.* **2006**, *11*, 13. [CrossRef]

95. Becker, E.; Jahn, T. Umrisse einer kritischen Theorie gesellschaftlicher Naturverhältnisse. In *Kritische Theorie der Technik und der Natur*; Böhme, G., Manzei, A., Eds.; Wilhelm Fink: München, Germany, 2003; pp. 91–112. English translation available online: http://www.isoe.de/ftp/darmstadttext_engl.pdf (accessed on 3 April 2017).

96. Lovbrand, E.; Pielke, R.; Beck, S. A Democracy Paradox in Studies of Science and Technology. *Sci. Technol. Hum. Values* **2011**, *36*, 474–496. [CrossRef]

97. Seidl, R.; Brand, F.S.; Stauffacher, M.; Krutli, P.; Le, Q.B.; Sporri, A.; Meylan, G.; Moser, C.; Gonzalez, M.B.; Scholz, R.W. Science with society in the anthropocene. *AMBIO* **2013**, *42*, 5–12. [CrossRef] [PubMed]

98. Aufenvenne, P.; Egner, H.; von Elverfeldt, K. On climate change research, the crisis of science and second-order science. *Construct. Found.* **2014**, *10*, 120–129.

99. Vanderstraeten, R. Observing Systems: A Cybernetic Perspective on System/Environment Relations. *J. Theory Soc. Behav.* **2001**, *31*, 297–311. [CrossRef]

100. Salmon, G.; Charbonnier, P. The two ontological pluralisms of French anthropology. *J. R. Anthropol. Inst.* **2014**, *20*, 567–573. [CrossRef]

*sustainability*

MDPI

*Article*

# Challenges for Social-Ecological Transformations: Contributions from Social and Political Ecology

Christoph Görg [1,*], Ulrich Brand [2], Helmut Haberl [1], Diana Hummel [3], Thomas Jahn [3] and Stefan Liehr [3]

[1] Institute of Social Ecology, 1070 Vienna, Alpen-Adria Universitaet Klagenfurt, 9020 Klagenfurt, Austria; helmut.haberl@aau.at

[2] Department of Political Science, University of Vienna, 1090 Vienna, Austria; ulrich.brand@univie.ac.at

[3] ISOE—Institute for Social-Ecological Research, 60486 Frankfurt/M., Germany; hummel@isoe.de (H.H.); jahn@isoe.de (T.J.); liehr@isoe.de (S.L.)

* Correspondence: christoph.goerg@aau.at; Tel.: +43-(0)1-522-4000-403

Received: 6 February 2017; Accepted: 4 June 2017; Published: 26 June 2017

**Abstract:** Transformation has become a major topic of sustainability research. This opens up new perspectives, but at the same time, runs the danger to convert into a new critical orthodoxy which narrows down analytical perspectives. Most research is committed towards a political-strategic approach towards transformation. This focus, however, clashes with ongoing transformation processes towards un-sustainability. The paper presents cornerstones of an integrative approach to social-ecological transformations (SET), which builds upon empirical work and conceptual considerations from Social Ecology and Political Ecology. We argue that a critical understanding of the challenges for societal transformations can be advanced by focusing on the interdependencies between societies and the natural environment. This starting point provides a more realistic understanding of the societal and biophysical constraints of sustainability transformations by emphasising the crisis-driven and contested character of the appropriation of nature and the power relations involved. Moreover, it pursues a transdisciplinary mode of research, decisive for adequately understanding any strategy for transformations towards sustainability. Such a conceptual approach of SET is supposed to better integrate the analytical, normative and political-strategic dimension of transformation research. We use the examples of global land use patterns, neo-extractivism in Latin America and the global water crisis to clarify our approach.

**Keywords:** social-ecological transformation; societal relations to nature; social ecology; political ecology; land use; resource-extractivism; water crisis; transdisciplinarity

---

## 1. Introduction

Transformation has become a major topic of sustainability research. The terminology indicates a shift both in the focus of research and in the understanding of the real scale of the challenges contemporary societies are facing. The consequences of this shift for policy-making are far reaching. In the last decades, incremental environmental governance was considered the cornerstone of a pragmatic policy approach. This approach, however, is increasingly criticised as being insufficient for coping with problems like climate change, biodiversity loss, resource depletion, food security or social inequalities. Consequently, the quest for a "societal transformations towards sustainability" [1–13] or even for a "Great Transformation" [14] emerged as a guiding theme.

This shift, on the one hand, opens up new perspectives. On the other hand, it runs the risk of narrowing down the scope of research and the corridor of possible action, leading to what has been termed "a new critical orthodoxy" [15]. The latter is characterised by a tension between the call for a comprehensive societal transformation and a strong trust in existing institutions like state, market, science, technology, and (Western) knowledge [15].

Without any doubt, transformations towards sustainability necessarily involve political-strategic aspects, including the capability to intervene into ongoing socio-political, economic, institutional and technological transformation processes, as well as providing transformative knowledge in various different political-administrative settings. However, there is a tendency towards more political-strategic projections at the expense of rigorous analytical approaches, capable of bridge transformation requirements and transformation strategies. What is needed, therefore, is a more solid understanding of those dominant societal dynamics that hinder a transformation towards sustainability. This presupposes considering a variety of interacting and conflicting transformation processes. Whereas some of these intentionally aim at sustainability (for example, the German "Energiewende"), others pursue different targets (for example, geopolitical strategies to secure resource access). Moreover, some of these transformations are the outcome of policies and structural societal conditions and thus unintended (and perhaps difficult or impossible to influence) by those actors that focus on deliberate, intended transformations towards sustainability [5,16], but perhaps influenced by the interests of powerful societal actors. Thus, conflicting and to some degree antagonistic societal processes and the power relations involved have to be considered. Only by explicitly addressing a plurality of conflicting transformation processes, a better analytic understanding can be achieved, which offers a more realistic approach for strategic interventions [12].

Besides the analytical and political-strategic dimensions, a third, rather normative dimension of the transformation debate remains unclear: What is a desirable, fair and feasible future for global societies? Sustainability transformations need adequate political strategies, but can only be successful if grounded in a robust analytical approach and a legitimate normative perspective. Drawing on the three-dimensioned sustainability discourse [17], we argue in this paper that a better integration of analytical perspectives on ongoing transformations of societal relations to nature and of normative considerations of what may constitute a desirable goal of global transformations towards sustainability is needed to improve the political-strategical aspirations of transformation research.

Several conceptual and methodological challenges prevent an integrated, i.e., analytical, normative and political-strategic understanding of social-ecological transformations (SET). In this paper, we outline a conceptual approach that aims at meeting these challenges, based on Social Ecology (SE) and Political Ecology (PE) (we are aware, that the use of the term Social-Ecological Transformation is not limited to SE and PE and we do not proclaim to have an ownership. Moreover, in such an article, it is not possible to present in detail SE and PE that are—in themselves—ramified. We refer to our work at the Institute for Social-Ecological Research in Frankfurt [18,19], the Institute for Social Ecology in Vienna [20,21] and the Department of Political Science at the University of Vienna [22,23], see also [24]. We are also aware that several other "transition" concepts exist in various scientific communities which, despite some similarities, differ in important respects from our concept of transformation. In ecology, for example, the concept of "ecological transition" or "regime shifts" denotes "substantial, long-lasting reorganizations of complex systems, such as ecosystems" [25]. In cultural ecology, the concept of ecological transition was used to describe sociocultural trends towards population growth, increasing mechanisation and growing exploitation of natural resources [26]. The "transition management" approach [27,28] asks for the processes how social innovations in niches are upscaled to regimes and landscapes. Both SE and PE refer to the concept of societal relations to nature (SRN), in German Gesellschaftliche Naturverhältnisse. However, the English translation of the German term Gesellschaftliche Naturverhältnisse is an issue: several translations are possible and in the use like societal nature relations. However, if we refer in the following to conceptual considerations how this concept may improve sustainability research, we explicitly use the term societal relations to nature (SRN), being aware that there are (minor) differences in the exact meaning even within the institutions involved. See for an elaborated understanding of the concept in sustainability research: [17], for the history of the concept in the German tradition of critical theory: [29]). Both SE and PE emphasizes the interactions between and co-constitution of environmental conditions and societal dynamics including inequalities and power relations. They consider the societal roots of the ecological crisis as part of a "multiple crisis" [30] or as "a crisis of societal relations to nature" [19]—a complex constellation, that is centred on dominant societal patterns of production and living (at a material as well as symbolic level), their ecological implications, and controversies about how to respond scientifically and politically. Both SE and PE stress the relevance of shaping SRN for every society and they offer several insights about how a crisis of SRN develops, e.g., regarding the societal metabolism of industrial societies that depend on fossil fuel (see below).

The aim of this paper is to demonstrate why such a point of departure can help to advance an integrated understanding of the challenges for societal transformations. For this purpose, we discuss three key assets of a combined SE and PE approach. First, by focusing on SRN, SE provides a more realistic way of understanding the societal and biophysical constraints of sustainably transformations at the interplay between several spatial and temporal scales (see Section 3). Second, PE emphasises the conflict-driven and contested character of the appropriation of nature and thus the power and domination-shaped character of SRN. With that, PE counters the often unreflected understanding of governance and political steering within the transformation debate by analysing actor constellations, the structural conditions of policy making and power relations (see Section 4). Third, both SE and PE pursue a transdisciplinary mode of research (TD; concerning SEC, in particular local studies but also Health Studies are important areas of TD research, see: [31–33]; ISOE follows TD as a rule, and the TD approach is based on own research on transdisciplinary concepts and methods, see [34,35], see also Section 5 of this paper for more details). TD is, at its core, about the co-production and integration of different forms of knowledge [25]. Such a mode of research is decisive for adequately understanding any strategy for transformations towards sustainability (see Section 5). Taking advantage of these assets, the conceptual approach of social-ecological transformations is supposed to help investigate problematic and non-sustainable structures and processes (analytical dimension), to contribute to transformations towards sustainability on the level of action and decision-making processes (political-strategic dimension) and analysing what are societally desirable and at the same time achievable conditions and ends (normative dimension).

In the following chapter, we discuss in more detail why a critical understanding of transformation is required. In Sections 3–5, we elaborate on the three assets of the combined SE and PE approach. In each of these sections, we try to demonstrate the relevance of their respective conceptual contributions by discussing particular examples from empirical research. Our aim here is merely to illustrate our arguments. We do not claim that the chosen examples can provide enough empirical evidence for actually proving the solidity of our approach—this is the object of further research. To conclude, we outline the contours of a critical and integrative approach to social-ecological transformations.

## 2. Transformations towards Sustainability—Strengths and Weaknesses of the Current Debate

Compared with the discussion about sustainable development that has evolved since the publication of the Brundtland-Report in 1987 and the Rio-Conference on Environment and Development in 1992, the current debate on transformations towards sustainability marks a progress [36]. The quickly growing relevance of the transformation concept is fuelled by the acknowledgement of a multiple crisis, including a crisis of SRN, which cannot be dealt with by environmental policy alone or by changes within other separated policy fields. In reverse, it questions the established institutional responses and its interplay [22,37,38]. However, as Nalau and Handmer argue, the discussion around transformative change is still emerging and it is not clear as to what transformation means, how it can be evaluated, and how the conceptions of transformation fit within the current understanding of dealing with policy problems in practice. [10] (p. 349). Despite the fuzziness of the concept, they conclude that transformation can be understood as a "fundamental shift that questions and challenges values and routine practices and changes prior perspectives employed to rationalise decisions and pathways" [5,10] (pp. 350, 668).

There is, however, much ambiguity and disagreement about the meaning and function of the concept. First of all, as the terms "sustainability transformation" or "transformations towards sustainability" (often equated with the aim of a "low-carbon society") indicate, most research is committed to a normative or political-strategic approach to transformation. Analytical clarity is often superseded by visionary and strategic orientations as the quote from Nalau and Handmer suggests (for more details [12]). Undoubtedly, visionary and strategic claims are making the debate so important. Yet they might run the risk of downplaying the socio-economic, political, cultural, and subjective dimensions that are deeply inscribed into SRN. For instance, we currently observe the rise of authoritarian rule and authoritarian neoliberalism as well as of right-wing parties [39,40] as a response to the financial crisis of 2008. Our observation is, however, that these tendencies and their root causes are not properly addressed in the transformation debate. Such tendencies deepen the crisis of SRN as they go hand in hand with climate change scepticism, the promotion of unconventional fossil fuels, or increased mining activities even in ecologically sensitive areas. The motivation here often is to protect a certain mode of living against the ramifications of the crisis and against a profound social-ecological transformation. The societal causes of non-sustainable dynamics remain, however, in the debate often rather opaque (see below).

Second, further challenges within the transformation debate emerge around the question of how to address biophysical constraints of societal development. Concepts such as "planetary boundaries" [41] or the "Anthropocene" [42,43] aim to operationalize those constraints, but they have several limitations. Beside limits of scientific certainty about the precise extent of thresholds (such as biodiversity loss), the spatial and temporal scale as much as the interplay of scientific analysis and societal appraisements (including cultural perceptions) represent major challenges. Scientific analyses are decisive to determine potential thresholds of (global) environmental change and resource use. However, boundaries have to be defined within societal and political processes, as they are necessarily linked to normative values. In other words, the biophysical constraints of societal development can only be determined with a transdisciplinary approach (see below). Moreover, the interaction between human activities (e.g., concerning land use) and natural processes (e.g., the services ecosystems provide or the contribution of land use to global warming) are crucial to determine the constraints and the option space for further resource use (e.g., concerning the options for feeding the world or for bioenergy use; [44]). These interactions are different at global or local scales or in short-term or long-term perspectives—but all these spatial and temporal scales are important to define the biophysical constraints of a desirable sustainable future.

If, for example, the boundaries for dangerous consequences of climate change are only addressed at a global scale, their impact on particular regions are neglected as much as the inequalities of certain modes of production or living across and within countries worldwide (as expressed with the term imperial mode of living, [13]). Thus, natural science analyses on the biophysical thresholds must be connected to social science analyses, dealing with both the existing societal demand of resources and the options and implications for desired futures. If the critical interactions between societal and natural processes and structures herein are neglected, a preference for large-scale technical solutions (e.g., for geoengineering as part of the "Good Anthropocene"; [45]) and a rejection of deep-rooted social-ecological transformations might be the political consequence.

Third, there is much ambiguity within the debate concerning the subjects or drivers, objects, scope and pace of such a transformation. Whereas some scholars argue for more technical or social innovations, emerging from niches in an unplanned transition process [46], others argue for an important role of the state [10,14]. But to address the subjects or drivers of transformation more profoundly requires analysing existing interest structures and power relations, hegemonic constellations and the very structure of the state and its steering capacity at various scales.

When it comes to the "objects" of transformation, it is analytically and politically highly relevant to identify what needs to be transformed. At the descriptive level, the global systems considered as overused reference objects include the global climate system, fertile land, soil or the oceans as resource sources or as the deposits of natural elements (prominently, [41,47]). Others point at "drivers" of change such as demographic trends, the globalisation of production, trade and financial markets, resource intensive modes of production and living, urbanisation, industrialisation in countries of the Global South, increasing resource use and prices, technological progress and digitalisation [14,48]. The multi-level perspective [28] focuses on stabilised lock-ins and path dependencies of existing socio-technological systems like energy, transport, housing or agro-food systems which are constituted by technologies, markets, policies, user practices and cultural meanings. This approach assumes that radical innovation emerges in niches and is pushed by dedicated actors. Innovation in niches might become relevant at the regime or even landscape level. By referring to the SRN concept, we argue that a more precise understanding of the objects of transformation is required in order to understand the "sustainability of unsustainability" and the rootedness of unsustainability in everyday practices [11,19,49,50]. It is neither an environmental crisis nor an interplay of multiple societal crises, but the interplay of both: a comprehensive crisis of the interactions between societal and biophysical processes (we will come back to this point in Section 4).

Moreover, there is a tension between a global and long-term transformation process and a plurality of transformation pathways at lower spatial, temporal and social scales marked by a variety of individual and societal actions conflicting to some degree with each other. Despite the globalist view on problems (i.e., planetary boundaries), the implicitly privileged scale of transformation seems to be the regional or national scale (in some cases the EU level). Even the Sustainable Development Goals or the Paris Agreement of the FCCC are to be implemented by nation states or national governments, respectively. This has to do with the existing political and economic conditions and reflects frustration with global governance approaches: beside the processes of the internationalisation of the state [51,52] and the emergence of global binding rules, the national political system is the one where binding decisions are taken and where the financial, legal and knowledge resources prevail. The economic system and the dominant economic groups are partly internationalised, but, at the same time, they stick to the national scale in the sense that the conditions of production, strategic resources, and compromises with workers and trade unions are still linked to the national scale. Recent crisis politics are a clear indicator for this claim. The "own" (national) business and growth remain crucial for politics.

A further tension is related to temporal scales. Most of the debate on transformations towards sustainability is concerned with the urgency of far reaching transformations in the face of accelerating climate change. From the perspective of the global climate system, there is strong evidence that a delay of actions will increase impact and costs and that the window of opportunity to avoid dangerous climate change is closing fast [53]. At the same time, the debate about the Anthropocene opens up the discussion to time scales beyond institutional capacities and human imagination [54]. Moreover, ongoing processes to mitigate climate change (e.g., the quest for technological "solutions" like nuclear power or bioenergy with carbon capture and sequestration, abbreviated BECCS, [55]) may represent a threat for sustainability not only for some further generations but for millions of years (e.g., concerning nuclear waste; [56]). Thus, short term requirements and long-term impacts are conflicting and must be balanced in policy making. However, existing economic and political institutions, focusing mainly on short-term returns in terms of money or political power, are incapable to address such long-term, interwoven and conflicting time scales. Thus, realistic transformation approaches need to balance visionary and future oriented strategies with an analysis of the constraints for future anticipation and long-term policy-making (including scientific uncertainties).

Towards a Critical Concept of Social-Ecological Transformation

To sum up, current research on transformations towards sustainability needs a better understanding of ongoing transformation processes and a better integration of long-term and short-term perspectives as well as on large-scale and regional or local-scale transformation approaches. What is largely missing in the current transformation debate are analyses that focus in more depth on the interactions between globalized societies and the natural environment, analysing resource use patterns and its social implications in terms of global inequalities as much as its impact on global ecosystems without denying local (including everyday), regional and national scales of problems and action (see below). A critical concept of social-ecological transformations points at a better understanding of the social-ecological dimensions of current transformation processes. This includes a better understanding of scale interactions, i.e., global, regional and local processes, and the systemic processes as much as the actor constellations and power relations involved. What is decisive is a better integration of analytical, normative and strategic dimensions of SET research and to focus more systematically on the challenges for shaping interwoven and conflicting transformation processes and their implication for SRN—i.e., a critical concept of SET. In the following, only some elements of such a critical concept of transformations will be discussed, focusing on conceptual achievements and empirical results provided from Social and Political Ecology over the last couple of years. Thus, the paper will not follow a "conventional" structure but it will mix up conceptual considerations with analytical discussions and empirical results.

### 3. Sociometabolic Transitions and Social-Ecological Constraints

It is one of the major achievements of the current debate that it emphasizes the interplay of several dimensions of resource use and environmental problems—from climate change and biodiversity loss and other environmental issues, up to resource use patterns and related socioeconomic crises. As discussed above, the debate could benefit from a better integration of long-term and short-term processes and of different spatial scales. Here, the Viennese tradition of Social Ecology offers to better integrate these dimensions. For several decades, this school of thought focuses on the interactions between societal processes and its biophysical environment. Major concepts elaborated in this regard are the concepts of societal metabolism and sociometabolic regimes [57,58]. Similar to the notion of societal metabolism coined long ago by Karl Marx [59], societies are addressed not as pure communication systems and not as to be separated from their biophysical foundations. At the contrary, the historical perspective focuses on the dynamic pattern of the appropriation and transformation of "nature" and its repercussion on societal change. Several concepts (important in the regard are concepts like the colonisation of nature, i.e., the transformations of natural ecosystems to increase yields; [57], operationalised in the human appropriation of net primary production or HANPP indicator [60], Long-term Socioecological Research [61,62] and Socio-Natural sites [63], and in particular, the concept of metabolic regimes and the transition between different regimes [64]) offers an important dimension of SETs often neglected within transformation research. Focusing in particular on large-scale and long-term implications of energy and resource use, sociometabolic regimes are marked by a certain energy system and related basic technologies, which constrains the option space for societal developments [58]. Hunter-gatherer societies, agrarian societies and industrial societies are the major types of sociometabolic regimes [65]. However, not only are there significant variations within each of these types, most existing societies include some features of several if not all of these regimes. For example, hunting and agriculture go hand in hand in many traditional societies, which are at the same time in contact with industrial society and use some tools and resources "imported" from there. In many parts of the world, the transition process between agrarian and industrial metabolic regimes are still ongoing, and about half of the world population is thought to still live rather in agrarian than in industrial society.

Three major messages can be distilled from research on sociometabolic transitions for the current debate, dealing with biophysical constraints: first, the industrial regime is in itself unstable and thus must be considered as a transformation society, due to its dependency on limited fossil energy sources [66]. The transition from biomass to fossil fuels as a main source of energy did not reduce society's biomass input; indeed, agrarian and industrial societies use roughly the same amount of biomass per capita and year (40–70 GJ/cap/year; [58]. The role of biomass, and hence of land use, changed fundamentally, however. While biomass represented quantitatively almost the entire source of primary energy available for all purposes in agrarian society, it largely served as the basis for food supply and some specific product groups (clothing, timber for construction and furniture, paper, etc.) in industrialized society. The factor 3–5 growth in primary energy supply associated with transitions from agrarian to industrial society was largely met by adding fossil fuels (and later large-scale hydropower, nuclear energy or renewables, which however still only represent relatively minor inputs today) to society's resource base. This allowed to overcome energetic limitations of agrarian societies (which ultimately were always limitations in access to land) and enabled enormous growth in resource use, and concomitantly economic activity and population density during the transition from agrarian to industrial society. However, this created new sustainability problems related both to the availability of resources and to the end-products of their use, e.g., climate change resulting from fossil fuel combustion.

From this perspective, the need for a "great transformation" and the search for a new energy system [61] is nothing astonishing or suddenly emerging. In contrast, the societal ignorance towards a limited resource base and the "limits to growth" needs to be explained. The crisis of SRN are thus strongly related to the societal inability and political unwillingness to take these limitations into account. From a critical perspective, this crisis is not only a crisis of resource use, but an expression of the societal constraints to shape their basic relations with nature, its energy source and the environmental implication of its resource use [29].

Secondly, the debate on biophysical constraints of societal development can be improved by starting from an analysis of sociometabolic regimes. From a social-ecological perspective, constraints are nothing given and fixed in an untouched nature or anchored in a biophysical system as such; there are always interactions between biophysical systems (e.g., ecosystems, but also the global climate system) and human interventions (e.g., the colonisation of ecosystems, but also fossil energies) that create the constraints [67]. The global climate system will probably function also beyond the Holocene—but the repercussions on human societies are increasingly problematic. From the beginning, systemic thresholds and tipping points must be analysed in a way that these interactions are acknowledged. Seen from these interactions of society and nature, it must be asked which characteristics within contemporary societies are responsible for their persistence in an unsustainable development pathway, i.e., their structural characteristics but also its power relations. However, these characteristics are not only anchored in the deep-rooted structural conditions of capitalism, as sometimes argued by ecological Marxists [68]. More relevant are specific modes of economic growth in certain phases of capitalism, in particular within Fordism, where resource use accelerates. Different transition pathways can be analysed within capitalism in more detail to assess the potential for successful interventions and more sustainable development pathways. For example, the "Great Acceleration" after WW2 [43] is marked by totally different resource use patterns at different times in different world regions [69]—and these differences may provide a starting point to define alternative transformation strategies.

Thirdly, the global scale and the long-term perspective is absolutely necessary for any serious analysis of current SETs. Neither the long-term feasibility—and thus biophysical constraints—nor questions of justice and (global) inequalities can be ignored for any transformation towards sustainability that take this term seriously. As conflicts over access and control of natural resources are becoming more prevalent in the future, resource fairness and justice becomes a major topic of SETs, both between and within countries [70]. Several concepts exist to analyse such large scale and long term perspectives, e.g., unequal ecological exchange between countries [69], and the unequal carbon footprint of households within China [71].

Biophysical Constraints of Land Use from a Social-Ecological Perspective

It is one of the characteristics of the current debate that land and land use becomes an increasingly important topic. High expectations on future economic options of land use (e.g., bio-economy and renewable energies) as much as dependencies and constraints are intensively discussed. Thus, land is a good example for the complexity of biophysical constraints and social-ecological interactions. Intuitively it seems obvious that land area represents a biophysical boundary. Its size is well known and largely invariable. Most human activities, among those some that are indispensable for survival, such as food production, require land. The earth's land mass amounts to 149 million km². Further, ~12% of this land area is covered permanently by ice and snow, and hence only ~130 million km² of land is potentially usable. Also, ~75% of the ice-free area are already used for infrastructure, housing, cropping, livestock grazing and forestry, although with widely varying intensity [72,73]. Most of the remaining ~25% is dry, rocky, steep or cold, and hence unproductive. Only the last pristine forests (~5–7% of the ice-free land) represent a reserve of fertile land—but using them would entail huge ecological costs such as carbon or biodiversity losses. Almost all additional production from land will hence entail either land-use competition or intensification of land use [74]. Hence, one might be tempted to think that it should be rather straightforward to define planetary boundaries [41] related to land, for example by calculating the "human appropriation of net primary production" or HANPP, i.e., the fraction of potential biomass productivity of land already used by humans, which has been estimated to amount to ~25% in the year 2000 [74]. However, empirical research has shown that in the past it has been possible to raise land productivity by large margins: In the last century, HANPP roughly doubled, but world population quadrupled and economic output grew 17-fold (but to some degree with negative side effects on biodiversity and regulating services; [75]). The transition from biomass to fossil fuels as society's main source of energy played a big role for this decoupling, as it helped in raising yields, e.g., through synthetic fertilizers, increased harvest indices of main crop plants (e.g., improved corn/shoot ratios of cereals) and almost limitless draught power from diesel-driven tractors [76]. Modelling for 2050 shows that further growth of food, fibre and bioenergy production is possible even without deforestation [44], but of course there exist important costs and trade-offs. For example, moving toward organic agriculture will require larger cropland areas and still provide less animal-product calories than the business-as-usual scenario. Sacrificing yield increases in order to reduce environmental pressures from intensive agriculture will reduce potentials to use land for carbon sequestration or biodiversity conservation, except if consumption in terms of overall volume and the fraction of animal products in diets is reduced accordingly. Thus, future options for SET need to consider option spaces for land use and its implications. Trade-offs and land use conflicts are unavoidable and current struggles on large scale land acquisitions ("landgrabbing") must be addressed properly: the power relations and dominant societal interests involved require a democratisation of SRN [77].

## 4. Political Dimensions of Social-Ecological Transformations

The challenges of shaping SRN are at the heart of a critical concept of SET [19,78]. This unavoidably leads to questions of politics in a wider sense. Concerning the political, there are two blind spots within the debate about sustainability transformations. The first one is a certain equation of "politics" with the state or governments. Despite much talk about "governance" as participation of various stakeholders, governments and the state are seen as the centre of the political. They are addressed as more or less unitary actors, responsible for the dealing with manifold problems.

However, analytical as well as political-strategic approaches to social-ecological transformation need to consider the inherent conflictive character of the dominant and intended alternative forms of the appropriation of nature and related societal nature relations. The conflicts can be tamed, compromises installed and even broad consensus over dominant societal nature relations and dealing with environmental problems can be reached (beyond the scope of this article is the political-ecological insight that historically-specific societal relations to nature, like those during Fordism, and provisioning systems, like auto-mobility as the predominant mode for mobility, can become for a certain period hegemonic [22,51,78]). What is mostly ignored, however, are the root causes of the ecological and the multiple crisis: the specific constellation of powerful economic actors in line with their political allies, capable of imposing their interests in the colonising and valorisation of nature. This is linked to power-driven discourses and the contested construction of the very meaning of ecological problems and crises. The ecological crisis has inherently a bio-physical and material but also a symbolic-discursive dimension [19,52,79].

Against the dominant quest for better cooperation, far-reaching sustainability transformations require conflictive strategies and actions mainly against dominant economic and political actors [51]. Moreover, for far reaching transformations of Northern modes of living, we need new mechanisms to integrate more or less large parts of the population into political processes and new institutions able to question the existing mode of production and living—i.e., a democratisation of political and social life [53]. In contrast, the mentioned "new critical orthodoxy" of sustainability transformations seems to trust very much in existing political and economic institutions and actors.

A second blind spot within the transformation debate is an under-determined understanding of political steering and the state themselves (we are aware of the fact that the "political" is much more comprehensive, including the public, civil society and even the site of production, consumption and the private. However, in the paper we focus on the state as a central instance of the political). A good example is the already mentioned overview by Nalau and Handmer with a specific focus on the interlinkages between sustainability transformations and policies. "Transformation has recently emerged as a suggested approach to manage change in societies given the increasing complexity of policy problems. ... well-planned and facilitated transformation calls for a careful consideration of what exactly needs to be changed and how" [10] (p. 355). The latter part of the quote motivates also our approach. However, we see the management perspective as reductionist when it mainly consists of a call for "new regulatory frameworks" or—in other contributions—a "strong and activating state" [14]. Here, politics is equated with public policies. In the debate on social-ecological transformation, policymakers—and behind them governments or states—are often assumed to be interested in handling collective problems, and hence in creating general welfare.

Instead, we argue, that beside the focus on policies (e.g., certain environmental measures) the very structures of polity (i.e., institutionalised forms of policies) and of politics (i.e., actors and conflicts about structures and political strategies) needs to be transformed towards sustainability. Again, to achieve this we need an adequate understanding of the state and the political.

Historical-materialist social, state and governance theory made important contributions to PE (for historical-materialist state theory in general see [80–84]; for the linking with PE see [37,52,78,85–91]). The analytical challenge is to conceptualise the state not only as a potential motor of sustainability transformations—this is important enough and dealt with in literature on the "green state" [92] or "environmental state" [93,94]. Beyond this it is key to understand how the state is deeply linked to un-sustainable modes of production and living, its links to dominant or even hegemonic social practices and rationalities, values, and discourses and how it became historically and still is crucial in the "generalisation" (Verallgemeinerung) of the fossilist metabolism.

In line with most state theoretical approaches, we understand the state as a specific materialised social institution that creates collectively binding decisions. Moreover, mainly the national state disposes over specific means to exercise a legitimate monopoly of the use of coercion (cf. on the recent debate about the internationalisation of the state: [52,95]. However, and in contrast to many other approaches, historical-materialist state theory considers the state not as a neutral entity, nor to be a mere instrument of capital or dominant social forces, but as a social relation. Therefore, the structures and actions of the state and modes of governance cannot be explained by themselves but rather through the consideration of social forces, practices and discourses, the (changing) societal context as well as the contested functions or tasks of the state in societal reproduction, e.g., the reproduction of existing societal nature relations. The latter implies that the state mainly secures and stabilises existing social relations like the social division of labour (along class, gender, and race, and also internationally); private property of the means of production and the private appropriation of the results of social production and the production of nature. Therefore, the institutional materiality of the state has to be understood against the background of the capitalist mode of production.

When it comes to this institutional materiality, it is a matter of fact that ecologically unsustainable societal structures and processes are deeply rooted in the state apparatus, its personnel and rules, their methods of functioning and their knowledge, and their modes and practices. As Nicos Poulantzas famously put it: the state can be understood as "a specific material condensation of a given relationship of forces" [80] (p. 73). This points at the co-constitutive character of society and state.

Moreover, the relational perspective considers the state as a "strategic field and process of intersecting power networks" [80] (p. 136) where—especially under more or less democratic conditions—different societal and political forces try to promote their interests, norms and values. Social-ecological conflicts are fought out and forces in favour of sustainability transformations act also on this strategic field that is asymmetrically structured and the conflicting actors pursue their strategies in alliances with state personnel and under specific rules and conditions (e.g., as "growth acceleration laws" in times of economic crises or selective environmental laws that don't affect economic interests). In that sense, the state is crucial to deal with manifold societal, economic and political conflicts and to facilitate the creation of consensus through stabilised and shifting relations of forces and compromises with its means of force, law and regulations, discourses and legitimacy, and material and immaterial resources. Hence, the state maps out the multiple terrains of struggle in the relations of production, through labour laws, education processes etc. In that sense, the state is crucial in giving interests and constellations of forces certain durability, in organising compromises and alliances as well as possible hegemony.

Of utmost importance for a political-strategic transformation perspective is the fact that the state as a materialised institution develops contradictions, tensions, and explicit struggles between societal forces—within the power bloc or beyond—it also takes the form of contradictions between different apparatuses and branches [84]. Bob Jessop [96] (p. 364) proposed that societal practices and forces need to be able to develop and pursue hegemonic projects that potentially become state projects. Those projects might create a certain unity of the highly heterogeneous state and its policies.

Dynamics of Resource Extractivism as Powerful Global Social-Ecological Transformations

A relational and political ecology understanding of politics and the state can be clarified by looking at the recent dynamic in Latin America. In the context of the worldwide raw-materials boom in the first decade of the 21st century, the question of the opportunities and limits of raw-materials-based development has moved to the forefront of political and scientific debates. In particular, this issue is being discussed intensively and controversially in Latin America. Development paths based on the production, extraction and export of raw materials and natural products—including agricultural and forestall ones—with the goal of reducing poverty and social inequality by means of enhanced export revenues and their distribution, have been analysed and criticised under the terms "extractivism" and "neo-extractivism" [97–100]; applying the concept to other regions, [77,101]. This was not at all new, but due to the historically unseen rise of prices for raw materials since 2003–2004, governments had an enormous space of action. This becomes obvious in 2017 as we can see in many countries that the downturn of the oil price since 2014 puts the distributive policies of this model in danger.

The so-called pink tide in Latin America with such an emphasis on distribution started when progressive governments came into power. This was expressed through the electoral victories of several left wing presidents since Hugo Chávez in Venezuela in 1998. However, they wanted to go beyond distributional politics. In principle, all governments wanted to reduce the dependency from the world market. Also, in the Andean countries there existed conflicting projects about the dominant and desirable forms of the appropriation of nature. Indigenous struggles and broad anti-neoliberal alliances led in Bolivia and Ecuador to progressive governments and the development of new constitutions. They came into force in 2009 in Bolivia and in 2008 in Ecuador, respectively, and proclaimed a harmonious relationship between society and nature. For the first time in history, the Ecuadorian constitution acknowledges in its article 72 the "rights of nature". However, in recent years their politics resulted in many respects in political frustration of many progressive social forces [100,102–105].

Analyses of the global resource boom are often undertaken from a PE perspective. This is especially the case when it comes to the role of politics and the state. Given the economic problems due to falling resource prices and demand and after the victories of right-wing candidates in countries like Argentina, the parliamentary victory of the right in Venezuela and the impeachment of the progressive Brazilian president, one of the intensively discussed questions in Latin America is: why were the governments not able—and in many cases even not willing—to reduce in an historically exceptional situation the dependency from the world market and foster certain forms of industrialisation and an internal market?

From the outlined political ecology perspective, it is accurate, in such world regions as Latin America, to characterise the state historically as "extractivist state" and currently as "neo-extractivist state". This might elucidate the social, and in fact political, rooting of "extractivist" projects formulated by manifold socio-economic and political actors—especially national and transnational corporations in the mining, fossil fuel and agricultural sector—and secured by international constellations, i.e., investment into resource extraction and demand for natural resources. However, the state is not only the executive instrument of dominant national and international groups or classes interested in resource extraction, but also is prepared to ignore, or, if resistance emerges, suppress other groups or classes. Although it is that, too, often enough; rather, it may also act as a mediator between interests and a "strategic terrain" that is dominated by powerful forces. Therefore, it will not be neutral but privilege certain interests over others [104,106,107]. Particularly, a hegemony-theoretical perspective oriented toward Antonio Gramsci might elucidate the fact that the development model of neo-extractivism has also ingrained itself into the mode of living of wage-earners, especially that of the Latin American urban middle classes. Resource use conflicts are denied, actors who oppose official politics are coopted, ignored or suppressed. A reflection of dominant and problematic resource use patterns were—and still are—not at all part of political debates; critical research is under pressure.

It is important to note that political conjunctures might change for shorter or longer moments social power relations as it was the case after 2000 with a certain political weakening of the bourgeoisie, especially in countries like Bolivia or Ecuador. However, a critical reflection on the state—especially the post-colonial state—helps us to look at the deeply rooted state structures, the bureaucratic links to the bourgeoisie, ongoing dependency from international conjunctures and to avoid a confusion between a change of government with the transformation of the state.

Given the scope of our paper, we can learn from recent developments in Latin America even more. It is obvious that the short-term perspective of governments, the extractivist industry and the beneficiaries of the price boom prevails by large any long-term perspective that is mainly formulated by indigenous peoples, local farmers and their associations—supported by critical intellectuals—who experience the negative impacts of neo-extractivism at first hand. Concerning the multi-scalar character of many dynamics, neo-extractivism shows that the mode of development in particular Latin American countries is linked to the consolidation of resource-intensive modes of production and living in the global North, the economic rise of countries with "emerging markets," and the resulting growing global demand for resources. Moreover, high prices in the raw-materials sector as the basis of neo-extractivism are not only due to any rise in demand, but also to the discovery of their suitability as a field of investment for overaccumulated capital, which might be called the "financialisation" of nature [23]. This means that not only the analysis of patterns of resource use and the processes and structures linked to them need to consider socio-economic and political dynamics elsewhere. This is also the case for the political-strategic and normative orientations at sustainability transformations.

Against governmental discourses and promises, during the boom of resource prices and related state income is was not at all clear whether the neo-extractivist development dynamics lead to a reduction of inequality and poverty at all (cf. [108,109]). However, neo-extractivism became dominant and even hegemonic because distributional politics towards the masses without questioning social structures enables societal compromises and fulfilled the socially dominant imaginary of "progress", but took place at the cost of nature.

## 5. A Transdisciplinary Approach to Social-Ecological Transformations—The Example of the Global Water Crisis

The topic 'water', with all its aspects of management, conservation, use and cross-sectoral linkages is a good example for illustrating why a transdisciplinary approach is central to a critical concept of SET. In this chapter, we start with a description of key characteristics of the global water crisis, its historical development, and endeavours for solving it. Using the concepts of regulation and transformation of SRN, we then take an analytical perspective on this process. We argue here that the ongoing transformation faces a variety of challenges and that an open, creative, and transdisciplinary research process is needed in order to shape this transformation in the sense of SET.

Water is essential for human welfare and healthy ecosystems. Today, however, it is highly under pressure up and beyond critical thresholds, leading to limitations or even breakdowns of entire social-ecological systems [75]. Since the early debates on regional water crises [110,111], a broad agreement on the existence and even aggravation of a global water crisis has developed [112–116]. This global water crisis is characterized by complex problems, which involve the interplay between local and regional scales and their underlying dynamics (ibid.). In particular, these problems refer to the issues of water availability including overuse, water pollution, and access to water. If left untackled, they challenge ongoing endeavours for bringing water, sanitation and food to the people and preserving the integrity of ecosystems. The emergence of such problems is usually not caused by only one, but by multiple interacting factors, which relate to how water is abstracted, made available, allocated, used, and finally released back into the environment.

Historically, the formation of the consensus that insufficient water availability is primarily not based in physical scarcity but in how water is managed was accompanied by the perception of an increasingly globalised crisis. This triggered the development of integrated approaches towards a more holistic management of water. The first regionalised approaches of co-ordinated management were brought on the international agenda at the International Water Conference in Rio del Plata (1977). The Conferences in Dublin and Rio de Janeiro in 1992 then marked milestones for the reanimation of already existing approaches for an Integrated Water Resources Management (IWRM) [117]—their arguably limited success in solving the global water crisis notwithstanding. Legislative processes in Europe (EU-WFD 2000) and normative processes of the United Nations (MDG 2000, SDG 2015) followed. Finally, scientific analyses of the role of water in various conflict situations [118] and of global water flows [119] made it eventually clear that the global dimension of water and in particular the societal impacts on water cannot be ignored anymore.

From the SRN perspective, the global water crisis addresses complex patterns of relations between natural water resources and social actors like households, farmers, enterprises, and suppliers [120]. In the Frankfurt approach to SE these patterns are also referred to as "first-order regulations" or "patterns of regulation" [19]. On a higher level, the manifestations of the crisis mentioned above are embedded in overarching structures and their dynamics: policies, multilateral treaties, and globalised markets are intentionally created second-order regulations, which influence how water is managed. They are defined by power relations, perceptions of rising social inequality, and global change processes like urbanisation and climate change. Such second-order regulations are also described as "modes of regulation" in SE (ibid.).

This shortened description of the development of the global water crisis suggests how signs of a looming crisis can cause intended and unintended transformation processes. IWRM, for example, was developed as a response to the crisis discourse and laid out guidelines for transformations towards a sustainable water management (with a focus on balancing the interdependencies between water and other resources). In the European Union the adoption of the EU Water Framework Directive (EU-WFD) in 2000 carried this concept into environmental legislation—even though the legislative process did not explicitly refer to any crisis discourse. The Millennium and in particular the Sustainable Development Goals (MDG and SDG) have put water also on their agendas. Moreover, the rights to water and sanitation have been recognized as human rights in 2010.

IWRM and the EU-WFD were important triggers for transformations towards sustainable water management because they have a direct impact on how knowledge about solutions is generated and how they are brought into practice. They serve as valuable boundary objects for different actors and disciplines, respectively as an important sectorial legal framework. Nevertheless, partly competing sectors like agriculture, industry, and energy production tried to react to water shortages by adaption and also by structural changes such as building reservoirs and implementing long-distance supply schemes or new technologies for intensified water abstraction. As a consequence, the risk of overuse increased, problems were not solved but shifted, and new conflicts emerged. The close connection between the agricultural and water sectors, for example, led to conflicting developments as demonstrated by the case of subsidies for certain water-intense agricultural products, which counteract measures for more efficient water uses or against water scarcity [121]. Water pricing with environmental levies had unintended effects and the implementation of drip irrigation or other water efficiency technologies in agriculture allowed an extension of irrigated land, causing the desired reduction of agricultural water use to fail [122]. Cities started to privatise their water supply and sanitation for improved economic efficiency but with the risk of higher barriers for interaction with other sectors [123,124]. This can be seen as part of commodification and commercialisation of water use. Often, these transformative actions were not coordinated and were driven by specific, isolated interests without an overall perspective. Thus, a truly transdisciplinary approach to the global water crisis is required, able for developing such a perspective.

The challenge of transformations towards sustainable water management is to move from partial solutions to an integrated and thus more sustainable balance of regulations concerning the competition on water with less unintended side-effects and critical trade-offs. In order to achieve that, we need to look at the issue of regulation of water-related SRN from a historical perspective. Many of today's patterns and modes of regulations developed over centuries and thus incorporate partly obsolete conditions. Therefore, old principles of water regulation—which are still strong drivers for long-term developments and action—may not hold anymore for the 21st century. An example is the shift from the principle 'one water for all purposes' to the more differentiated view of 'water of different quality for different purposes' [125,126] including the understanding of wastewater as a resource. Politicians and the international community are currently joining forces in order to achieve a consensus about this paradigm shift [127]. What becomes apparent is that it opens a space for innovative patterns of regulation. The differentiated management of blue, green and grey water flows along with corresponding alternative technological, infrastructural and socio-economic solutions can serve as an example here (ibid.). Furthermore, Goal 6 of the SDG addresses water prominently, has strong cross-linkages to several other Goals like food, health, energy, cities, climate change and biodiversity, and relates also to water and sanitation as human rights. The SDG call on developed and developing countries to take integrated action and apply holistic thinking depending on their specific needs. These developments pave the ground for a social-ecological transformation of water management. They are, however, not the answer for everything as the issue of groundwater shows: Groundwater is neglected in the SDG despite its huge importance for water supply and its critical condition in many regions worldwide. All these things considered, sustainable water management approaches will show a higher complexity than previous ones because of the close coupling of scales and sectors. Cities, for example, are important drivers of development and in case future urban water supply and sanitation take the shift of paradigms seriously, infrastructures, resource flows and their governance will be transformed within the cities and in relation to the hinterland. Finally, relevant stakeholders have to become much more part of feedback processes in water management. This leads to new participatory structures, which are more adaptive to ongoing dynamics.

Shaping the transformations towards a sustainable water management is an open and creative process. This means that, in addition to empirical studies, new knowledge needs to be produced by differentiating, critically assessing and (re-)integrating what we already know. The combination of well-known technologies for innovative solutions in the case of greywater use and circular water economy is an example for this. The knowledge to be integrated is plentiful: natural science models and data from geohydrology, for example, have to be combined with an improved understanding of water governance regimes and institutions, innovative technological options and their requirements, and socio-economic and cultural practices of direct and indirect water users. Heuristic and analytical approaches like the concept of social-ecological systems (SES) as complex, adaptive systems [19,128] and a theoretical and conceptual foundation of 'regulation' [129] allow for the analytical decomposition of problems and the innovative (re-)composition of solutions. They also help to make new ideas like the water-energy-food nexus (FAO 2014) analytically accessible and to open up a participatory research agenda, which extends the knowledge base by incorporating extra-scientific knowledge [130].

Transdisciplinary research, which includes interdisciplinary cooperation, is able to capture the complex and multi-dimensional character of the global water crisis. This mode of research aims to produce the three types of knowledge [131] that are necessary to link the analytical, political-strategic, and normative dimensions of social-ecological transformations: system knowledge for a better understanding of the structures and processes that fuel the global water crisis, orientation knowledge about the requirements and standards of a future sustainable water management, and transformation knowledge on how the process towards a sustainable water management can be shaped. In doing so, transdisciplinary research adopts a (self)reflexive and (self)critical attitude: it routinely scrutinises its own procedures, methods, and practices of knowledge production as well as the different new roles for scientists that come along with it [131,132].

## 6. Conclusions and Outlook: A Critical Approach towards Social-Ecological Transformations

This paper argues that the current debate on transformations towards sustainability can be improved by a critical, inter- and transdisciplinary approach to social-ecological transformations, based on conceptual and empirical achievements from SE and PE. As shown above, several analytical challenges can be addressed building on existing work. An integrative perspective that aligns analytical, normative and strategic dimensions is at the centre of our proposal. For sure, this integrated perspective can only be developed within a wider research community, due to the broad array of competences required.

First, a focus on resource use patterns and their implications on ecosystems (especially biodiversity) and food, biomass production and water are important elements in this regard. It can be demonstrated that current societies are in themselves unstable and crisis driven "societies in transformation" and that it is a historically open question whether existing resource use patterns can be re-regulated in a sustainable way.

Second, resource use patterns evolve over long periods and are based on as well as stabilisers of power relations and hegemonic constellations. We argue that a relational perspective is the differentia specifica for a critical concept of transformation. According to Marx, modes of production correspond to the 'relations of production—relations which human beings enter into during the process of social life, in the creation of their social life' [133]. The concept focuses on structures and processes by means of which society organises its material foundations (i.e., its metabolism with nature), socioeconomically, politically, culturally, and subjectively. It identifies dominant societal structures and processes and their necessarily contradictory and crisis driven reproduction. In that sense, resource use patterns might become hegemonic. However, in times of crisis or catastrophes or at particular scales—often the local one where those who are negatively affected by certain patterns live—existing SRN can be contested and shaped. The degree of shaping—a smooth modernisation or a profound transformation—depends on power relations and strategies, on feasible alternatives and biophysical constraints.

Third, inter- and transdisciplinary research had developed conceptual approaches and empirical methods to address the complex dynamics of SRN. However, the challenges of SETs are going further and require additional conceptual and empirical work to contribute to an improved strategical approach for sustainability transformations. Therefore, a critical understanding of current transformations towards unsustainability must be developed, that allows for improved strategies towards sustainability.

Fourth, this critical approach requires a consideration of the interplay of change and persistence, critical developments, ruptures and discontinuities, instead of simply linear developments. A major challenge is the interplay of several overlapping transformation processes, both intended and unintended, that requires an analytical perspective comprehensive enough to address the societal context (including the barriers for SET) and the side effects of certain processes while not neglecting the case specifics. It should also give a better picture of the potentials and obstacles of initiatives and proposals for sustainability transformations, i.e., the political-strategic and normative dimension.

Fifth, from a social-ecological and political-ecological perspective, SET always occurs: SRN are always regulated and only temporarily stabilized and the modes of regulations create the causes for further transformations. This is not a minor statement because it makes the respective analyses sensitive for the fact that the dominant tendencies or "grammars" of transformation need to be reflected [15]. Thus, from our perspective the question is not whether SRN will transform but what dominant tendencies or "grammars" stand behind such transformations. We assume three strong tendencies ("grammars") that structure the industrial and fossilist mode of production and regulation of SRN: one such a grammar is the colonising of nature or land taking, a tendency that exists throughout history and is shaped by societal power relations and domination. "Nature" is increasingly shaped by human activities, whereas global societies are increasingly affected by repercussions and crises tendencies, but in an unequal way in the Global South and the Global North. Beyond the pure economic rationality of capitalism, this grammar is deeply anchored within the dominant dualistic European pattern of understanding and its belief in the domination of nature [134] (cf. [135]). Therefore, it requires a plurality of worldviews and knowledge types, open for alternatives to the dominant modes of regulation of SRN.

The second is the capitalist grammar of capital accumulation, the growth imperative and the predominance of the production of surplus values over the production of use values. The former goes hand in hand with the valorisation and overexploitation of nature (and the work force) and forms certain modes of regulation [19]. In that sense, political economy is "political ecology's bread-and-butter" [90] (p. 343) and crucial for our approach, too. This does not mean that all societal relations are structured along capitalist imperatives; they might co-exist with subsistent, solidary and cooperative forms of production and living. Moreover, non-paid care work is decisive to organise the material and symbolic reproduction of societies and, hence, the appropriation of nature [136]. In fact, different modes of regulations do not merely co-exist and are often co-constitutive but also conflicting with each other [51]. Thus, the destructive "logic" of the valorisation of nature for capital accumulation is not without alternatives. In fact, these tendencies are contested, counter-tendencies can evolve—and do evolve—due to severe crises and social struggles of actors that intend to impede the destructive tendencies of dominant regulations of societal relations to nature. We assume that such struggles are decisive for alternative resource patterns and that a closer analysis of such resource use patterns in space and time may provide a starting point to define alternative transformation strategies (see above Section 3)—but this of course needs further investigation. Moreover, broader analytical lenses are required to understand societal and socio-ecological dynamics, e.g., a more comprehensive understanding of the economy and societal reproduction that goes beyond the formal market and money economy and towards considerations of non-monetarised forms of production and labour and related societal nature relations [136].

The third tendency or "grammar" is related to our strictly multi-scalar perspective that does not lose out of sight the global. We argued that—despite the relevance of the local level and in particular local struggles—the claims for sustainability transformations are, in principle, global but in fact they refer mainly to national or regional scales. The national scale, however, is dominant due to the density of the national political systems (compared to the international one) and the dominance of strategies of competitiveness that are mainly pursued at the national scale. This is the basis that capitalist development occurs unevenly, both in space and in time. It also installs a powerful mechanism or tendency to externalise the negative preconditions and consequences of production and consumption to other regions. Biesecker and Hofmeister [136] call this a constant and absent "shadow of externalisation" that drives societal dynamics in certain regions, makes life more attractive at the cost of the living conditions in other regions. These complex processes of externalisation are a cornerstone of constant social-ecological transformations and need to be considered and changed by any project of sustainability transformations.

**Acknowledgments:** Funding by the Austrian Science Funds (FWF), project No. P29130-G27, is gratefully acknowledged. This research contributes to the Global Land Programme (http://www.futureearth.org/projects/glp-global-land-programme). We also thank two anonym reviewers for their comments and suggestions.

**Author Contributions:** Christoph Görg and Ulrich Brand as lead authors contributed equally to the paper, the other authors are listed in alphabetical order. The whole paper was drafted and all sections were commented by all authors. Major contributions are in the following order: Ulrich Brand and Christoph Görg have written Sections 1, 2, 4 and 6, Helmut Haberl and Christoph Görg have written Section 3 and Diana Hummel, Thomas Jahn and Stefan Liehr have written Section 5.

**Conflicts of Interest:** The authors declare no conflict of interest.

## References

1. Hackmann, H.; St. Clair, A.L. *Transformative Cornerstones of Social Science Research for Global Change*; ISSC: Paris, France, 2012.
2. JPI CLIMATE. *Strategic Research Agenda for the Joint Programming Initiative "Connecting Climate Knowledge for Europe"*; JPI CLIMATE: Helsinki, Finland, 2011.
3. Westley, F.; Olsson, P.; Folke, C.; Homer-Dixon, T.; Vredenburg, H.; Loorbach, D.; Thompson, J.; Nilsson, M.; Lambin, E.; Sendzimir, J.; et al. Tipping Toward Sustainability. Emerging Pathways of Transformation. *AMBIO* **2011**, *40*, 762–780. [CrossRef] [PubMed]
4. Leach, M.; Rockström, J.; Raskin, P.; Scoones, I.; Stirling, A.C.; Smith, A.; Thompson, J.; Millstone, E.; Ely, A.; Arond, E.; et al. Transforming Innovation for Sustainability. *Ecol. Soc.* **2012**, *17*, 11. [CrossRef]
5. O'Brien, K. Global environmental change II: From adaptation to deliberate transformation. *Prog. Hum. Geogr.* **2012**, *36*, 667–676. [CrossRef]
6. Frantzeskaki, N.; Loorbach, D.; Meadowcroft, J. Governing societal transitions to sustainability. *Int. J. Sustain. Dev.* **2012**, *15*, 19–36. [CrossRef]
7. Muradian, R.; Walter, M.; Martinez-Alier, J. Hegemonic transitions and global shifts in social metabolism: Implications for resource-rich countries. Introduction to the special section. *Glob. Environ. Chang.* **2012**, *22*, 559–567. [CrossRef]
8. Brie, M. *Futuring. Perspektiven der Transformation im Kapitalismus über ihn Hinaus*; Brie, M., Ed.; Westfälisches Dampfboot: Münster, Germany, 2014.
9. Fischer-Kowalski, M.; Hausknost, D. *Large Scale Societal Transitions in the Past*; Social Ecology Working Paper No 152; Institute of Social Ecology: Vienna, Austria, 2014.
10. Nalau, J.; Handmer, J. When is transformation a viable policy alternative? *Environ. Sci. Policy* **2015**, *54*, 349–356. [CrossRef]
11. Jonas, M. Transition or Transformation? A Plea for the Praxeological Approach of Radical Societal Change. In *Praxeological Political Analysis*; Jonas, M., Littig, B., Eds.; Routledge: Oxford, UK; New York, NY, USA, 2017; pp. 116–133.
12. Brand, U. How to get out of the multiple crisis? Towards a critical theory of social-ecological transformation. *Environ. Values* **2016**, *25*, 503–525. [CrossRef]
13. Brand, U.; Wissen, M. Social-Ecological Transformation. In *International Encyclopedia of Geography*; Castree, N., Goodchild, M.F., Kobayashi, A., Liu, W., Marston, R.A., Eds.; Wiley-Blackwell: Hoboken, NJ, USA, 2017.
14. WBGU. *World in Transition. A Social Contract for Sustainability*; WBGU: Berlin, Germany, 2011.
15. Brand, U. "Transformation" as a New Critical Orthodoxy: The Strategic Use of the Term "Transformation" does not Prevent Multiple Crises. *GAIA Ecol. Perspect. Sci. Soc.* **2016**, *25*, 23–27. [CrossRef]
16. Pelling, M. *Adaptation to Climate Change: From Resilience to Transformation*; Routledge: London, UK, 2010.

17. Jahn, T. Theory of Sustainability? Considerations on a Basic Understanding of "Sustainability Science". In *Theories of Sustainable Development*; Enders, J.C., Remig, M., Eds.; Routledge: London, UK; New York, NY, USA, 2015; pp. 30–42.

18. Becker, E.; Jahn, T. *Soziale Ökologie. Grundzüge einer Wissenschaft von den Gesellschaftlichen Naturverhältnissen*; Becker, E., Jahn, T., Eds.; Institut für Sozial-Ökologische Forschung Campus Verlag: Frankfurt, Germany; New York, NY, USA, 2006.

19. Hummel, D.; Jahn, T.; Keil, F.; Liehr, S.; Stieß, I. Social Ecology as Transdisciplinary Science—Conceptualizing, Analyzing, and Shaping Societal Relations to Nature. *Sustainability* **2017**. submitted.

20. Fischer-Kowalski, M.; Weisz, H. Society as hybrid between material and symbolic realms. *Adv. Hum. Ecol.* **1999**, *8*, 215–251.

21. Haberl, H.; Fischer-Kowalski, M.; Krausmann, F.; Winiwarter, V. *Social Ecology. Society-Nature Relations across Time and Space*; Haberl, H., Fischer-Kowalski, M., Krausmann, F., Winiwarter, V., Eds.; Human-Environment Interactions; Springer: Dordrecht, The Netherlands, 2016.

22. Brand, U.; Görg, C. Regimes in Global Environmental Governance and the Internationalization of the State: The Case of Biodiversity Politics. *Int. J. Soc. Sci. Stud.* **2013**, *1*, 110–122. [CrossRef]

23. Brand, U.; Wissen, M. Financialisation of Nature as Crisis Strategy. *J. Entwicklungspolitik* **2014**, *30*, 16–45. [CrossRef]

24. Kramm, J.; Schaffartzik, A.; Pichler, M.; Zimmermann, M. Social ecology—State of the art and future prospects. *Sustainability* **2017**. submitted.

25. Carpenter, S.R.; Brock, W.A. Rising variance: A leading indicator of ecological transition: Variance and ecological transition. *Ecol. Lett.* **2006**, *9*, 311–318. [CrossRef] [PubMed]

26. Bennett, J.W. *The Ecological Transition: Cultural Anthropology and Human Adaptation*; Pergamon Press: New York, NY, USA, 1976.

27. Loorbach, D. Transition Management for Sustainable Development: A Prescriptive, Complexity-Based Governance Framework. *Governance* **2010**, *23*, 161–183. [CrossRef]

28. Geels, F.W. Ontologies, socio-technical transitions (to sustainability), and the multi-level perspective. *Res. Policy* **2010**, *39*, 495–510. [CrossRef]

29. Görg, C. Shaping Relationships with Nature—Adaptation to Climate Change as a Challenge for Society. *ERDE J. Geogr. Soc. Berl.* **2011**, *142*, 411–428.

30. Chomsky, N. The multiple crises of neoliberal capitalism and the need for a global working class response. *Int. Soc. Rev.* **2016**, *101*. Available online: http://isreview.org/issue/101/multiple-crises-neoliberal-capitalism-and-need-global-working-class-response (accessed on 8 June 2017).

31. Haas, W. Method précis: Transdisciplinary Research. In *Social Ecology, Society-Nature Relations across Time and Space*; Springer: Dordrecht, The Netherlands, 2016; pp. 555–557.

32. Weisz, H.; Haas, W. Health through Socioecological Lenses. In *Social Ecology, Society-Nature Relations across Time and Space*; Springer: Dordrecht, The Netherlands, 2016; pp. 559–576.

33. Petridis, P.; Fischer-Kowalski, M. Island Sustainability: The Case of Samothraki. In *Social Ecology, Society-Nature Relations across Time and Space*; Springer: Dordrecht, The Netherlands, 2016; pp. 543–554.

34. Bergmann, M.; Jahn, T.; Knobloch, T.; Krohn, W.; Pohl, C.; Schramm, E. *Methods for Transdisciplinary Research. A Primer for Practice*; Campus Verlag: Frankfurt, Germany; New York, NY, USA, 2012.

35. Jahn, T.; Keil, F. An actor-specific guideline for quality assurance in transdisciplinary research. *Futures* **2015**, *65*, 195–208. [CrossRef]

36. Brand, U.; Brunnengräber, A.; Andresen, S.; Driessen, P.; Haberl, H.; Hausknost, D.; Helgenberger, S.; Hollaender, K.; Læssøe, J.; Oberthür, S.; et al. Debating transformation in multiple crises. In *World Social Science Report 2013: Changing Global Environments*; United Nations Educational, Scientific and Cultural Organization; Organisation for Economic Co-operation and Development; International Social Science Council: Paris, France, 2013; pp. 480–484.

37. Park, J.; Conca, K.; Finger, M. *The Crisis of Global Environmental Governance: Towards a New Political Economy of Sustainability*; Park, J., Conca, K., Finger, M., Eds.; Taylor & Francis: London, UK; New York, NY, USA, 2008.

38. Newell, P. *Globalization and the Environment: Capitalism, Ecology and Power*; John Wiley & Sons: Hoboken, NJ, USA, 2013.

39. Bruff, I. The Rise of Authoritarian Neoliberalism. *Rethink. Marx.* **2014**, *26*, 113–129. [CrossRef]

40. Oberndorfer, L. From new constitutionalism to authoritarian constitutionalism—New Economic Governance and the state of European democracy. In *Asymmetric Crisis in Europe and Possible Futures*; Jäger, J., Springler, E., Eds.; Routledge: Milton Park, UK; New York, NY, USA, 2015; pp. 186–207.

41. Rockström, J.; Steffen, W.; Noone, K.; Persson, Å.; Chapin, F.S.I.; Lambin, E.; Lenton, T.; Scheffer, M.; Folke, C.; Schellnhuber, H.J.; et al. Planetary Boundaries: Exploring the Safe Operating Space for Humanity. *Ecol. Soc.* **2009**, 14. [CrossRef]

42. Crutzen, P.J. Geology of mankind. *Nature* **2002**, *415*, 23. [CrossRef] [PubMed]

43. Steffen, W.; Broadgate, W.; Deutsch, L.; Gaffney, O.; Ludwig, C. The trajectory of the Anthropocene: The Great Acceleration. *Anthr. Rev.* **2015**, *2*, 81–98. [CrossRef]

44. Erb, K.-H.; Lauk, C.; Kastner, T.; Mayer, A.; Theurl, M.C.; Haberl, H. Exploring the biophysical option space for feeding the world without deforestation. *Nat. Commun.* **2016**, *7*, 11382. [CrossRef] [PubMed]

45. Stirling, A. Time to Rei(g)n Back the Anthropocene? Available online: http://steps-centre.org/2015/blog/time-to-reign-back-the-anthropocene/ (accessed on 3 February 2017).

46. Grin, J.; Rotmans, J.; Schot, J. *Transitions to Sustainable Development: New Directions in the Study of Long Term Transformative Change*; Routledge: New York, NY, USA, 2010.

47. United Nations Environment Programme (UNEP). *Towards a Green Economy: Pathways to Sustainable Development and Poverty Eradication*; United Nations Environment Programme, Ed.; UNEP: Nairobi, Kenya, 2011.

48. German Bundestag. *Final Report of the Expert Commission on Growth, Well-Being and Quality of Life*; German Bundestag: Berlin, Germany, 2013.

49. Blühdorn, I. Post-capitalism, post-growth, post-consumerism? Eco-political hopes beyond sustainability. *Glob. Discourse* **2017**, *7*, 42–61. [CrossRef]

50. Schultz, I.; Stieß, I. Linking Sustainable Consumption to Everyday Life. A social-ecological approach to consumption research. In *Proceedings: Perspectives on Radical Changes to Sustainable Consumption and Production (SCP)*; Tukker, A., Charter, M., Vezzoli, C., Eds.; Greenleaf Publishing Ltd.: Sheffield, UK, 2008; pp. 288–300.

51. Brand, U.; Görg, C.; Hirsch, J.; Wissen, M. *Conflicts in Environmental Regulation and the Internationalisation of the State: Contested Terrains*; Routledge: London, UK; New York, NY, USA, 2010.

52. Brand, U.; Görg, C.; Wissen, M. Second-Order Condensations of Societal Power Relations: Environmental Politics and the Internationalization of the State from a Neo-Poulantzian Perspective. *Antipode* **2011**, *43*, 149–175. [CrossRef]

53. Intergovernmental Panel on Climate Change (IPCC). *Climate Change 2013: The Physical Science Basis. Summary for Policymakers*; Intergovernmental Panel on Climate Change: Geneva, Switzerland, 2013.

54. Görg, C. Zwischen Tagesgeschäft und Erdgeschichte: Die unterschiedlichen Zeitskalen in der Debatte um das Anthropozän. *GAIA Ecol. Perspect. Sci. Soc.* **2016**, *25*, 9–13. [CrossRef]

55. Gasser, T.; Guivarch, C.; Tachiiri, K.; Jones, C.D.; Ciais, P. Negative emissions physically needed to keep global warming below 2 °C. *Nat. Commun.* **2015**, *6*, 7958. [CrossRef] [PubMed]

56. Brunnengräber, A.; Görg, C. Nuclear waste in the Anthropocene. Uncertainties and unforeseeable time scales in the disposal of nuclear waste. *GAIA Ecol. Perspect. Sci. Soc.* **2017**, *26*, 96–99.

57. Fischer-Kowalski, M.; Erb, K.-H. Core Concepts and Heuristics. In *Social Ecology: Society-Nature Relations across Time and Space*; Haberl, H., Fischer-Kowalski, M., Krausmann, F., Winiwarter, V., Eds.; Springer International Publishing: New York, NY, USA, 2016; pp. 29–61.

58. Krausmann, F. Transitions in Sociometabolic Regimes throughout Human History. In *Social Ecology: Society-Nature Relations across Time and Space*; Haberl, H., Fischer-Kowalski, M., Krausmann, F., Winiwarter, V., Eds.; Springer International Publishing: New York, NY, USA, 2016; pp. 63–92.

59. Marx, K. Das Kapital Bd. 1. In *Marx-Engels-Werke 23*; quoted from the German edition; Dietz: Berlin, Germany; International Publishers: New York, NY, USA, 1996.

60. Haberl, H.; Erb, K.-H.; Krausmann, F. Human Appropriation of Net Primary Production: Patterns, Trends, and Planetary Boundaries. *Annu. Rev. Environ. Resour.* **2014**, *39*, 363–391. [CrossRef]

61. Haberl, H.; Fischer-Kowalski, M.; Krausmann, F.; Martinez-Alier, J.; Winiwarter, V. A socio-metabolic transition towards sustainability? Challenges for another Great Transformation. *Sustain. Dev.* **2011**, *19*, 1–14. [CrossRef]

62. Singh, S.J.; Haberl, H.; Chertow, M.; Mirtl, M.; Schmid, M. *Long Term Socio-Ecological Research*; Springer: Dordrecht, The Netherlands; Heidelberg, Germany; London, UK; New York, NY, USA, 2013.

63. Winiwarter, V.; Schmid, M.; Dressel, G. Looking at half a millennium of co-existence: The Danube in Vienna as a socio-natural site. *Water Hist.* **2013**, *5*, 101–119. [CrossRef]

64. Fischer-Kowalski, M.; Rotmans, J. Conceptualizing, Observing, and Influencing Social–Ecological Transitions. *Ecol. Soc.* **2009**, *14*. [CrossRef]

65. Fischer-Kowalski, M.; Haberl, H. Metabolism and colonization. Modes of production and the physical exchange between societies and nature. *Innov. Eur. J. Soc. Sci. Res.* **1993**, *6*, 415–442. [CrossRef]

66. Sieferle, R.P. *Das Ende der Fläche: Zum Gesellschaftlichen Stoffwechsel der Industrialisierung*; Böhlau: Köln, Germany; Weimar, Germany; Wien, Austria, 2006.

67. Haberl, H.; Erb, K.-H. Land as a planetary boundary—A socioecological perspective. In *Handbook of Growth and Sustainability*; Victor, P., Dolter, B., Eds.; Edward Elgar: Cheltenham, UK, 2017.

68. Moore, J. *The Capitalocene—Part II: Abstract Social Nature and the Limits to Capital*; Fernand Braudel Center and Department of Sociology, Binghamton University: Binghamton, NY, USA, 2014.

69. Schaffartzik, A.; Wiedenhofer, D.; Eisenmenger, N. Raw Material Equivalents: The Challenges of Accounting for Sustainability in a Globalized World. *Sustainability* **2015**, *7*, 5345–5370. [CrossRef]

70. Pichler, M.; Staritz, C.; Küblböck, K.; Plank, C.; Raza, W.; Peyré, F.R. *Fairness and Justice in Natural Resource Politics*; Routledge: New York, NY, USA, 2016.

71. Wiedenhofer, D.; Guan, D.; Liu, Z.; Meng, J.; Zhang, N.; Wei, Y.-M. Unequal household carbon footprints in China. *Nat. Clim. Chang.* **2016**, *7*, 75–80. [CrossRef]

72. Erb, K.-H.; Gaube, V.; Krausmann, F.; Plutzar, C.; Bondeau, A.; Haberl, H. A comprehensive global 5 min resolution land-use data set for the year 2000 consistent with national census data. *J. Land Use Sci.* **2007**, *2*, 191–224. [CrossRef]

73. Ellis, E.C.; Kaplan, J.O.; Fuller, D.Q.; Vavrus, S.; Goldewijk, K.K.; Verburg, P.H. Used planet: A global history. *Proc. Natl. Acad. Sci. USA* **2013**, *110*, 7978–7985. [CrossRef] [PubMed]

74. Haberl, H. Competition for land: A sociometabolic perspective. *Ecol. Econ.* **2015**, *119*, 424–431. [CrossRef]

75. Millenium Ecosystem Assessment (MEA). *Ecosystems and Human Well-Being: Desertification Synthesis*; World Resources Institute: Washington, DC, USA, 2005.

76. Krausmann, F. Milk, Manure, and Muscle Power. Livestock and the Transformation of Preindustrial Agriculture in Central Europe. *Hum. Ecol.* **2004**, *32*, 735–772. [CrossRef]

77. Pichler, M. "People, Planet & Profit": Consumer-Oriented Hegemony and Power Relations in Palm Oil and Agrofuel Certification. *J. Environ. Dev.* **2014**, *22*, 370–390.

78. Görg, C. *Regulation der Naturverhältnisse*; Westfälisches Dampfboot: Münster, Germany, 2003.

79. Adger, W.N.; Benjaminsen, T.A.; Brown, K.; Svarstad, H. Advancing a Political Ecology of Global Environmental Discourses. *Dev. Chang.* **2001**, *32*, 681–715. [CrossRef]

80. Poulantzas, N. *State, Power, Socialism*; NLB: London, UK, 1980.

81. Jessop, B. *State Power. A Strategic-Relational Approach*; Polity Press: Cambridge, UK, 2007.

82. Sauer, B.; Wöhl, S. Feminist Perspectives on the Internationalization of the State. *Antipode* **2011**, *43*, 108–128. [CrossRef]

83. Gallas, A.; Bretthauer, L.; Kannankulam, J.; Stützle, I. *Reading Poulantzas*; Gallas, A., Bretthauer, L., Kannankulam, J., Stützle, I., Eds.; Merlin Press: London, UK, 2011.

84. Brand, U.; Görg, C. Post-Fordist governance of nature. The internationalization of the state and the case of genetic resources: A Neo-Poulantzian Perspective. *Rev. Int. Polit. Econ.* **2008**, *15*, 567–589. [CrossRef]

85. Peet, R.; Watts, M. *Liberation Ecologies: Environment, Development, Social Movements*, 2nd ed.; Peet, R., Watts, M., Eds.; Routledge: London, UK; New York, NY, USA, 2004.

86. Robinson, W.I. *Latin America and Global Capitalism: A Critical Globalization Perspective*; John Hopkins University Press: Baltimore, MD, USA, 2008.

87. Robbins, P. *Political Ecology*; Blackwell: Malden, MA, USA; Oxford, UK; Carlton, Australia, 2004.

88. Newell, P. The political economy of global environmental governance. *Rev. Int. Stud.* **2008**, *34*, 507–529. [CrossRef]

89. Forsyth, T. *Critical Political Ecology: The Politics of Environmental Science*; Routledge: London, UK; New York, NY, USA, 2003.

90. Mann, G. Should political ecology be Marxist? A case for Gramsci's historical materialism. *Geoforum* **2009**, *40*, 335–344. [CrossRef]

91. Wissen, M. *Gesellschaftliche Naturverhältnisse in der Internationalisierung des Staates*; Westfälisches Dampfboot: Münster, Germany, 2011.

92. Bäckstrand, K.; Kronsell, A. *Rethinking the Green State: Environmental Governance Towards Climate and Sustainability Transitions*; Bäckstrand, K., Kronsell, A., Eds.; Routledge: Abingdon, UK, 2015.

93. Duit, A.; Feindt, P.H.; Meadowcroft, J. Greening Leviathan: The rise of the environmental state? *Environ. Politics* **2016**, *25*, 1–23. [CrossRef]

94. Sommerer, T.; Lim, S. The environmental state as a model for the world? An analysis of policy repertoires in 37 countries. *Environ. Politics* **2016**, *25*, 92–115. [CrossRef]

95. Brand, U.; Wissen, M. Strategies of a green economy, contours of a green capitalism. In *The International Political Economy of Production*; van der Pijl, K., Ed.; Edward Elgar: Cheltenham, UK, 2015; pp. 508–523.

96. Jessop, B. *State Theory: Putting the Capitalist State in Its Place*; Polity Press: Cambridge, UK, 1990.

97. Gudynas, E. Alcances y contenidos de las transiciones al Post-Extractivismo. *Ecuad. Debate* **2011**, *82*, 61–79.

98. Svampa, M. Resource Extractivism and Alternatives: Latin American Perspectives on Development. *J. Entwicklungspolitik* **2012**, *28*, 43–73. [CrossRef]

99. Burchardt, H.-J.; Dietz, K. (Neo-)extractivism—A new challenge for development theory from Latin America. *Third World Q.* **2014**, *35*, 468–486. [CrossRef]

100. Brand, U.; Dietz, K.; Lang, M. Neo-Extractivism in Latin America—One side of a new phase of global capitalist dynamics. *Rev. Cienc. Política Bogotá* **2016**, *11*, 125–159. [CrossRef]

101. Bridge, G. Global production networks and the extractive sector: Governing resource-based development. *J. Econ. Geogr.* **2008**, *8*, 389–419. [CrossRef]

102. Tapia, L. *El Estado de Derecho Como Tiranía*; CIDES-UMSA: La Paz, Bolivia, 2011.

103. Lander, E. The State in the Current Processes of Change in Latin America. *J. Entwicklungspolitik* **2012**, *28*, 74–94. [CrossRef]

104. Alcoreza, R.P. *Cartografías Histórico-Políticas. Extractivismo, Dependencia y Colonialidad. Dinámicas Moleculares*; Wordpress: La Paz, Bolivia, 2014.

105. Alcoreza, R.P. *El Conservadurismo de los Gobiernos Progresistas*; Wordpress: La Paz, Bolivia, 2015.

106. Rey, M.T. *Estado y Marxismo. Un Siglo y Medio de Debates*; Prometeo Libros Editorial: Buenos Aires, Argentina, 2007.

107. Camacho, O.V. Paths for Good Living: The Bolivian Constitutional Process. *J. Entwicklungspolitik* **2012**, *28*, 95–117. [CrossRef]

108. Lander, E.; Arze, C.; Gómez, J.; Ospina, P.; Álvarez, V. *Promesas en su Laberinto: Cambios y Continuidades en los Gobiernos Progresistas de América Latina*; CEDLA; IEE; CIM: La Paz, Bolivia; Quito, Ecuador; Caracas, Venezuela, 2013.

109. Veltmeyer, H. The Political Economy of Natural Resource Extraction: A New Model or Extractive Imperialism? *Can. J. Dev. Stud.* **2013**, *34*, 79–95. [CrossRef]

110. Warne, W.E. Water crisis is present. *Nat. Resour. J.* **1969**, *9*, 53–62.

111. Adams, M.E. Nile Water: A Crisis Postponed? Hydropolitics of the Nile Valley. *Econ. Dev. Cult. Chang.* **1983**, *31*, 639–643. [CrossRef]

112. Biswas, A.K. Deafness to global water crisis: Causes and risks. *Ambio* **1998**, *27*, 492–493.

113. Famiglietti, J.S. The global groundwater crisis. *Nat. Clim. Chang.* **2014**, *4*, 945–948. [CrossRef]

114. Boelens, R.A.; Crow, B.; Dill, B.; Lu, F.; Ocampo-Rader, C.; Zwarteveen, M.Z. Santa Cruz Declaration on the Global Water Crisis. *Water Int.* **2014**, *39*, 246–261. [CrossRef]

115. Srinivasan, V.; Lambin, E.F.; Gorelick, S.M.; Thompson, B.H.; Rozelle, S. The nature and causes of the global water crisis: Syndromes from a meta-analysis of coupled human-water studies. *Water Resour. Res.* **2012**, *48*, W10516. [CrossRef]

116. United Nations Development Programme (UNDP). *Human Development Report 2006—Beyond Scarcity: Power, Poverty and the Global Water Crisis*; United Nations Development Programme: New York, NY, USA, 2006.

117. Global Water Partnership (GWP). *Integrated Water Resources Management*; Technical Report No. 4; Technology Advisory Committee: Stockholm, Sweden, 2000.

118. Gleick, P.H. Water and Conflict: Fresh Water Resources and International Security. *Int. Secur.* **1993**, *18*, 79–112. [CrossRef]

119. Hoekstra, A.Y. *The Global Dimension of Water Governance: Nine Reasons for Global Arrangements in order to Cope with Local Water Problems*; UNESCO-IHE Institute for Water Education: Delft, The Netherlands, 2006.

120. Kluge, T. Integrated Water Resources Management (IWRM) as a Key for Sustainable Development. In *The Economics of Global Environmental Change. International Cooperation for Sustainability*; Cogoy, M., Steininger, K.W., Eds.; Edward Elgar Publishing: Celtenham, UK; Northampton, MA, USA, 2007; pp. 134–154.

121. Hummel, D.; Kluge, T.; Liehr, S. Globale Handelsströme Virtuellen Wassers. Eins—Entwicklungspolitik Information Nord-Süd, Dossier Virtuelles Wasser. *Fachzeitschrift* **2006**, *22*, VII–VIII.

122. Schramm, E.; Sattary, E. *Scenarios for Closed Basin Water Management in the Zayandeh Rud Catchment*; ISOE-Materialien Soziale Ökologie; ISOE—Institute for Social-Ecological Research: Frankfurt-Main, Germany, 2014.

123. Barraqué, B. *Urban Water Conflicts: UNESCO-IHP*; Barraqué, B., Ed.; Urban Water Series; Taylor & Francis: Paris, France, 2011.

124. Wissen, M.; Naumann, M. A New Logic of Infrastructure Supply: The Commercialization of Water and the Transformation of Urban Governance in Germany. *Soc. Justice* **2006**, *33*, 20–37.

125. Kluge, T. Umgang mit Wasser als Hydraulischer Maschinerie. Kollektive Hintergrundvorstellungen zur Mechanisierung des Wassers. In *Wasser (Anlässlich des Internationalen Kongresses 1998)*; Wienand: Köln, Germany, 2000.

126. Wilderer, P.A.; Schreff, D. Decentralized and centralized wastewater management: A challenge for technology developers. *Water Sci. Technol.* **2000**, *41*, 1–8.

127. Kluge, T. Wasserwiederverwendung—Eine Schlüsseltechnologie zwischen technischer und sozialer Innovation. In *Mit Abwasserbehandlung Zukunft Gestalten: 88. Darmstädter Seminar Abwassertechnik*; Verein zur Förderung des Instituts IWAR der TU Darmstadt, Ed.; Schriftenreihe WAR: Darmstadt, Germany, 2016; pp. 43–54.

128. Liehr, S.; Röhrig, J.; Mehring, M.; Kluge, T. Addressing Water Challenges in Central Northern Namibia: How the Social-Ecological Systems Concept Can Guide Research and Implementation. *Sustainability* **2017**, submitted.

129. Kluge, T.; Liehr, S.; Schramm, E. *Strukturveränderungen und Neue Verfahren in der Ressourcenregulation*; ISOE-Diskussionspapiere; ISOE-Institut für Sozial-Ökologische Forschung: Frankfurt am Main, Germany, 2007; Volume 27.

130. Winker, M.; Schramm, E.; Schulz, O.; Zimmermann, M.; Liehr, S. Integrated water research and how it can help address the challenges faced by Germany's water sector. *Environ. Earth Sci.* **2016**, *75*, 1226. [CrossRef]

131. Jahn, T.; Bergmann, M.; Keil, F. Transdisciplinarity: Between mainstreaming and marginalization. *Ecol. Econ.* **2012**, *79*, 1–10. [CrossRef]

132. Jahn, T. Wissenschaft für eine nachhaltige Entwicklung braucht eine kritische OrientierungSustainability Science Requires a Critical Orientation. *GAIA Ecol. Perspect. Sci. Soc.* **2013**, *22*, 29–33. [CrossRef]

133. Marx, K. Das Kapital Bd. 3. In *Marx-Engels-Werke 25*, 3rd ed.; Quoted from the German edition; Dietz: Berlin, Germany; International Publishers: New York, NY, USA, 1998.

134. Horkheimer, M.; Adorno, T.W. Dialektik der Aufklärung. In *Gesammelte Schriften*; Horkheimer, M., Ed.; Fischer: Frankfurt am Main, Germany, 1987; Volume 5.

135. Görg, C. Societal Relationships with Nature: A Dialectical Approach to Environmental Politics. In *Critical Ecologies. The Frankfurt School and Contemporary Environmental Crises*; Biro, A., Ed.; University of Toronto Press: Toronto, ON, Canada, 2011; pp. 43–72.

136. Biesecker, A.; Hofmeister, S. (Re)productivity: Sustainable relations both between society and nature and between the genders. *Ecol. Econ.* **2010**, *69*, 1703–1711. [CrossRef]

*sustainability*

MDPI

*Article*

# How the Social-Ecological Systems Concept Can Guide Transdisciplinary Research and Implementation: Addressing Water Challenges in Central Northern Namibia

Stefan Liehr [1,2,*], Julia Röhrig [3], Marion Mehring [1,2] and Thomas Kluge [1]

[1] ISOE—Institute for Social-Ecological Research, Hamburger Allee 45, 60486 Frankfurt am Main, Germany; mehring@isoe.de (M.M.); kluge@isoe.de (T.K.)
[2] Senckenberg Biodiversity and Climate Research Centre (BiK-F), Senckenberganlage 25, 60325 Frankfurt am Main, Germany
[3] German Aerospace Center (DLR), Linder Höhe, 51147 Cologne, Germany; julia.roehrig@dlr.de
* Correspondence: liehr@isoe.de; Tel.: +49-69-7076919-36

Received: 6 February 2017; Accepted: 16 June 2017; Published: 26 June 2017

**Abstract:** Research aimed at contributing to the further development of integrated water resources management needs to tackle complex challenges at the interface of nature and society. A case study in the Cuvelai-Etosha Basin in Namibia has shown how semi-arid conditions coinciding with high population density and urbanisation present a risk to people's livelihoods and ecosystem health. In order to increase water security and promote sustainable water management, there is a requirement for problem-oriented research approaches combined with a new way of thinking about water in order to generate evidence-based, adapted solutions. Transdisciplinary research in particular addresses this issue by focusing on the problems that arise when society interacts with nature. This article presents the implementation of a transdisciplinary research approach in the above-mentioned case study. The concept of social-ecological systems (SES) plays a key role in operationalising the transdisciplinary research process. Application of the SES concept helps to outline the problem by defining the epistemic object, as well as structure the research process itself in terms of formulating research questions and developing the research design. It is argued here that the SES concept is not merely useful, but also necessary for guiding transdisciplinary sustainability research and implementation. The study from Namibia clearly demonstrates that the introduction of technological innovations such as rainwater and floodwater harvesting plants requires a social-ecological perspective. In particular this means considering questions around knowledge, practices and institutions related to water resources management and includes various societal innovations alongside technologies on the agenda.

**Keywords:** Cuvelai-Etosha Basin; savannah ecosystems; ecosystem services; integrated water resources management; rainwater and floodwater harvesting; social-ecological systems; transdisciplinary research

## 1. Introduction

Many semi-arid regions in the world are facing problems such as water scarcity, infertile land, the impacts of climate change, population growth, urbanisation processes and inadequate water supply and sanitation [1–4]. Water-related problems are manifold and complex in how they are manifested in different sectors and at different levels in society [5]. In order to improve water security and address inequality and injustice, there needs to be improved understanding of the key societal and ecological elements and drivers of these complex systems as well as their interrelations, and their (self-regulatory) capacity needs to be strengthened [5]. It is therefore a challenge to develop evidence-based solutions that increase the resilience of these systems [6] and pave the way for broader replication in future in the region and beyond.

Fundamental research projects and straightforward development cooperation are pushed to their limits in the face of these challenges. The answer is to shift the focus towards linked 'research and development' approaches, often referred to as 'research for development' [7]. This is particularly the case if potential solutions and their implementation are linked to innovations such as specific technological developments and new forms of management strategies and practices. Suitable approaches need to fulfil specific requirements such as the integration of different disciplines and the linking of science with society in order to allow an analysis of the complex interplay between social and ecological processes and structures and offer feasible solutions for sustainable resource management [8,9].

Transdisciplinary research is able to meet this challenge. In particular it addresses problems with societal relevance at the science-practice interface and therefore necessitates the involvement of stakeholders and relevant scientific disciplines [8,10]. The concept of social-ecological systems (SES) is very well suited to transdisciplinary research for sustainability because these systems can be understood as concrete epistemic objects in the real world of spatio-temporal phenomena [11]. These phenomena are viewed and conceptualised differently by various authors, e.g., as 'human-nature relations' by Glaser [11,12] or as 'societal relations to nature' by Becker [13] and Hummel et al. [14]. SES support the structuring of real-world problems on a clear conceptual foundation, thereby taking into account the aforementioned interplay between society and nature which is a typical issue that can be addressed by transdisciplinary approaches. When applied to case studies, SES are usually understood to be provisioning systems, with "provisioning" being used in a wide sense to include any material or even symbolic benefit that society can draw from natural resources [14].

In this paper, the implementation of rainwater and floodwater harvesting (RFWH) in the Cuvelai-Etosha Basin (CEB) in central northern Namibia served as a case study for applying the SES concept in a transdisciplinary research setting. The case study is part of a larger transdisciplinary research and development project designed to last for around ten years. The CEB is a typical example of semi-arid conditions in which there is limited availability and inefficient use of water [9,15]. Thus, the growing population is exposed to physical water scarcity [15] which is aggravated by inadequate socio-economic capacities and challenged by climate change [16]. The aim of implementing RFWH is to improve water availability and thus enhance the ability to buffer water fluctuations. Furthermore, RFWH is designed to increase food self-sufficiency as a cross-sectoral impact. Taken together, RFWH has considerable potential to contribute to increased sustainability and the enhanced wellbeing of people in the region [17,18].

The exceptional duration of the project and its funding framework along with water harvesting as a highly relevant strategy for improving water security offered the ideal prerequisites for a systemised reflection on the implementation of all the steps in the transdisciplinary research approach. It offered new insights from practice into how transdisciplinarity and the SES concept are interlinked, and the consequences of this on the formulation of a research question and the design of research tasks on the ground.

This article starts with a presentation of the case study and the project setting, then outlines the conceptual and methodological approach of transdisciplinarity and the SES concept, before showing how this shapes and performs the implementation of RFWH. The applicability and limits of this approach are discussed and conclusions presented from a decade of project work.

## 2. The Namibian Case Study

### 2.1. Case Study Area

The CEB is the southern part of the transboundary Cuvelai catchment, which is shared by Angola and Namibia (Figure 1). The CEB covers an area of 34,723 km$^2$, representing 4.2% of the national territory. The climate in the CEB is classified as semi-arid, with one rainy season between October and March. Average annual rainfall varies between 200 mm in the southwest and 600 mm in the northeast [15]. The annual evaporation rates of approximately 2600 mm are extremely high [19]. Very variable rainfall and contrasting extremes result in either too much or too little water [9]. The climatic and topographic conditions in the CEB mean that the region lacks permanent water sources. Instead, the basin is characterised by a system of ephemeral rivers and small water courses (known as oshanas) that drain water from Angola into the Etosha Salt Pan in the south. The pan is located within the Etosha National Park, and water flow is crucial for the park's ecosystem and its high biodiversity. As the topography is flat and rainfall events are intensive, the basin is regularly affected by floods. The water quality of the oshana water degrades rapidly after the rainy season due to the high rate of evaporation and the uncontrolled use of water by humans and animals. Climate change is likely to pose additional challenges to securing sufficient water for ecosystems and people [20,21].

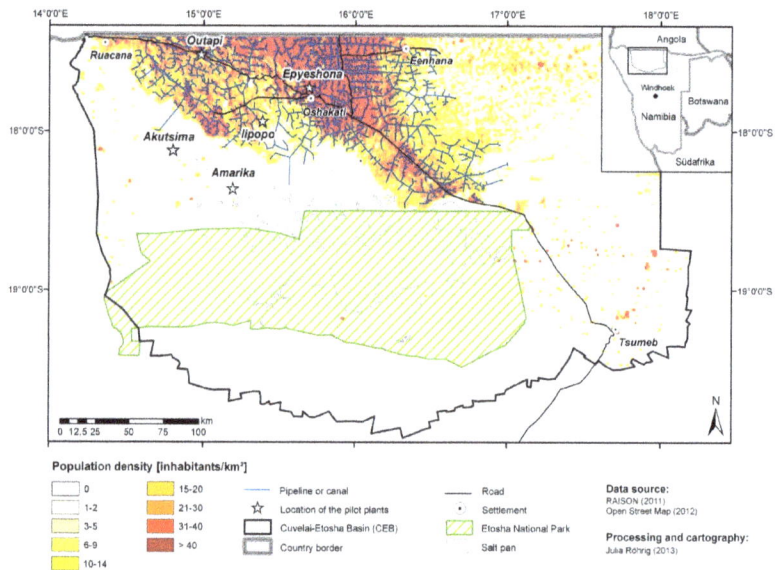

**Figure 1.** The study area in central northern Namibia and the location of CuveWaters pilot plants for all implemented technologies. The pilot sites for rainwater and floodwater harvesting are at Epyeshona and Iipopo.

Water availability has influenced settlement activities, land use and livelihood strategies for decades. The basin is the most densely populated rural area in Namibia and has around 850,000 inhabitants [15], which is almost half the population. These circumstances complicate the essential task of meeting basic needs for livelihood security, such as water and food supply. Thus precarious living conditions for the population and a sensitive dependency on the dynamics of natural conditions coincide with an ecosystem characterised by exploitation and scarcity.

The technological solution for supplying drinking water to the communities is a long-distance canal and a pipeline system fed by the Kunene River, the border river between Angola and Namibia. This transboundary and interbasinal water transfer is one of the largest supply networks in Africa [19,22]. Where the supply network ends, people depend on water from hand-dug wells or seasonal surface water and rainfall for seasonal agriculture, livestock farming and household consumption [23,24]. Apart from this long-distance infrastructure, water availability is unreliable and its quality degrades rapidly, especially during the dry season. Rainwater and floodwater harvesting for domestic use does not systematically occur in the CEB. In the 1950s and 1960s, pump-storage dams were constructed for floodwater harvesting, but high evaporation and inadequate soil conditions brought these attempts to an end [25,26].

In the basin, livestock farming is traditionally the predominant source of livelihood. It is complemented by rain-fed agriculture, primarily based on pearl millet (known locally as "mahangu") and fishing during the wet season [15,27]. Rainfall is not typically stored in the project region, therefore no crops are planted during the dry season, which means the productivity level is fairly low. Food security in the basin is therefore highly dependent on food imports from South Africa. These common practices and a low level of education mean that there is limited knowledge about the hydrological cycle, water quality or irrigation farming. In contrast, there is wide-ranging and valuable indigenous knowledge about disaster prediction and coping mechanisms based on close observation of animals, vegetation, weather and celestial bodies.

After the country's independence, the Namibian government began to draw up reforms to improve knowledge and practices with the aim of sustainably managing the scarce water and land resources under the paradigm of Integrated Water Resources Management (IWRM) [16,28]. This process has resulted in various changes to the institutional landscape. New organisations have been set up at different levels as part of the IWRM implementation strategy and decentralisation process. One example is the Basin Management Committees (BMCs), which represent all the stakeholders in their respective basins or sub-basins. The establishment of BMCs began in 2003, but despite considerable demand to strengthen the involvement of stakeholders, the creation of further BMCs is somewhat behind schedule [28]. While key water policies are in harmony with IWRM principles, their enshrinement in law is long overdue, which has weakened regulatory practice, reduced institutional empowerment [28] and hampered reforms aimed at including the local population in the management and financing of the water supply. As for food supply, local food production is regarded as critical for food self-sufficiency and security, and large-scale irrigation projects are being promoted to improve the production level and generate independence from imports [29,30].

### 2.2. The CuveWaters Project

From 2004 to 2015, the international research and development project "CuveWaters: Integrated Water Resources Management in Central Northern Namibia (Cuvelai Basin) in the SADC-Region" investigated how adapted technology-based solutions could contribute to IWRM in the CEB. A transdisciplinary research approach was applied based on the integration of science, technology and society [31]. This approach is reflected in the project design and structure: its empirical and technical components are closely linked to integrative societal elements, providing the basis for adaptive problem-solving since interlinkage promotes social embedding of the technologies along with the active involvement of the institutional players and local population.

The funding programme of the German Federal Ministry for Education and Research (BMBF) comprised three main project phases and a preceding six-month exploratory project phase. Each transition to the next phase was accompanied by an evaluation of the previous results and a proposition for the future project concept.

During the exploratory project phase of CuveWaters, key stakeholders in Namibia and technology partners in Germany were identified. The common goal was to establish a multi-resource mix for water use that was designed "to improve the living conditions of people in the project region" [31] (p. 689). The underlying methodological approach and implementation process are described in more detail in [9]. 'Multi-resource mix' means that water from different sources is made available as a result of adapted technological solutions, and then used as drinking or irrigation water depending on its quality. Three technological solutions were identified with the Namibian partners and marked the start of the multi-resource mix: rainwater and floodwater harvesting, and groundwater desalination along with sanitation and water reuse.

After the exploratory project phase, its implementation began in 2006 in three main phases—the initial, pilot and diffusion phases—and was completed by the end of 2015 [31]. The first and second main project phases were an interdisciplinary endeavour involving stakeholders such as communities and administrations. Using what is known as a demand-responsive approach [32], the technological design was jointly developed, appropriate sites were identified and pilot plants were implemented. The third main project phase focused on handing over ownership, disseminating acquired knowledge and evaluating it in terms of its contribution to societal and scientific progress. This provided the basis for safeguarding implementation on the ground for a period beyond the end of the project. However scientists, industry partners and policy makers were also addressed as target groups in order to scale up efforts and transfer the technologies to other regions.

## 3. Conceptual and Methodological Background

### 3.1. The Transdisciplinary Research Approach

Transdisciplinary research approaches are increasingly being adopted in research on resource management problems because they appear to be very well suited to addressing the complexity of pressing problems [8,9,33,34]. In the CuveWaters project, the approach of Jahn et al. [8] was applied. This approach shows a clear theoretical foundation [14] and exhibits a well developed implementation process based on a history of project experiences [8]. According to [8], three consecutive phases can be distinguished in an ideal transdisciplinary research process:

- Phase 1: formation of a common research object (problem transformation)
- Phase 2: production of new knowledge (interdisciplinary integration)
- Phase 3: transdisciplinary integration (evaluation of new knowledge for its contribution to societal and scientific progress).

Phase 1 comprises the sequence of two transformations, from the societal problem to a boundary object and then to an epistemic object from which research questions can be derived. Boundary objects have a loose structure and aim "to accommodate individual perspectives and meanings while at the same time maintaining an identity that is recognized by all parties involved" [8] (p. 5). In contrast, epistemic objects are much more structured and have a well-defined meaning and use [35]. SES are typically examples of an epistemic object [8]. In phase 2, the roles of researchers and stakeholders are clarified and an integration concept for the research designed and implemented. Disciplinary action and interdisciplinarity are crucial in this process and are understood as integral parts of transdisciplinarity. If an SES has been formulated in phase 1, this can guide and structure research design and implementation. Finally, phase 3 concerns the assessment of integrated results and provides products for science and society. The focus is therefore on contributions to both societal and scientific progress. All three phases were applied in the present case study.

The three transdisciplinary phases correspond to the project phases of the case study but are not strictly identical. The exploratory project phase relates to phase 1 of the transdisciplinary process, the first and second main project phases relate to different stages of succession in phase 2, while the third and final main project phase relates to phase 3, with a smooth transition between these phases.

### 3.2. The Social-Ecological Systems (SES) Concept

The SES concept formalises the relationships between nature and society [12]. On this basis, it can serve as a starting point for operationalising transdisciplinary research by creating a link to research practice [11,14]. The SES concept is presented as part of the Frankfurt Social Ecology and the results section below uses the case study to demonstrate how this concept can guide and structure a transdisciplinary project.

In recent decades, the SES concept has become increasingly important in the international discourse on sustainability science [35–39]. It is used to describe, analyse and model human-nature interactions. In the discourse on the SES concept, both the Stockholm-based Resilience Alliance and the work of Elinor Ostrom have played a leading role. Research by the Resilience Alliance mainly focuses on the adaptive management of ecosystems from a resilience perspective [37,40]. Ostrom and her colleagues conceive of SES as a general framework with which to analyse institutions and governance systems, and then apply this framework in particular to the area of common-pool resources [41] or questions of the systems' robustness [42]. However, more recent studies link and integrate the concepts of SES and ecosystem services (ESS) [12,43–46]. This development allows mutual reinforcement when it comes to systematically conceptualising their benefits to society and reveals the underlying structures and processes that drive it. The various types of services within the ESS concept also extend the potential to interlink different areas of critical human-nature interactions, such as water, food and energy supply.

In this article, the latter discourse is referred to in the integration of the SES and ESS concepts, and the conceptual work published by [12,43,47,48]. The above human-nature interactions are interpreted as 'societal relations to nature' [49,50], in accordance with the formal research programme of the Frankfurt Social Ecology [14]. SES are understood to be the translation of the theoretical framework into research practice [35]. As part of the Frankfurt Social Ecology research programme, the concept of SES [48] was interlinked with the concept of ESS back in 2008, and 'management' was explicitly integrated to address the societal influence on nature. Both management and ESS form a dynamic regulatory cycle of complex interactions between actors and ecosystem functions (Figure 2). Together they allow an integrated analysis of how ecosystem-based benefits support society under changing conditions such as modified management strategies, but also climate change or biodiversity loss, as a function of the underlying research question.

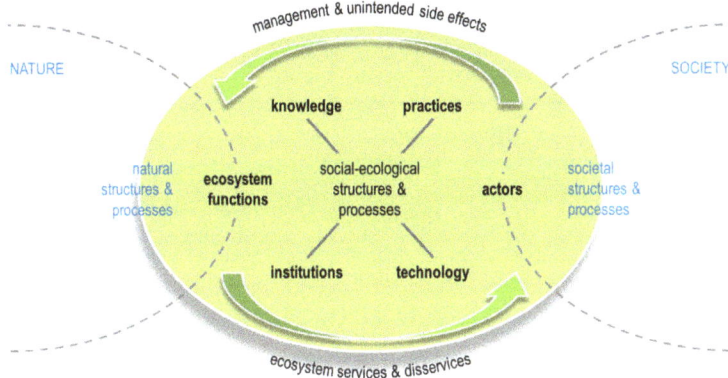

**Figure 2.** The concept of social-ecological systems (SES) [12].

SES (green circle) are defined as hybrid and emergent systems that are nested in both society and nature. In contrast to other concepts [39], SES here are not understood as a mere additive overlap of the natural and societal sphere. Although the relevant natural and societal structures and processes are part of SES, an additional hybrid social-ecological sphere is at the centre of SES [43,48]. This hybrid sphere is interpreted as an emergent outcome of highly nested interactions between various components over space and time; it shows its own specific social-ecological structures and processes which cannot be attributed to the natural or societal sphere or merely to the area in which they overlap. Nonetheless, SES also draw on those societal (social, economic, cultural and political) and natural (biological, geological, chemical and physical) structures and processes that are relevant for the considered focus of research. The general nonlinearity of SES, along with the multitude of interacting structural components and feedback processes, gives rise to their characterisation as complex, adaptive systems.

'Social-ecological structures and processes' determine the analytical core of SES. Examples are processes such as seeding, irrigation, harvesting, ploughing and extracting water, or structures such as gardens and waterholes. A further analysis of these structures and processes is offered in early works by Hummel et al. [48]. They identify knowledge, practices, institutions and technology as mediating dimensions (also called contextual factors) between society and nature. These dimensions are referred to here and their specification refined with regard to a clearer emphasis of their hybrid character. In this context, 'knowledge' comprises all forms of understanding by scientists and non-scientists referring to nature-society interrelations. Societal 'practices' represent routinised types and patterns of behaviour regarding the use of ESS in their material or symbolic relevance. The term 'institutions' addresses the fields of economy, politics, law and culture that create a regulatory system of rules for action with regard to the use of ESS. 'Technology' refers to all man-made material structures and developments that are designed and applied in order to interfere with ecosystem functions with the aim of making specific services available for use [43,47] (modified).

The key components of SES are actors and ecosystem functions. 'Actors' are defined here as persons and groups of persons who influence management and, subsequently, ecosystem functions with their actions, or who are affected by changes in ecosystem functions, services or disservices. The influencing and receiving quality of the interaction can vary for each actor. Actors can come from certain functional groups of society such as consumers, traders, farmers, resource managers and policy makers with respective interests and motivations. In addition, scientists can also be actors if it is assumed that the knowledge produced by research interferes with the activities of the other actors. 'Ecosystem functions' meanwhile are defined as the capacity of natural processes and components to provide goods and services that satisfy human needs, directly and indirectly [51]. They are a generic part of the natural dynamics and, as mentioned above, are subject to changes caused by societal management actions.

Finally, 'ecosystem services' comprise the benefits to society derived from ecosystems [51–53]. 'Ecosystem disservices' are interpreted as the opposite of ESS in the sense that they cause harm to society (e.g., financial or health risks and food insecurity). They are therefore perceived as negative factors for human wellbeing [54]. Actors influence the system either directly or indirectly via intended 'management' activities or their 'unintended side effects'.

## 4. Results—Transdisciplinary Implementation of Water Harvesting from an SES Perspective

This section refers to the technology of rainwater and floodwater harvesting to illustrate the practical implementation of the above-mentioned transdisciplinary research process. Here, the SES concept is applied in its heuristic and analytical functions to structure both the problem and the research process.

*4.1. Phase 1: Formation of a Common Research Object (Problem Transformation)*

4.1.1. RFWH as the Boundary Object

The exploratory project phase presented the first opportunity to develop and strengthen a network of contacts with key Namibian stakeholders. Discussions took place with representatives from the Namibian Ministry for Agriculture, Water and Forestry (MAWF), the Desert Research Foundation of Namibia (DRFN), Oshakati Town Council, Oshikoto Regional Council, potential users and other experts in Namibia and Germany. These discussions led to the identification and consolidation of rainwater harvesting and sub-surface water storage at household and community level as technologies of a so-called multi-resource mix. At the end of the first main phase of the project, sub-surface water storage became floodwater harvesting: modifications to the technical concept necessitated a change in name and henceforth only the term 'floodwater harvesting' will be used. Since the management and use of rainwater and floodwater have very similar characteristics, they were treated and analysed together. Important arguments for the joint agreement on RFWH as the common object of work were: (i) the potential to make use of an unused water resource, (ii) its low complexity, (iii) the combination with gardening and thus its contribution to food self-sufficiency and income generation, and finally (iv) the high interest of potential users.

At this stage, RFWH enabled cooperation between the heterogeneous group of actors and therefore could be interpreted as the boundary object of subsequent research.

### 4.1.2. The SES 'Small-Scale Food Production System' as the Epistemic Object

At the early stage of the exploratory project phase, RFWH included plans for the use of the harvested water for gardening. Therefore, RFWH can be understood as a small-scale system within an overarching system for coupled water and food provision. The aim of introducing harvesting and storage of rainwater and floodwater for all-season gardening was to create a completely new opportunity for the productive and efficient use of local water and land resources. However, it was not intended for RFWH to substitute other forms of supply, but rather to generate an additional source of decentralised production and income. This technology was not widely available in the region at the time and was to be implemented with local material and capacities. Considerable potential for widespread implementation and diffusion was therefore expected.

The application of the SES concept, which commenced in the exploratory project phase and was subsequently continued, embedded the principal technological idea of RFWH (the boundary object) into a broader and structured context. The systems perspective on the coupling of water and land management practised in RFWH shifted the focus to the ecosystem service of 'food' and triggered an investigation from an integrative perspective (Figure 3). The SES concept allowed a stronger, more structured epistemic object to be created.

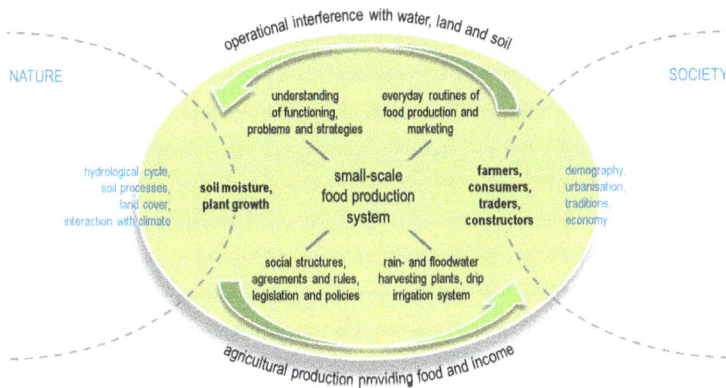

**Figure 3.** Application of the concept of social-ecological systems (SES) to the given case study of rainwater and floodwater harvesting as part of a small-scale food production system. At this level, SES serve as an epistemic object for research and can be understood as provisioning systems.

From a systems perspective, household farmers are primary 'actors'. Furthermore, food consumers, traders and construction workers also play a central role in generating the demand for locally produced food, as well as for income and labour. The key 'ecosystem functions' are water storage in the soil and primary plant production. They depend on numerous processes that regulate nutrient and water fluxes and determine biomass production. Actors and ecosystem functions are interlinked in a feedback loop. The primary demand for food drives water and land 'management', which in turn influences ecosystem functions and generates agricultural products as the key 'ecosystem service'. Consequences of management interventions include changes in the natural structures and processes such as the hydrological cycle (e.g., water buffering and reduction in run-off), soil processes (e.g., moisture and nutrient distribution) and land cover (e.g., cultivation of crops and land consumption of harvesting and storage components). Potentially 'unintended side effects' of management with possible adverse effects on ecology could be the introduction of new vegetable plants as invasive species in the region, the application of pesticides that are harmful to human or ecosystem health, soil leaching and reduced groundwater recharge due to water retention. Beyond the scope of this study, there are other long-term side effects of intensive and widespread land management, such as changes to local weather systems or landslides. Subsequent benefits of the key 'food' ecosystem service are primarily nutrition, health, income and labour. Managing the ecosystem and receiving its services further generates cultural benefits with regard to knowledge acquisition and environmental education, as well as the adaptation and enhancement of cultural farming traditions. 'Ecosystem disservices' may occur if, for example, new pests are introduced and distributed as a result of gardening activities.

The four mediating dimensions will be explained and exemplified below with regard to their role in the design of an integrated concept for research.

### 4.1.3. Defining the Research Question

Understanding RFWH as a boundary object and applying the SES concept helped define the epistemic object and identify the research question 'How do the two harvesting technologies need to be adapted to and embedded within the complex social-ecological context in order to provide the basis for the diffusion of a sustainable and complementary source of water, food and income?' This pivotal research question entailed further research questions such as: What are the particular needs of people when it comes to designing rainwater and floodwater harvesting plants? What forms of governance are appropriate in socio-cultural and institutional settings? What suitable ways might there be to develop the capacities of people and institutions? What economic perspectives exist based on cost-benefit considerations and options for financing? What are the potential adverse ecological effects?

### 4.2. Phase 2: Production of New Knowledge (Interdisciplinary Integration)

#### 4.2.1. Clarification of the Roles of Researchers and Stakeholders

Before starting the actual work, it was important to clarify the respective roles and responsibilities of the researchers and stakeholders. The role of researchers was to put the transdisciplinary approach into practice on the basis of a sound methodology. To achieve this goal, they had to secure the participation of stakeholders in the research, planning and implementation phases. This demanded a high level of integration without compromising the efforts of the various disciplines. Stakeholders performed the role of knowledge holders, experts and supporters. They had the task of ensuring that RFWH was implemented taking into consideration relevant policies and questions of governance, everyday practices and restrictions, and socio-economic and cultural conditions and constraints. In an advanced phase of the project, ownership of the pilot plants had to be adopted by a Namibian stakeholder. A lengthy process preceded the assignment of this very specific role. During this process, it was jointly decided that the communities themselves, or more precisely the traditional authority represented by the headman, would take over responsibility for the facilities. The Agricultural Extension Services of the MAWF were assigned the task of ensuring that the users received assistance. Finally, the northern campus of the University of Namibia (UNAM) in Ongwediva took over the role as the centre for capacity development and technical studies on RFWH.

#### 4.2.2. Design of an Integration Concept for Research

As a result of phase 1, which involved applying the SES concept in order to define the epistemic object, it became evident that the implementation of reliable rainwater and floodwater harvesting 'technology' had to include knowledge, practices and institutions in order to guarantee sustainability (Figures 2 and 3).

Although water harvesting is a relatively simple technology, those building it require specific 'knowledge' (know-how), as do the farmers who operate and maintain the facilities, including the gardens, in order to generate yield and profit. For instance, they need to know the optimal times for sowing, harvesting and selling. The respective actors have to understand how the plant is structured and how it functions, as well as its interactions with external factors. The actors need to possess specific problem-solving skills, which means that after understanding the problem, they are able to weigh up solution strategies and alternatives, and act on the basis of the decisions made.

Knowledge paves the way for new 'practices' and is therefore an integral part of stakeholders' actions and everyday routines. At the same time, practices contribute to long-term experiences and the deeper understanding of system dynamics. It becomes evident that practices predominantly address the question of 'how', while knowledge addresses the question of 'why'. Practices mainly cover the areas of water collection and storage, gardening and marketing. These areas of activity are right at the interface of society and nature, and therefore notably exhibit their immanent hybrid character: they are triggered by societal demands for benefits and their respective ESS, they are strongly influenced by social, economic and institutional considerations, and they have an impact on the water cycle, soil structure and land cover.

'Institutions' are relevant in order to sustainably embed implementation, safeguard the results and initiate early and effective diffusion processes within a favourable environment [55]. When selecting suitable locations for and designing such facilities, the active involvement of local and regional stakeholders and their institutional background is essential to meet local demand and ensure all-round commitment. Particularly in community-orientated approaches, institutions play a decisive role in safeguarding compliance with norms and rules, securing adherence to legislation and policies, or establishing suitable communication strategies to resolve conflicts. In the Namibian case of RFWH, the traditional authorities represented by the headmen took ownership of the communal harvesting plants. Farmers' groups and traditional authorities signed a mutual agreement that set out the rules and responsibilities. On the ministerial side, the Extension Officers of the Directorate of Agricultural Production, Engineering and Extension Services (DAPEES) at the Ministry of Agriculture, Water and Forestry (MAWF) had the task of providing regular assistance on farming and financing issues, including possible subsidies, and training opportunities. IWRM-related institutions such as the BMC were still in the development phase as far as the development of structures and capacities was concerned. Those institutions were involved but not yet able to play a strong role in RFWH. Instead, other non-IWRM-related local institutions such as the Rural Development Centres (RDC), Constituency Councils and Agriculture Development Centres (ADC) came to the fore, acting as promoters, supporters and sponsors since RFWH in combination with gardening serves their interests with respect to community development. On an overarching regional and national level, institutions needed to be involved in the process in order to develop standards and funding models to further support scaling up the processes. They were also required to embed the new technological concepts in strategies and policies for water, agriculture and food production, climate change adaptation and questions of property rights.

This had consequences on the design of the research process with respect to the areas of core activities. All the four mediating dimensions explained above needed to be addressed and integrated. They therefore constituted four areas of activities for research. Furthermore, transdisciplinarity with the methodologically guided involvement of stakeholders in a participation process was chosen to constitute a fifth core activity. Finally, the performance of empirical studies with respect to different, more discipline-based tasks was an additional core activity in the project. Thus, the following six core activities structured the research of CuveWaters:

- technological development
- knowledge management
- capacity development
- governance and institutionalisation
- participation

- empirical studies.

These core activities were strongly interlinked since participation, for example, contributed to all the other activities. The structure in Table 1 represents this interlinking.

4.2.3. Implementation of the Research Process

The core activities comprised a rich repertoire of methods and tasks (Table 1). Here, integration became evident since the core activities not only referred to their own corresponding mediating dimension within the SES concept, but also contributed to the other dimensions. This is the core structural element of integration in the research process.

**Table 1.** Integration concept for research and repertoire methods and tasks showing how the core activities of the transdisciplinary research process contributed to the mediating dimensions of a sustainable implementation from an SES perspective.

| Core Activities | Mediating Dimensions | | | |
| --- | --- | --- | --- | --- |
| | Knowledge | Practices | Institutions | Technology |
| Technological development | Generation of knowledge about technological solutions (engineering sciences) | Provision of information about requirements for construction, operation and maintenance | Provision of information about technical specifications for planning and decision-making (e.g., costs, production capacities) | Construction of pilot plants, adaptation and optimisation of technologies to the specific social-ecological conditions |
| Knowledge management | Development of manuals, planning instruments and toolkits; public relations; drafting of scientific publications | Provision of knowledge-based instruments for planning and decision-making; development of informative toolkits for operational practice | Provision of knowledge-based instruments for planning and decision-making; development and transfer of implementation concepts | Development and implementation of analytical tools for monitoring and interpreting the plants' technical status |
| Capacity development | Training and knowledge transfer (e.g., manuals and toolkits); academic training with lectures and summer school, scientific theses & internships | Training for improved skills of farmers and constructors; supervision of scientific theses & internships to improve IWRM practice; community health clubs | Training for institutions (train-the-trainer) for ownership, implementation and operation, scaling up | Training for technical service providers and technical staff |
| Governance and institutionalisation | Stakeholder and policy analysis; literature review; development of implementation concepts including recommendations for future implementation and scaling up | Establishment of responsibilities for training programmes (e.g., for plant construction, small-scale gardening, marketing) | Continuous communication and cooperation; institutional embedding incl. ownership, implementation support and legal safeguarding; development of financing models | Development of adapted technical standards; initialisation of standardisation (e.g., tank materials, safety requirements) |
| Participation | Community/farmer workshops; user involvement in monitoring & evaluation | Community/farmer workshops; demand-responsive approach | Involvement of institutional representatives; demand-responsive approach | Adaptation of pilot plant designs; pilot construction under real-life conditions with local workers |
| Empirical studies | Monitoring & evaluation; social-ecological impact assessment; financial analyses and economic impact studies | Optimisation of practices (e.g., fertiliser use, crop selection, efficient irrigation); studies on socio-cultural perspectives on water | Adaptation of institutions (e.g., responsibilities, rights, operation); analysis for substantiated arguments (e.g., economy) | Optimisation of technological components |

As part of the 'technological development', different rainwater and floodwater pilot plants were established to investigate whether the technologies were feasible options for improving livelihoods in Namibia on the basis of scientifically derived evidence. Rainwater harvesting plants have been in operation at Epyeshona in the peri-urban surroundings of the town of Oshakati since 2010 [32,56,57]. The criteria for the plant design were a mean precipitation of 470 mm/a and, as a result of the socio-empirical survey, a demand for gardening space amounting to 150 to 230 m$^2$ per farmer, as well as the availability of catchment areas for rainfall harvesting. Two technical and organisational options were tested with regard to their potential to meet people's needs: at household level, three pilot plants harvested rainwater from roof catchments and stored it in tanks (30 m$^3$) for use in household gardens, while at a community level, rainwater was harvested from a ground catchment and a greenhouse roof, stored in different reservoirs (in total 200 m$^3$) and used for irrigation by six households in a joint management initiative. Given the variability in rainfall, these designs should have allowed gardening with full storage capacity in approximately three out of four years. In-depth analysis of different gardening options using a tank flow model and an irrigation water model showed that even the negative impacts of climate change could be offset partly or completely by specific adaptation measures [58].

Floodwater harvesting was implemented at Iipopo in 2012 [57] based on similar assumptions regarding the gardening area demand per farmer and with the constraint that only a marginal amount of discharge from the oshana should be captured so as not to compromise its regular flow. Here, water from the oshana was pumped into different storage reservoirs (in total 400 m$^3$) during the rainy season, when the water quality is at its best. The stored water was used for irrigation by up to ten households on a cooperative basis. All the plants had a water-efficient drip-irrigation system installed. Despite experience acquired in the 1950s and 1960s with what are known as Stengel dams (storage capacities between 25,000 and 100,000 m$^3$), the project's rainwater and floodwater harvesting technologies were new and innovative in the region.

In order to share and disseminate the experiences and results from the project, different activities were conducted under the umbrella of 'knowledge management'. This project component comprised in particular the development of instruments to support decision-making and planning processes. By developing such tools, the project wanted to ensure that the essential results and experiences of the project were made available to decision-makers or planners in a practical and suitable way beyond the duration of the project [59]. In particular, a toolkit for rainwater harvesting was designed jointly with Namibian partners. The final version of the toolkit consisted of applied information for all phases in the planning, construction and operation of the technology.

The aim of the academic and non-academic 'capacity development' activities was to improve and transfer knowledge and practices at all levels. The project therefore ran about 30 training courses for rainwater and floodwater harvesting and subsequent gardening. Farmers were trained to operate and maintain the facilities beyond the end of the project, but also to regulate social conflicts within the new setting of community management. People were trained in the construction of harvesting and gardening facilities, which was also an investment in working skills and job creation.

The 'governance and institutionalisation' side of the project comprised a stakeholder analysis, a review of relevant political documents and a policy analysis. Such analyses together with continuous communication and exchange provided strong back-up for the incorporation of the project activities into current regional and national planning activities. In the long term, it supports the political and legal safeguarding of users and could foster future dissemination processes [56]. By strengthening local food production, the project contributed to blueprints such as Namibia's Vision 2030 [29] and Millennium Development Goals No. 1 and 7 [60].

Concepts that are purely technologically sophisticated can easily clash with users' socio-cultural needs and practices. 'Participation' is essential to meet existing needs and to embed new technologies. Different activities ensured that stakeholders were actively involved in the planning, implementation and evaluation processes, as well as in operational structures. Therefore, a new participatory method—the demand-responsive approach—was jointly developed with the Namibian partners (for details see [32,61]).

The project's 'empirical studies' project generated valid and fundamental information and data to identify impacts on SES as a whole, evaluate their sustainability and further adapt the technologies. Efficient and feasible pilot plants were an essential basis for facilitating their adoption by Namibian partners and were therefore crucial to dissemination. For optimisation and evaluation, a comprehensive socio-technical monitoring programme was developed with the communities at every project site. For instance, cost-benefit analyses at household level identified 'ferro-cement' to be the most cost-efficient material for tank construction [56]. Furthermore, empirical studies were crucial in addressing questions raised by stakeholders or institutions about how to adapt implementation in future. They also identified the risks and uncertainties for all parties, particularly regarding economic considerations such as investment, operation and maintenance costs.

*4.3. Phase 3: Transdisciplinary Integration (Evaluation of New Knowledge for Its Contribution to Societal and Scientific Progress)*

What contributions were made to societal and scientific progress? The main achievements were the successful implementation of RFWH, the adoption of responsibilities as part of the ownership process, and the establishment of everyday use of this technology. Appropriate storage volumes allowed small-scale gardening throughout the year, enriched and diversified the farmers' diet, and allowed income generation by selling the harvest at nearby markets [56]. Jobs in plant construction and gardening activities could be created and a stimulus for the regional economy generated [56]. Reduced dependence on food imports from South Africa was also expected. Additionally, the research and development process also provided new opportunities to acquire self-confidence and social acceptance. Apart from water and food security, the project therefore addressed an improvement in the socio-economic and health situation and, furthermore, the enhancement of knowledge and practices relating to natural resources management including gardening [32].

All the mediating dimensions played their part in this achievement: the technological development of a variety of solutions for different conditions and forms of application [62,63], the establishment of ownership and governance structures [64], training and other instruments for capacity development such as the RFWH toolkit [65], studies on the impact of alternative gardening variants and the contribution to adapting to climate change [58]. A significant proportion of the references used were clearly non-academic, which is a typical and often underrated (societal) outcome of transdisciplinary research.

The acceptance and pioneering nature of this technology—as important signs of societal progress— were impressively reflected in the term 'Green Villages', which was given to the pilot sites of Epyeshona and Iipopo by the Namibians themselves. They also used the term as a brand in order to boost the sales of their harvest at market. The respective communities became model sites for other communities and additional rainwater harvesting plants have since been implemented, initiated by single households and by Namibian institutions. The project supported these additional implementations via indirect consultation only. As also stated by Vreugdenhil et al. [55], the flexibility of the existing pilot schemes was a critical requirement for adapting the RFWH technology to different fields of application and varying conditions. As mentioned above, one of these additional plants has been implemented at the UNAM northern campus, funded by the Gesellschaft für Internationale Zusammenarbeit (GIZ), and is being used for academic education and technical studies.

A technology toolkit for RFWH was developed that contains information for planning and detailed technical manuals for construction, including materials and tools lists [65]. The toolkit is intended for technical experts (e.g., Agricultural Extension Offices and Rural Development Centres) and supports future capacity development such as training and knowledge dissemination. It also ensures the sustainability of RFWH because the pilot plants have increasingly become role models for further dissemination.

At the interface with policy, outcomes and recommendations addressed the cross-sectoral benefits of RFWH, the reduced dependency on climate variations and effects of climate change, as well as the need for state or donor-financed training and dissemination centres [18,56]. The latter created a setting that promotes small-scale water harvesting and gardening initiatives and enables them to be diffused in a self-organised manner [18]. However critical points were also mentioned. For example, the initial investment required may be not affordable by considerable parts of the population, and regional, national or even international financing options should be considered for the promotion of RFWH.

The pivotal research question of how water-harvesting technologies need to be adapted and embedded were answered by these outcomes. Solutions of a flexible technical design (rainwater or floodwater, aboveground or underground tanks or alternatively ponds, different catchment areas and storage volumes, variants of tank material) and a flexible management approach (individual or community) were key in terms of technology. However, just as important in the process was the close involvement of users and the responsible institutions guided by the methodology [32]. At the same time, the development of a demand-responsive approach together with the Namibian partners and the deliberations on the application of the SES concept for bringing transdisciplinarity into practice contributed to the scientific progress that was made.

## 5. Discussion

This study demonstrated how the transdisciplinary research concept developed by Jahn et al. [8] could be applied to a case study concerning the implementation of RFWH in central-northern Namibia. Here, the application of the SES concept played a crucial role. The SES concept facilitated the formulation of the epistemic object for research by structuring the problem situation (and boundary object). The concept then served as a basis for the research design and thus guided a solution-oriented research process. The transdisciplinary approach and SES concept were closely interlinked and complemented one other. While the transdisciplinary approach was manifested in a strong procedural influence on the research due to the respective phases and steps inherent in it, the application of the SES concept strongly guided the systemic perspective, the structure of tasks, the methodological repertoire and its integrative dimension, and disciplinary composition of the project.

The implementation of such an ideal transdisciplinary research process is still an exception rather than the rule. One major reason is that this kind of research process is quite time consuming and funding schemes are usually still not adequately adapted to the requirements of this mode of research. The case study presented here began with an open concept, and only the intense involvement of the stakeholders in a series of workshops and consultations allowed the demand for RFWH to be identified and concrete socio-technical design of suitable variants of RFWH in combination with gardening to be developed. The subsequent implementation on the ground was followed by extensive measures around capacity development in the construction, operation and marketing of the products, advice and support to the pilot facilities, the establishment of community management schemes and strategies for conflict solution that were new to the people in the area. The associated changes in the communities and for other stakeholders were judged to be new and essential structures and processes, thus the introduction of RFWH as a technological innovation worked hand in hand with the respective innovations in society. Finally, the adoption of ownership took a considerable time before and after the mere technical implementation. Funding conditions often do not allow the thorough implementation of all steps of the whole transdisciplinary process and involve the risk of projects struggling with long-term sustainability if they are constrained to parts of the process. Thus, an adaptation to the funding frameworks is the first critical factor for the complete operationalisation and success of transdisciplinary approaches.

The present case study demonstrated that the SES concept played a central role in the operationalisation of transdisciplinarity. The SES concept used here originated from the Frankfurt Social Ecology [14]. The current literature on SES is diverse and the underlying goals, disciplinary backgrounds, addressed scales and levels of differentiation vary considerably [66]. However, the authors argue that this diversity is important in order to reflect the complexity of research questions and problems. The concept presented in this article builds on a balanced view of the mediating dimension of knowledge, practices, technology and institutions, and highlights their key role in the dynamics between actors and ecosystem functions, with management and ESS as symmetric counterparts. Concepts of human wellbeing or the valuation of ESS can be integrated as further analytical differentiations for the unspecified societal structures and processes behind the actors' component. Those concepts can also be related to the four mediating dimensions, e.g., the valuation of ESS is interlinked, *inter alia* with knowledge and institutions. In contrast, Ostrom [36] focuses strongly on questions of governance and introduces a framework with multi-tier variables for a thorough analytical perspective. Authors such as Bennett et al. [67,68] and García-Llorente et al. [69] place ESS at the centre of SES. They shift the focus to the interrelation and trade-offs of ESS as a function of societal drivers [68,69] or at the valuation of ESS [46]. Finally and critically, it depends on the given social-ecological issue and research question as to which conceptual approach is most suitable for structuring and guiding the transdisciplinary research process.

As discussed, the presented approach provided guidance for the research process towards a combination of technological and social innovations. Another issue of considerable importance was the challenge of intercultural integration and collaboration. Different knowledge backgrounds, forms of communication, conflict management, institutional structures, perceptions of water and nature, value systems and traditions exist in parallel when jointly working on RFWH as the common boundary object. In this regard, the strong participation of stakeholders and consideration of local demands and views about the issues supported the project's ambition to address this challenge. As an example, a team of Namibian experts, project partners and advisors analysed the heterogeneous perspectives of water held by the three groups of Namibian-German project facilitators, community members and local authorities. They found that the perceptions differ—even between Namibian stakeholders—with regard to the questions of the 'infinity of water resources' and 'water as an economic good', while the roles of women and participation were not disputed [70]. This reinforced efforts in knowledge integration and transfer or participation, and influenced the themes and priorities of subsequent activities such as community workshops and studies. In general, the broad repertoire of methods and tasks represent the wide perspective of the transdisciplinary approach, which includes awareness of a project's intercultural endeavours in the context of research and development.

## 6. Conclusions

The application of the SES concept in this study shows that innovations in technologies alone are not sufficient to drive a system's overall sustainability. To make a system more sustainable in the long run, the introduction of new technologies must be accompanied by appropriate developments in knowledge, practices and institutions, all of which are crucial to the system. Located at the interface between the societal and natural spheres, these dimensions of action display a hybrid character. They are an expression of the complexity of the system, which needs to be taken into account when structuring and conducting the research (and development) process. Thus the results of this study also show that the technological activities must be supplemented by various accompanying activities, such as knowledge management, capacity development, governance and institutionalisation, participation, and empirical studies. These activities represent and address the aforementioned dimensions of the SES concept in action and further principal tasks of research. It becomes apparent that innovations in technologies are underpinned by corresponding changes in society, more specifically by social innovations.

The study further demonstrates how the SES concept helps to structure transdisciplinary research. The results from the Namibian case study highlight that the SES concept serves as an epistemic object, but also guides the design of the research process. It has an integrative function in terms of permitting analysis of different (basic) societal relations to nature in areas such as nutrition, land use or labour and production. In the case study, the provision of water and food was approached by relating resource management and ESS to their societal benefits, actors and ecosystem functions.

Furthermore, linking the SES concept with the ESS concept explicitly addresses the dynamic feedback loop between actors and the ecosystem functions of SES, and relates it to a clear conceptual framework. It also captures a broader system-related perspective for the ESS concept and thus allows a more comprehensive analysis of drivers and responses to changes in ESS.

Additionally, the system approach allows generalities to be deduced and also helps generate new knowledge about the question of transformation and how to shape the process towards sustainable water management, as presented here. Both are important for the dissemination of integrated solutions. However, the transferability of the pilot plants remains a challenging task and still carries uncertainties. The promotion of scaling up processes such as the development of funding models, the evaluation of all dimensions that are crucial to the respective SES, and reflections on the degree of adaptation required for transfer to other contexts are vital for long-lasting success beyond the duration of the project.

Finally, conducting and implementing transdisciplinary research requires relevant funding schemes. If the goal of the sustainable and long-lasting success of such a research process is to be taken seriously, there is an urgent need to adapt current research funding because transdisciplinary research is much more time consuming than fundamental research. The case study presented here, which lasted around ten years and included a funded exploratory project phase, is still the exception rather than the rule.

**Acknowledgments:** This study was funded by Germany's Federal Ministry of Education and Research (BMBF) (CuveWaters project, grant numbers 0330766A, 033L001A, 033W014A) and the 'LOEWE' (Landes-Offensive zur Entwicklung Wissenschaftlich-ökonomischer Exzellenz) research funding programme of Hessen's Federal Ministry of Higher Education, Research, and the Arts (BiK-F). Special thanks to the CuveWaters project partners in Namibia and Germany, in particular to Alexander Jokisch from the Technische Universität Darmstadt, for discussing details on harvesting technologies. We also thank our colleagues at ISOE, Diana Hummel, Thomas Jahn, Alexandra Lux, Robert Lütkemeier and Engelbert Schramm, for their conceptual input to SES. We are very grateful to Maxine Demharter, Heidi Kemp and Claire Tarring for proofreading this work. We would like to thank our colleagues from BiK-F, Karen Hahn, Katja Heubach, Thomas Hickler, Lasse Loft and Uwe Zajonz, for the discussions on earlier versions of the SES concept. Finally, we express our gratitude to the reviewers for their very helpful and constructive comments.

**Author Contributions:** Stefan Liehr and Julia Röhrig wrote the introduction and the sections on the Namibian case study and results. Stefan Liehr and Marion Mehring formulated the conceptual and methodological background. Stefan Liehr discussed the results. Thomas Kluge, Stefan Liehr and Julia Röhrig elaborated the conclusions. The revision was conducted by Stefan Liehr.

**Conflicts of Interest:** The authors declare no conflict of interest.

# References

1. Abdulla, F.; Eshtawi, T.; Assaf, H. Assessment of the Impact of Potential Climate Change on the Water Balance of a Semi-arid Watershed. *Water Resour. Manag.* **2009**, *23*, 2051–2068. [CrossRef]
2. Warren, A.; Sud, Y.C.; Rozanov, B. The future of deserts. *J. Arid Environ.* **1996**, *32*, 75–89. [CrossRef]
3. Fang, C.; Bao, C.; Huang, J. Management Implications to Water Resources Constraint Force on Socio-economic System in Rapid Urbanization: A Case Study of the Hexi Corridor, NW China. *Water Resour. Manag.* **2007**, *21*, 1613–1633. [CrossRef]
4. Falkenmark, M. Adapting to climate change: Towards societal water security in dry-climate countries. *Int. J. Water Resour. Dev.* **2013**, *29*, 123–136. [CrossRef]
5. Boelens, R.A.; Crow, B.; Dill, B.; Lu, F.; Ocampo-Rader, C.; Zwarteveen, M.Z. Santa Cruz Declaration on the Global Water Crisis. *Water Int.* **2014**, *39*, 246–261. [CrossRef]
6. Rockström, J.; Falkenmark, M.; Allan, T.; Folke, C.; Gordon, L.; Jägerskog, A.; Kummu, M.; Lannerstad, M.; Meybeck, M.; Molden, D.; et al. The unfolding water drama in the Anthropocene: Towards a resilience based perspective on water for global sustainability. *Ecohydrology* **2014**, *7*, 1249–1261. [CrossRef]
7. Maselli, D.; Lys, J.-A.; Schmid, J. *Improving Impacts of Research Partnerships*; Geographica Bernensia: Berne, Switzerland, 2006.
8. Jahn, T.; Bergmann, M.; Keil, F. Transdisciplinarity: Between mainstreaming and marginalization. *Ecol. Econ.* **2012**, *79*, 1–10. [CrossRef]
9. Kluge, T.; Liehr, S.; Lux, A.; Moser, P.; Niemann, S.; Umlauf, N.; Urban, W. IWRM concept for the Cuvelai Basin in northern Namibia. *Phys. Chem. Earth Parts A/B/C* **2008**, *33*, 48–55. [CrossRef]
10. Deppisch, S.; Hasibovic, S. Social-ecological resilience thinking as a bridging concept in transdisciplinary research on climate-change adaptation. *Nat. Hazards* **2013**, *67*, 117–127. [CrossRef]

11. Glaser, M. The Social Dimension in Ecosystem Management: Strength and Weaknesses of Human-Nature Mind Maps. *Hum. Ecol. Rev.* **2006**, *13*, 122–142.

12. Mehring, M.; Bernard, B.; Hummel, D.; Liehr, S.; Lux, A. Halting Biodiversity Loss: How Social-Ecological Biodiversity Research Makes a Difference. *Int. J. Biodivers. Sci. Ecosyst. Serv. Manag.* **2017**, *13*, 172–180. [CrossRef]

13. Becker, E. Problem Transformation in Transdisciplinary Research. In *Unity in Knowledge in Transdisciplinary Research for Sustainability. Encyclopedia of Life Support Systems (EOLSS)*; Hirsch Hadorn, G., Ed.; Developed under the Auspices of UNESCO: Oxford, UK, 2002/2008.

14. Hummel, D.; Jahn, T.; Keil, F.; Liehr, S.; Stieß, I. Social Ecology as Critical, Transdisciplinary Science—Conceptualizing, Analyzing, and Shaping Societal Relations to Nature. *Sustainability* **2017**, accepted.

15. Mendelsohn, J.; Jarvis, A.; Robertson, T. *A Profile and Atlas of the Cuvelai-Etosha Basin*; RAISON & Gondwana Collection: Windhoek, Namibia, 2013.

16. MAWF. *Development of an Integrated Water Resources Management Plan for Namibia—National Water Development Strategy and Action Plan*; Theme Report No. 8; MAWF: Windhoek, Namibia, 2010.

17. Langridge, R.; Christian-Smith, J.; Lohse, K.A. Access and Resilience: Analyzing the Construction of Social Resilience to the Threat of Water Scarcity. *Ecol. Soc.* **2006**, *11*, 18. [CrossRef]

18. Liehr, S.; Schulz, O.; Kluge, T.; Jokisch, A. *Water Security and Climate Adaptation through Storage and Reuse*; ISOE Policy Brief No. 1; ISOE: Frankfurt am Main, Germany, 2015.

19. Heyns, P. The Namibian perspective on regional collaboration in the joint development of international water resources. *Int. J. Water Resour. Dev.* **1995**, *11*, 467–492. [CrossRef]

20. Shongwe, M.E.; van Oldenborgh, G.J.; van den Hurk, B.; de Boer, B.; Coelho, C.A.; van Aalst, M.K. Projected changes in mean and extreme precipitation in Africa under global warming. Part I: Southern Africa. *J. Clim.* **2009**, *22*, 3819–3837. [CrossRef]

21. Bernstein, L.; Bosch, P.; Canziani, O.; Chen, Z.; Christ, R.; Davidson, O.; Hare, W.; Huq, S.; Karoly, D.; Kattsov, V. *Climate Change 2007: Synthesis Report. Contribution of Working Groups I, II and III to the Fourth Assessment Report of the Intergovernmental Panel on Climate Change*; IPCC: Geneva, Switzerland, 2007.

22. Liehr, S. Driving forces and future development paths of central northern Namibia. In *Exploring Sustainability Science: A Southern African Perspective*; Burnes, M., Weaver, A., Eds.; African SUN MeDIA: Stellenbosch, South Africa, 2008; pp. 431–467.

23. Sturm, M.; Zimmermann, M.; Schütz, K.; Urban, W.; Hartung, H. Rainwater harvesting as an alternative water resource in rural sites in central northern Namibia. *Phys. Chem. Earth Parts A/B/C* **2009**, *34*, 776–785. [CrossRef]

24. Government of the Republic of Namibia (GRN). *Initial National Communication to the United Framework Convention on Climate Change*; Country Study Coordinated and Prepared by the Directorate of Environmental Affairs of the Ministry of Environment and Tourism (MET), 2002. Available online: http://unfccc.int/resource/docs/natc/namnc1.pdf (accessed on 20 June 2017).

25. Stengel, H.W. *Wasserwirtschaft Waterwese Water affairs in S.W.A.*; Afrika-Verlag Der Kreis: Windhoek, Südwestafrika, 1963.

26. Mendelsohn, J.; el Obeid, S.; Roberts, C. *A Profile of North-Central Namibia*; Gamsberg Macmillan Publishers: Windhoek, Namibia, 2000.

27. Polak, M.; Liehr, S. Theoretical reflections about the analysis of water governance in coupled social-ecological systems. In *Water Governance—Challenges in Africa: Hydro-Optimism or Hydro-Pessimism*; Anne, I., Ed.; Peter Lang AG—Internationaler Verlag der Wissenschaften: Bern, Switzerland; Berlin, Germany; Bruxelles, Belgium; Frankfurt am Main, Germany; New York, NY, USA; Oxford, UK; Wien, Austria, 2012; pp. 65–80.

28. MAWF. *Integrated Water Resources Management Plan for Namibia*; MAWF: Windhoek, Namibia, 2010.

29. Government of the Republic of Namibia (GRN). *Namibia Vision 2030. Policy Framework for Long-Term National Development*; GRN: Windhoek, Namibia, 2004.

30. National Planning Commission (NPC). *Namibia's Fourth National Development Plan (NDP4) 2012/13 to 2016/17*; NPC: Windhoek, Namibia, 2012. Available online: http://www.gov.na/documents/10181/14226/NDP4_Main_Document.pdf (accessed on 20 June 2017).

31. Liehr, S.; Brenda, M.; Cornel, P.; Deffner, J.; Felmeden, J.; Jokisch, A.; Kluge, T.; Müller, K.; Röhrig, J.; Stibitz, V.; et al. From the Concept to the Tap—Integrated Water Resources Management in Northern Namibia. In *Integrated Water Resources Management: Concept, Research and Implementation*; Borchardt, D., Bogardi, J.J., Ibisch Ralf, B., Eds.; Springer: Cham, Switzerland, 2016; pp. 683–717.

32. Zimmermann, M.; Jokisch, A.; Deffner, J.; Brenda, M.; Urban, W. Stakeholder participation and capacity development during the implementation of rainwater harvesting pilot plants in central northern Namibia. *Water Sci. Technol.* **2012**, *12*, 540. [CrossRef]

33. Schneider, F. Approaching water stress in the Alps: transdisciplinary coproduction of systems, target and transformation knowledge. In *Managing Alpine Future II—Inspire and Drive Sustainable Mountain Regions, Proceedings of the Innsbruck Conference, Innsbruck, Austria, 21–23 November 2011*; Borsdorf, A., Stötter, J., Veulliet, E., Eds.; Verlag der Österreichischen Akademie der Wissenschaften: Vienna, Austria; pp. 107–117.

34. Dewulf, A.; Brugnach, H.; Ingram, H.; Termeer, C. The co-production of knowledge about water resources: Framing, uncertainty and climate change. In Proceedings of the 7th International Science Conference on the Human Dimensions of Global Environmental Change, UN Campus, Bonn, Germany, 26–30 April 2009; pp. 1–14.

35. Becker, E. Social-Ecological Systems as Epistemic Objects. In *Human-Nature Interactions in the Anthropocene: Potentials of Social-Ecological Systems Analysis*; Glaser, M., Krause, G., Ratter, B., Welp, M., Eds.; Routledge: London, UK, 2012; pp. 37–59.

36. Ostrom, E. A general framework for analyzing sustainability of social-ecological systems. *Science* **2009**, *325*, 419–422. [CrossRef] [PubMed]

37. Folke, C. Resilience: The emergence of a perspective for social–ecological systems analyses. *Glob. Environ. Chang.* **2006**, *16*, 253–267. [CrossRef]

38. Berkes, F. From Community-Based Resource Management to Complex Systems: The Scale Issue and Marine Commons. *Ecol. Soc.* **2006**, *11*, 45. [CrossRef]

39. Fischer-Kowalski, M.; Weisz, H. Society as Hybrid between Material and Symbolic Realms. Toward a Theoretical Framework of Society-Nature Interrelation. *Adv. Hum. Ecol.* **1999**, *8*, 215–251.

40. Walker, B.; Carpenter, S.; Anderies, J.; Abel, N.; Cumming, G.S.; Janssen, M.; Lebel, L.; Norberg, J.; Peterson, G.D.; Pritchard, R. Resilience Management in Social-ecological Systems: A Working Hypothesis for a Participatory Approach. *Conserv. Ecol.* **2002**, *6*, 14. [CrossRef]

41. Ostrom, E. A diagnostic approach for going beyond panaceas. *Proc. Natl. Acad. Sci. USA* **2007**, *104*, 15181–15187. [CrossRef] [PubMed]

42. Anderies, J.M.; Janssen, M.A.; Ostrom, E. A Framework to Analyse the Robustness of Social-ecological Systems from an Institutional Perspective. *Ecol. Soc.* **2004**, *9*, 18. [CrossRef]

43. Hummel, D.; Jahn, T.; Schramm, E. *Social-Ecological Analysis of Climate Induced Changes in Biodiversity—Outline of a Research Concept*; BiK-F Knowledge Flow Paper No. 11; ISOE: Frankfurt am Main, Germany, 2011.

44. Maes, J.; Teller, A.; Erhard, M.; Liquete, C.; Braat, L.; Berry, P.; Egoh, B.; Puydarrieux, P.; Fiorina, C.; Santos-Martín, F. *Mapping and Assessment of Ecosystems and Their Services An Analytical Framework for Ecosystem Assessments under Action 5 of the EU Biodiversity Strategy to 2020*; Technical Report; Publication office of the EU: Luxembourg, 2013.

45. Reyers, B.; Biggs, R.; Cumming, G.S.; Elmqvist, T.; Hejnowicz, A.P.; Polasky, S. Getting the measure of ecosystem services: A social–ecological approach. *Front. Ecol. Environ.* **2013**, *11*, 268–273. [CrossRef]

46. Martín-López, B.; Gómez-Baggethun, E.; García-Llorente, M.; Montes, C. Trade-offs across value-domains in ecosystem services assessment. *Ecol. Indic.* **2014**, *37*, 220–228. [CrossRef]

47. Hummel, D. *Population Dynamics and Supply Systems: A Transdisciplinary Approach*; Campus Verlag: Frankfurt am Main, Germany, 2008.

48. Hummel, D.; Hertler, C.; Niemann, S.; Lux, A.; Janowicz, C. *Versorgungssysteme als Gegenstand Sozial-Ökologischer Forschung: Ernährung und Wasser*; demons working paper No. 2; ISOE: Frankfurt am Main, Germany, 2004.

49. Becker, E.; Jahn, T.; Hummel, D. Gesellschaftliche Naturverhältnisse. In *Soziale Ökologie—Grundzüge einer Wissenschaft von den Gesellschaftlichen Naturverhältnissen*; Becker, E., Jahn, T., Eds.; Campus Verlag: Frankfurt am Main, Germany; New York, NY, USA, 2006; pp. 174–197.

50. Görg, C.; Brand, U.; Haberl, H.; Hummel, D.; Jahn, T.; Liehr, S. Challenges for Social-Ecological Transformations. Contributions from Social and Political Ecology. *Sustainability* **2017**, accepted.

51. De Groot, R.S.; Wilson, M.A.; Boumans, R.M. A typology for the classification, description and valuation of ecosystem functions, goods and services. *Ecol. Econ.* **2002**, *41*, 393–408. [CrossRef]

52. Sukhdev, P.; Wittmer, H.; Schröter-Schlaack, C.; Nesshöver, C.; Bishop, J.; ten Brink, P.; Gundimeda, H.; Kumar, P.; Simmons, B. *The Economics of Ecosystems and Biodiversity: Mainstreaming the Economics of Nature: A Synthesis of the Approach, Conclusions and Recommendations of TEEB*; TEEB: Geneva, Switzerland, 2010.

53. MEA. *Millennium Ecosystem Assessment—A Framework for Assessment*; Island Press: Washington, DC, USA, 2005.

54. Lyytimäki, J.; Sipilä, M. Hopping on one leg—The challenge of ecosystem disservices for urban green management. *Urban For. Urban Green.* **2009**, *8*, 309–315. [CrossRef]

55. Vreugdenhil, H.; Slinger, J.; Thissen, W.; Ker, R.P. Pilot Projects in Water Management. *Ecol. Soc.* **2010**, *15*, 13. [CrossRef]

56. Woltersdorf, L.; Jokisch, A.; Kluge, T. Benefits of rainwater harvesting for gardening and implications for future policy in Namibia. *Water Policy* **2014**, *16*, 124. [CrossRef]

57. Röhrig, J.; Schuldt-Baumgart, N.; Krug von Nidda, A. *Water Is Life—Omeya Ogo Omwenyo*; CuveWaters Report; ISOE: Frankfurt am Main, Germany, 2013.

58. Woltersdorf, L.; Liehr, S.; Döll, P. Rainwater Harvesting for Small-Holder Horticulture in Namibia: Design of Garden Variants and Assessment of Climate Change Impacts and Adaptation. *Water* **2015**, *7*, 1402–1421. [CrossRef]

59. Röhrig, J.; Liehr, S. How to provide and transmit project outcomes to support decision makers in the long run? Approach and instruments of the CuveWaters project. In Proceedings of the International Conference on Integrated Water Resources Management, Dresden, Germany, 12–13 October 2011.

60. National Planning Commission (NPC), Republic of Namibia. *Namibia 2013 Millennium Development Goals Interim Progress Report No. 4*; NPC: Windhoek, Namibia, 2013. Available online: http://www.na.undp.org/content/namibia/en/home/library/mdg/mdgsrep2013 (accessed on 20 June 2017).

61. Deffner, J.; Mazambani, C. *Participatory Empirical Research on Water and Sanitation Demand in Central Northern Namibia: A Method for Technology Development with a User Perspective*; CuveWaters Papers No. 7; ISOE: Frankfurt am Main, Germany, 2010.

62. Jokisch, A.; Schulz, O.; Kariuki, I.; Krug von Nidda, A.; Deffner, J.; Liehr, S.; Urban, W. *Rainwater Harvesting in Central-Northern Namibia*; Factsheet; ISOE: Frankfurt am Main, Germany, 2015.

63. Jokisch, A.; Schulz, O.; Kariuki, I.; Krug von Nidda, A.; Deffner, J.; Liehr, S.; Urban, W. *Floodwater Harvesting in Central-Northern Namibia*; Factsheet; ISOE: Frankfurt am Main, Germany, 2015.

64. Schulz, O.; Jokisch, A.; Deffner, J.; Woltersdorf, L.; Liehr, S.; Urban, W.; Kluge, T. *Rain- and Floodwater Harvesting—Implementation Concept*; Report; ISOE: Frankfurt am Main, Germany, 2015.

65. Schulz, O.; Jokisch, A.; Kariuki, I. *The Technical Toolkit for Rain- and Floodwater Harvesting (RFWH Toolkit)*, 2nd ed.; ISOE, TU Darmstadt: Frankfurt am Main, Germany, 2015. Available online: http://www.cuvewaters.net/Toolkits.112.0.html (accessed on 20 June 2017).

66. Binder, C.R.; Hinkel, J.; Bots, P.W.G.; Pahl-Wostl, C. Comparison of Frameworks for Analyzing Social-ecological Systems. *Ecol. Soc.* **2013**, *18*. [CrossRef]

67. Bennett, E.M.; Cramer, W.; Begossi, A.; Cundill, G.; Díaz, S.; Egoh, B.N.; Geijzendorffer, I.R.; Krug, C.B.; Lavorel, S.; Lazos, E.; et al. Linking biodiversity, ecosystem services, and human well-being: Three challenges for designing research for sustainability. *Curr. Opin. Environ. Sustain.* **2015**, *14*, 76–85. [CrossRef]

68. Bennett, E.M.; Peterson, G.D.; Gordon, L.J. Understanding relationships among multiple ecosystem services. *Ecol. Lett.* **2009**, *12*, 1394–1404. [CrossRef] [PubMed]

69. García-Llorente, M.; Iniesta-Arandia, I.; Willaarts, B.A.; Harrison, P.A.; Berry, P.; Bayo, M.d.M.; Castro, A.J.; Montes, C.; Martín-López, B. Biophysical and sociocultural factors underlying spatial trade-offs of ecosystem services in semiarid watersheds. *Ecol. Soc.* **2015**, *20*. [CrossRef]

70. Biggs, D.; Heyns, P.; Klintenberg, P.; Mazambani, C.; Nantanga, K.; Seely, M. *Bridging Perspectives on IWRM in the Cuvelai Basin*; CuveWaters Report; ISOE, DRFN: Frankfurt am Main, Germany; Windhoek, Namibia, 2008.

*sustainability*

MDPI

*Article*

# Ecosystem Services as a Boundary Concept: Arguments from Social Ecology

Christian Schleyer [1,*], Alexandra Lux [2,3], Marion Mehring [3,4] and Christoph Görg [5]

[1]  Section of International Agricultural Policy and Environmental Governance, University of Kassel, 37213 Witzenhausen, Germany

[2]  ISOE—Institute for Social-Ecological Research, Transdisciplinary Methods and Concepts, 60486 Frankfurt/M., Germany; lux@isoe.de

[3]  Senckenberg Biodiversity and Climate Research Centre (BiK-F), 60325 Frankfurt/M., Germany; mehring@isoe.de

[4]  ISOE—Institute for Social-Ecological Research, Biodiversity and People, 60486 Frankfurt/M., Germany

[5]  Institute of Social Ecology, 1070 Vienna, Alpen-Adria University, 9020 Klagenfurt, Austria; christoph.goerg@aau.at

*  Correspondence: schleyer@uni-kassel.de; Tel.: +49-179-11-66-474

Received: 4 February 2017; Accepted: 16 June 2017; Published: 26 June 2017

**Abstract:** Ecosystem services (ES) are defined as the interdependencies between society and nature. Despite several years of conceptual discussions, some challenges of the ES concept are far from being resolved. In particular, the usefulness of the concept for nature protection is questioned, and a strong critique is expressed concerning its contribution towards the neoliberal commodification of nature. This paper argues that these challenges can be addressed by dealing more carefully with ES as a boundary concept between different disciplines and between science and society. ES are neither about nature nor about human wellbeing, but about the mutual dependencies between nature and human wellbeing. These mutual interdependencies, however, create tensions and contradictions that manifest themselves in the boundary negotiations between different scientific disciplines *and* between science and society. This paper shows that approaches from Social Ecology can address these boundary negotiations and the power relations involved more explicitly. Finally, this implies the urgent need for more inter- and transdisciplinary collaboration in ES research. We conclude (1) that the social–ecological nature of ES must be elaborated more carefully while explicitly focussing on the interdependencies between nature and society; (2) to better implement inter- and transdisciplinary methods into ES research; and (3) that such ES research can—and to some extent already does—substantially enhance international research programmes such as Future Earth.

**Keywords:** ecosystem services; social–ecological systems; social ecology; transdisciplinary research; interdisciplinary research; colonisation; boundary concept

## 1. Introduction

The concept of Ecosystem Services (ES) has been created to address the interactions between nature and society. It describes the relevance of ecosystem functions for human wellbeing [1]. Moreover, the concept raises high expectations to inform decision-making on the interrelation and dependencies between societal and natural processes. Its scope is very broad, ranging from agricultural products to climate regulation and cultural services. Thus, it is dedicated to showcase trade-offs across various economic sectors, different fields of human wellbeing (from health and food to aesthetic perceptions) and related policy fields, and to reveal synergies among them while taking ecological limits seriously.

What is not yet adequately addressed is the specific character of the ES concept at the interplay of societal and natural processes (see [2]). The concept is neither about ecosystems, i.e., on biophysical processes as an object of natural science research, nor is it about the economic value and the goods provided by ecosystems, i.e., on socioeconomic valuations or cultural perceptions, analysed by economics and social sciences. Ecosystem services are about how both processes mutually depend on each other, i.e., about the interdependencies between society and nature. Such mutual interdependencies require a specific inter- and transdisciplinary research, which can be provided by Social Ecology [3–5].

The main objective of this article is to further the social–ecological perspective in order to strengthen the ES concept in the context of inter- and transdisciplinary research. In Section 2, we elaborate from a perspective of Social Ecology on what is meant by the notion that the ES concept is inherently social–ecological in nature. From there, we deal with it as a boundary concept between different disciplines and between science and society. An analysis from a social–ecological perspective serves also as a better starting point to explain the expectations of the concept and its application. For this reason, the theoretical underpinnings and practical challenges of ES as a boundary concept are introduced (Section 3). For implementing this social–ecological perspective, interdisciplinary (Section 4) and transdisciplinary (Section 5) collaboration are needed, focussing on the relationship between ES and the colonisation of nature. Finally, we showcase the requirements and implications for an application of the concept in social–ecological research.

## 2. The Ecosystem Services Concept from a Perspective of Social Ecology

After several years of research and conceptual discussions, there is still considerable disagreement in ES research and practice. Until now, several gaps of the ES concept have been scrutinised, most of them related to its anthropocentric and economic impetus (e.g., [6]). Much of the criticism calls for improvements in terminologies [7,8] and methods [9]. Whether the debate on ES deals with an agreed classification is also scrutinised [10–12], and there are calls for a more participatory approach towards ES research that takes issues of social perceptions and co-production seriously, especially concerning cultural services [13–15]. With respect to valuation, application of the ES concept is often criticised for its bias towards monetary valuation and the commodification of nature [16–18].

Starting from Social Ecology, however, offers a slightly different perspective on the concept. It emphasises that the ES concept is neither about ecosystems nor about societal goods and human wellbeing solely, but about the interdependencies between human wellbeing and nature [2,19,20]. Some of the challenges addressed in the current debates are related to the question of how to deal with this social–ecological character of ES. For example, research on ES assumes, usually implicitly, that ES are provided by ecosystems as natural processes independent from human influences. This assumption, however, ignores the role of human interventions (labour, technology, and capital) in improving the supply of ES (in particular, agricultural goods, but also forest products and cultural or regulating services) by transforming 'natural ecosystems' into human-modified cultural landscapes [21–25]. While one study shows that humans indeed often contribute to the maintenance and enhancement of ecosystems [26], another emphasises *"that increased use of manufactured and financial capital might deliver higher quantities of ecosystem services in the short-term but is often associated to several trade-offs in space and time"* [27] (p. 271). Such interventions, in reverse, are captured by the concept of colonisation, an important concept of Social Ecology [23,28] (see Section 4) and offer a different starting point for an analysis of the capacity of ecosystems to contribute to human wellbeing. Thus, a more careful treatment of this social–ecological character may substantively improve the application of the ES concept. In this regard, the paper builds upon inter- and transdisciplinary research and conceptual discussions conducted at the Institute of Social Ecology (SEC) in Vienna (Austria) and the ISOE—Institute for Social–Ecological Research in Frankfurt/Main (Germany) (see [5] in this issue).

The concrete potential for an improved uptake of the ES concept is shown in this paper by discussing fundamental issues of the scientific conceptualisation of ES at the boundaries between different scientific disciplines *and* between science and society. We emphasise that approaches from Social Ecology are well-suited to address the related boundary negotiations and the power relations involved. For this reason, we argue in this paper that inter- and transdisciplinary research is needed to improve the societal relevance of the ES concept and the link to human wellbeing. We also reflect on the potential discrepancy between a concept that inherently depends on inter- and transdisciplinary research approaches and 'thinking', as well as respective forms of collaboration, and their practical application.

## 3. Ecosystem Services as a Boundary Concept: Potential and Application

The ES concept can be considered as a boundary concept (e.g., [19,29,30]). A boundary concept enables researchers from different disciplines, policy makers, and other stakeholders to develop a common language [31], and respectively integrate and derive knowledge relevant to their field [32]. At the same time, the exact meaning and conceptualisation of such a boundary concept is contested both between different scientific disciplines, *and* between science and society; this will be determined in boundary negotiations [33]. Thus, the vagueness of the ES concept (e.g., [6]) is both a blessing and a curse. On the one hand, the vagueness is the outcome of continuous, complex, and demanding negotiations (boundary work) about, for example, how the ES concept is defined, how implementation and application is interpreted, which foci are set, and which tools are used. With respect to policy and decision-making, these negotiations are not only a scientific exercise but—to some degree—also linked to political or societal interests [34–36]. Moreover, due to its nature as a boundary concept, knowledge claims concerning ES from different scientific disciplines are linked to certain fields of policy-making and may potentially conflict with other fields (e.g., conservation biology linked to nature protection conflicts vs. agricultural research which emphasises provisioning services [37,38]). These differences may also cause ambiguities and tensions in the core meaning of the concept, concerning, for example, the strengths and limits of economic valuation methods (including the debate on how to define, quantify, and value ES [10–12]) or the need for integrated approaches of monetary and non-monetary valuation methods [39–41]. These tensions and ambiguities may also refer to the objectives involved (e.g., its anthropocentric bias) or different strategies to define the building blocks of the conceptual framework. The latter become apparent when comparing, for example, the steps and arrows of the Cascade Model in [12] (see Figure 1) with the Stairways Model in [21] (see Figure 2). Similarly, scientists, policy makers, and other stakeholders may impose very different normative values and their respective key components on application of the ES concept: sustainability (ecological, economic, social); justice (distributional, procedural, and recognition); and diversity (biological, cultural/linguistic, and institutional).

Looking back into the history of the ES concept, these boundary negotiations can be understood as major reasons for some conceptual ambiguities and the related scientific debates. The ES concept was revisited and elaborated further in the context of the Millennium Ecosystem Assessment [1], later during the international TEEB (The Economics of Ecosystems and Biodiversity) initiative [42], and in the IPBES (Intergovernmental Science-Policy Platform on Biodiversity and Ecosystem Services) context [43] and beyond. Over time, expectations were raised that the concept may improve the interface between science and society and the relevance of science for policy-making and the management of ecosystems [19]. Indeed, as pointed out in one study [32], the language of ES has often been deliberately used by scientists and environmentalists with the precise aim of tailoring ecological knowledge to users, particularly policy-makers and politicians. As those users often have economic and social concerns as their main priority, economic language is perceived as a powerful tool to convince policy- and other decision-makers outside the environmental policy domain. Used in this context, it is expected that the ES concept may substantially inform policy-makers and other stakeholders when drafting or revising policies.

From the beginning, however, there were concerns, in particular, about the limits of the economic 'language'. For example, the TEEB-approach failed to produce one single (economic) number to demonstrate the global relevance of biodiversity and ecosystem services. By contrast, it emphasised the need to contextualise the link between science and policy, and to develop specific approaches for certain audiences, for example, national and international policy-makers, but also regional and local policy-makers and businesses (see teebweb.org). During the negotiations of IPBES, the economic bias of the ES concept, and its application, became a matter of political debate and scientific concern. Because some countries challenged the ES concept as a perceived tool for the commodification of nature, the conceptual framework developed within IPBES explicitly emphasizes different and conflicting worldviews and asks for a translation between them [43]. Therefore, valuation of ecosystems and their services is increasingly recognised as a contested issue which needs careful consideration, due to the nature of ecosystem services as a boundary concept.

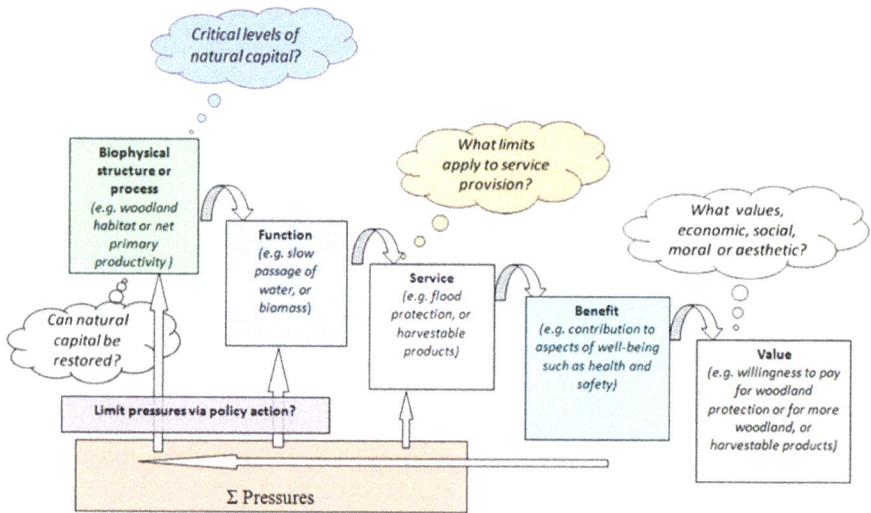

**Figure 1.** The ecosystem services Cascade Model [12].

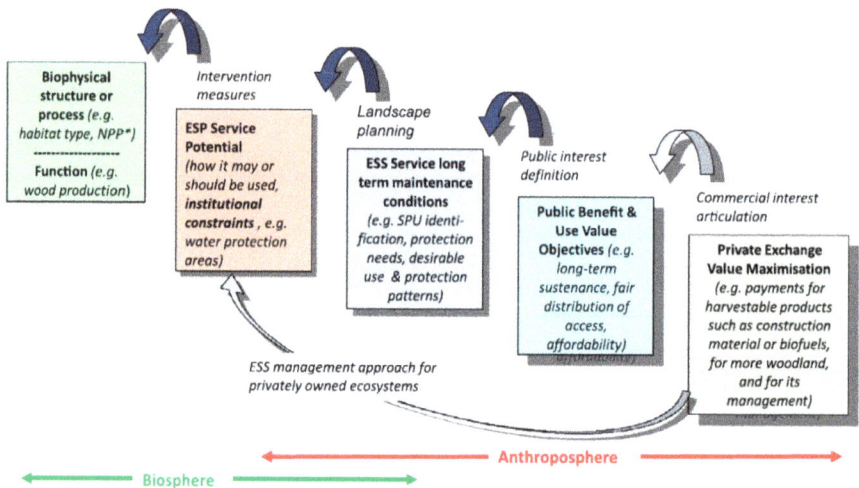

**Figure 2.** The ecosystem services Cascade Model, as adapted by [21], turning the cascade into stairways. NPP = Net Primary Production; ESP = Ecosystem Service Potential; ESS = Ecosystem Service.

Input from different scientific disciplines and non-scientific knowledge must be incorporated, and different purposes and different kinds of valuation must be considered. Further, different kinds of purpose may include different kinds of application (e.g., awareness raising, policy negotiations for mainstreaming ES, or monitoring of ecosystems and their state). Within these purposes, differing cultural perceptions and different kinds of valuation must be reflected carefully, especially concerning economic valuation tools within certain cultural contexts [14]. Within certain contexts, monetary values may not be perceived as legitimised measures or indicators (e.g., the economic value of a certain piece of land or a tree which is considered as sacred) or even insignificant (see [22] for an example resulting from extensive field work); these efforts may not be perceived as helpful, and instead, as undermining the foundations of certain cultures. Thus, to realise its potential as a tool at the interface between science, general society, and policy-making, the boundary work, which is often concealed within the knowledge claims of different disciplines, must be revealed and reflected carefully. More precisely, inter- and transdisciplinary research on ES should avoid oversimplified assumptions about the application of ecosystem valuation, and requires participatory mechanisms to integrate the often very diverse value attributions of stakeholders towards various ecosystem services [21].

After many years of discussion in the scientific domain, and a multitude of attempts to implement the ES concept (or parts of it) in practice, developing and establishing such an integrated approach of inter- and transdisciplinary research remains challenging; many investigations have failed to integrate biophysical and socio-cultural dimensions adequately, or suffered from inappropriate design of stakeholder involvement—if these challenges were addressed at all ([14,41,44]). Further, applying the ES concept on the ground comes along with very concrete and non-trivial knowledge needs on the part of different stakeholders, including methods and tools for ES assessment and valuation and for dealing with trade-offs made explicit in the process [31,45,46].

In summary, following the approach of ES as a boundary concept, it is possible to better understand the controversies and debates surrounding this concept both within science *and* between science and practice. This understanding is an important precondition for improving the applicability of the ES concept and its ability to address societal issues [47,48]. At the same time, the concept and its vagueness offer opportunities to address, and to some extent integrate, the perspectives of several policy fields while dealing with respective differences in meaning and definition. Against this background, once again, inter- and transdisciplinarity as integrated research modes come to the fore. They support inclusive processes that, inter alia, allow to reveal contested assumptions on the sustainable use of ecosystem services.

## 4. Social–Ecological Nature of Ecosystem Services: Challenges for Interdisciplinary Collaboration

The need for interdisciplinary collaboration on ES is widely acknowledged, at least in principle. The concept was developed over the last decade as a result of strict interdisciplinary collaboration between natural scientists, mostly from ecology, and social scientists, in particular from ecological economics (for the history of the concept, refer to [49]). It is no coincidence that the most prominently involved subdisciplines are those which are heterodoxies, or at least are open to cross-disciplinary collaboration, for example, ecological economics and conservation biology. Nevertheless, there is concern from several sides that interdisciplinary collaboration remains challenging. For example, Norgaard warned several years ago that the use of the ES concept may move from an eye-opener to a complexity blinder [41], when ignoring the huge conceptual challenges involved. Other challenges are related to certain parts of the ES concept, such as the intangible values of cultural services [14] and the tendency to express all kinds of values only, or preferably, in monetary terms [40]. For the purpose of this article, one challenge is particularly important and is often concealed in ES research: the social–ecological nature of ES, that represents a specific challenge for interdisciplinary research.

One of the major challenges for addressing the social–ecological nature of ES is the assumption that ES are provided directly by ecosystem functions, as suggested in the so-called ES Cascade Model developed by [12]. The Common International Classification of Ecosystem Services (CICES) was developed to provide a basis for an agreed classification system [50], and to avoid duplication and misrepresentation; the Cascade Model describes ES as a rather linear flow from ecosystems to humans. Biophysical structures and processes generate ecosystem functions, which in turn provide ES to humans benefitting from them, depending on their values (see Figure 1). Feed-back loops are included in this cascade, as pressures in ecosystems may induce policy responses. Moreover, several knowledge gaps are highlighted concerning the critical levels of ES provisioning, the restoration capacity of ecosystems, and the different kinds of valuation approaches involved, within and beyond economic values. Yet, the basic idea is a linear delivery of services and benefits, based on the capacity of ecosystems and their biophysical properties.

This basic idea can be challenged by at least two considerations: first, the very nature of ecosystems in most parts of the world, which are mostly culturally transformed landscapes, and which depend in their delivery of services very much on the modification through human activities [24,25]; and, second, the very nature of benefits, which in most cases need interventions in the form of labour and technology [22]. At first glance, the first argument does not seem to be decisive, as the same modification of ES delivery is also addressed by the feed-back loop of the Cascade Model. The difference, however, is characterised by the notion that the modifications of landscapes manifest themselves not only as pressures, but also as the supply of both provisioning services (e.g., production of food and feed, or provision of shelter) and regulating services (e.g., flood protection) increasing or appearing in the first place due to modification of a concrete landscape. Thus, maintaining the resulting landscape depended—and still depends—on human labour and technology, and the input of other resources such as fertilisers and knowledge. From a social–ecological perspective, it thus depends on continuing what we call the colonisation of nature [51].

The term 'colonisation of nature' was introduced to analyse the human interventions into ecosystems as part of the Neolithic revolution and the introduction of agriculture [40]. From this time onwards, the productivity of ecosystems increased significantly as did the proportion that humans appropriate from the net primary production of ecosystems. This is measured by the HANPP-index (Human Appropriation of Net Primary Production), the amount of Net Primary Production (NPP) used by societies. More precisely, *"the HANPP indicator tells about how much of the potentially available annual plant biomass production (NPP) is appropriated through human colonizing interventions, through deliberate changes of the land cover (for example the conversion of pristine forest to grassland or arable land, or the sealing of soils by construction of cities and roads), and/or through harvest"* [28] (p. 40). HANPP and embodied HANPP (eHanpp) in certain products such as food and feed stuff represents not only an important indicator for the relevance of 'nature' for societies, but also the other way around—for the constraints of this appropriation in the overall productivity of ecosystems [52]. As the colonisation of certain ecosystems differs due to the actual amount of technology and labour that certain societies are able to apply, it offers also a starting point for the analysis of the consequences of land use (changes) in ecological and social terms. Global inequalities as much as ecological consequences are closely linked, for example, to the amount of NPP in certain regions that are used or that can be used for beef production or other goals such as climate regulation.

From the perspective of Social Ecology, the interlinkages between ecosystems and benefits, or between nature and society, are conceptualised in a different way. The term 'colonisation of nature' clearly denotes *"the intended and sustained transformation of natural systems, by means of organized social interventions, for the purpose of improving their utility for society. [ . . . ] A colonizing intervention must both be causally effective in changing some biophysical condition; it must make a difference in the world of matter. Likewise, it must be culturally conceived of, organized and monitored; it must, 'make sense' in the world of communication"* [53] (p. 234). This concept acknowledges that the impact of societies on nature does not always undermine the functional requirements of ecosystems, as assumed in many nature conservation approaches, but often increases the capacity of an ecosystem for service provision. This does not mean, however, to ignore critical levels or constraints in the appropriation of nature. First, it denotes the dependencies of societies on ecosystems and the need to invest labour, resources, and money in their functioning. Moreover, limits exist, either in terms of available NPP or in terms of the loss of biodiversity and other negative side effects resulting from certain forms of appropriation (e.g., the use of fertilisers resulting in eutrophication of rivers and lakes). The former can be analysed using the HANPP-approach, whereas the latter needs additional analysis. As the original Cascade Model indicates, these questions can be addressed by the ES concept in general, but currently represent a major gap in research. In particular, the relationship between biodiversity loss and ES provision needs further investigation [54].

In addition, the work on 'social–ecological dynamics of ES' [2] point to a functional relation between ES provision and demand by society: nature and society exert mutual influence. A study from the Socotra Archipelago (Yemen) clearly illustrates that the provision of key ES (here: fish biomass for food production and household economies) is significantly correlated with the monsoon patterns, and consequently, with the fishing behaviour of the local communities. This seasonality in ES provision triggers complex adaption of local livelihood strategies. The local people adapted to this variation by, for example, developing trading activities with people from the island interior [2]. This study shows that the relation between ES and societal benefit cannot be explained by the Cascade Model [12] where ES only flow to society. It is rather the interlinkages—and in this case the dependency between ecosystems and benefits—that matter.

Another point is related to the first challenge, that service provision needs ongoing societal interventions in form of money, labour, and technology. To further elaborate on this intervention, the authors of [21] modified the Cascade Model further and introduced the concept of Ecosystem Service Potentials (ESP). The authors also merged the ecosystem structures and components, and the ecosystem functions into one category of ecosystem properties, representing the biophysical basis of service provision (see Figure 2). They argue that neither are the potentials determined by the functions, nor can they be assessed by analysing the services only: ESP are generated in complex social processes (e.g., increased commodification of nature or struggles about local commons), and these processes determine the kinds of services ultimately realised, or not (e.g., the loss of regulating services). Describing the process in a stepwise manner, first, an ESP must be recognised and the services the system could potentially provide must be identified (the authors call this 'use-value attribution'). The mobilised ESP can then be either directly consumed, or marketed as commodities. Thus, the benefits provided are either non-monetary benefits or monetary ones [22].

The difference between these benefits is decisive in terms of the societal processes which govern the appropriation of nature and, thus, enable a better understanding of the societal processes involved. Use value is the precondition that any good or service fits to human needs and creates a benefit. It is clearly linked to certain cultural value systems (e.g., in terms of what is perceived as a food, but also to cultural services such as spiritual values). In reverse, monetary values in capitalist societies are governed by exchange relations established by the (world) market, and thus, tend to ignore the relevance of ecosystems and nature for the actual benefits they provide. (It is an empirical question whether regional markets, not linked to the world market, still exist. In any case they are always threatened by global markets.) Use values which stimulate the overuse of resources can, of course, undermine the functional requirements of ecosystems. However, monetary or exchange values tend to ignore the biophysical requirements of ecosystems categorically and, thus, create a contradiction also called the 'metabolic rift' of capitalist societies [55–57]. Thus, this contradiction is the underlying cause why capitalism tends to destroy the productivity of land and soil [58,59]: it is not the intervention into nature (i.e., the colonisation of nature) but a specific kind of intervention as institutionalised into the capitalist market system that must be addressed by response strategies [60].

## 5. The Role of Integration in Ecosystem Services Research: The Need for a Transdisciplinary Approach

The sections above show that integration as the main challenge of ES as a boundary concept appears in different ways, particularly:

(a)   integrating scientific disciplines, i.e., the existing disciplinary knowledge as well as their modes of knowledge generation (methodologies);

(b)   integrating knowledge from different sources (scientific and non-scientific);

(c)   integrating different purposes for applying the ES concept (from assessment and awareness raising to policy advice and support for decision-making–which may all come together in a specific project, see, for example, the TEEB-studies [42]);

(d)   integrating different perceptions of ES (from 'nature's gifts' to commodities) by different actors in civil society, policy, science, etc.;

(e)   integrating perspectives of different action fields when it comes to take note of trade-offs between ES and their use or their conservation; and

(f)   integrating different types of knowledge such as system, orientation, and transformation knowledge [61] for each step of the ES models, either the Cascade or the Stairways Model, as each step involves highly political decisions (e.g., [35,36]. In particular, orientation knowledge (e.g., which ES should be protected?) and transformation knowledge (e.g., how can we achieve sustainable use of ES for everybody?) is needed (see also [29]).

The integration tasks are anything but trivial when it comes to analysing the state and (future) changes of ES, and developing options for their management. While coping with integration, the major critique of the ES concept becomes vital: addressing integration means, on the one hand, to practice boundary work on terminology, methods, value attributions, and so on. On the other hand, integration of different perceptions, perspectives, and ultimately, different knowledge systems, has to deal with the manifold relations between nature and society, as pointed out in Section 4.

If the aim of ES-related research is to produce results that are scientifically sound and informative for practitioners and stakeholders, it should be conducted in a problem-oriented, integrative, and target-oriented manner. In this regard, Hirsch Hadorn et al. [62] pointed out that these challenges cannot be fully covered by applied research approaches, when issues cross boundaries between disciplines and problem fields. Such complex issues can only be tackled with transdisciplinary approaches [61–63]. Thus, the potentials of a transdisciplinary research mode must also be considered in ES research that aims to understand the dynamic interrelations between biophysical structures, their perception as ES, and their translation into benefits and values.

Jahn and colleagues [61] (p. 8f.) state, that *"Transdisciplinarity is a critical and self-reflexive research approach that relates societal with scientific problems; it produces new knowledge by integrating different scientific and extra-scientific insights; its aim is to contribute to both societal and scientific progress; integration is the cognitive operation of establishing a novel, hitherto non-existent connection between the distinct epistemic, social-organizational, and communicative entities that make up the given problem context."* In these terms, Transdisciplinary Research (TDR) aims to enable mutual learning processes between science and society.

There are various frameworks for conducting transdisciplinary research. Here, we focus on the conceptual, ideal-typical model of transdisciplinarity proposed by the authors of [61]. This model's basic assumption is that *"developing solutions for societal problems always means and requires linking these problems to gaps in the existing bodies of knowledge, that is, to scientific problems. Seemingly self-evident, this proposition, however, allows us to conceptualize the contributions to societal and scientific progress as the two epistemic ends of a single research dynamic."* [61] (p. 4). This general argument is applicable to and highly relevant for ES research. As described above, the ES concept and its scientific pillars are contested–at least when it comes to conflicting worldviews and systems of values within. Additionally, the basis for science-driven assessments is not always perceived to be sufficient. Recent efforts such as IPBES, the TEEB-studies at different scales, or European knowledge brokers such as biodiversity knowledge (KNEU) or EKLIPSE—The European Mechanism for Supporting Better Decision Making on Biodiversity and Ecosystem Services, indicate a need for system knowledge in the field of biodiversity and ecosystem services. They should support a better understanding of processes and the interlinkages between ecosystems and society for the 'production' and 'consumption' of ES. Some years ago, the debate on the so-called 'post 2015-targets' for biodiversity showed that there is also a need for knowledge that supports defining objectives for conservation and sustainable management of ES [64]. In such a situation, the ways and means of reaching those objectives may be highly controversial. In the end, the call for a sound consideration of trade-offs within different policy strategies, or between various actions, suggests that transformation knowledge is needed [65].

Against this background, the problem-orientation of TDR is able to give new impulses to the ES concept as a boundary concept. In Figure 3, an exemplary TDR process for working with ES is illustrated. In the upper part of the figure, the main function of ES as boundary object comes to the fore: a boundary object is the starting point for formulating a transdisciplinary epistemic object that relates to a joint problem. Thus, it demands to be able to relate a real-world problem—for example, searching for ways to solve conflicts between agricultural production and conservation in a certain area—to scientific research questions, such as the causes and effects of changes in species distribution. If using ES as a boundary concept, a possible intersection in this example could be the analysis of drivers and pressures for ES changes. This example shows that in TDR—ideal-typically—a joint problem is constituted that originates neither solely in society nor solely in science (cf. [61] (p. 4ff.). With this starting point, new integrated knowledge can be produced in order to develop (partial) answers to the initial-societal and scientific questions (path in the middle of the Figure 3). Participatory elements in combination with strong methods for relating scientific and non-scientific knowledge, such as focus groups, stakeholder workshops, the Delphi method, or integration through the formation of interdisciplinary teams are at the core [66,67]. While multi- or interdisciplinary efforts focus mainly on the knowledge production and often result in scientific papers with limited outreach regarding the initial real-world problem, TDR foresees a further step: for serving society through research, applications, and communication, the transdisciplinary integration (lower part of the Figure 3) is crucial. The knowledge and the (partial) answers gained are to be evaluated against the background of the joint problem and the initial questions. Then, two kinds of 'products' can be derived: on the one hand, there is a need for products that are applicable in practice, for example, decision-support tools, guidelines, and learning videos; on the other hand, scientific contributions are derived that are innovative for the scientific praxis. Examples mentioned in Figure 3 (model of social–ecological systems, SES) show that those scientific contributions go beyond disciplinary discourses. This example is mentioned here, because ES are a kind of scientific eye-opener for a better understanding of the societal relations to nature [2,68].

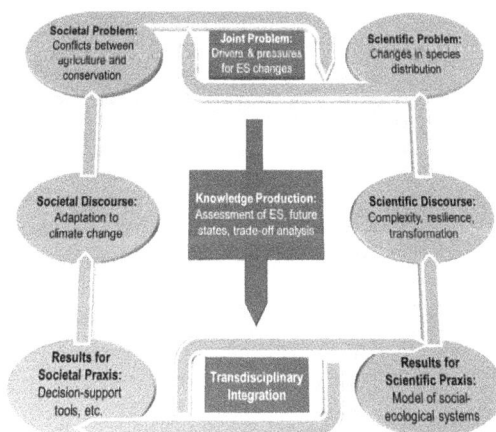

**Figure 3.** Framing ES research as a Transdisciplinary Research (TDR) process, based on the ideal-typical model for TDR by [61] (p. 5).

## 6. Rethinking Ecosystem Services Research: Implications and Requirements towards a Social–Ecological Application

In the final section, we provide and discuss implications of and requirements derived from the analysis of ES research presented above for the further development of social–ecological approaches. First, there is a need for identifying and applying empirical and analytical methods and approaches that foster a more systematic and comprehensive understanding of the interdependencies between nature and society. Current forms and applications of (ES or impact) assessments often fail to sufficiently account for those interdependencies. This is particularly true for capturing the dynamics of ES supply and demand [2]. In this regard, we see great potential for bridging the ES Cascade Model [12] and the Stairways Model [21]; to go beyond a linear model of service provision and to analyse more carefully the identification of ecosystem functions, services, and benefits in the light of respective societal processes (e.g., migration and urbanization) and the boundary negotiations involved. Monetary valuation, for example, can be strictly rejected by several cultures and this conflict between monetary and anti-monetary, not only *non*-monetary valuation, creates strong boundary negotiations which may lead to a refusal of the concept of ES itself [22]. The introduction of ESPs and the notion that their identification is essentially part of and the result of complex social processes more explicitly links the observed (dynamics in) ecosystem stocks and flows to political and societal dynamics. Here, transdisciplinary approaches are essential to identify and analyse the complex interlinkages between societal and natural processes. If we take TDR seriously, as illustrated in Figure 3, there is a need to integrate ES research more explicitly into wider frameworks or larger concepts such as SES (see [2], but also [65,69,70]) or socioeconomic metabolism [51,71]) in order to capture the intended and unintended activities from society on nature that might influence ES provisioning. This promises a better consideration of the social–ecological dynamics of ecosystem services, namely the functional relation between ES supply and demand influencing each other [2].

Moreover, a better link of ES research with other approaches may also improve their uptake within ongoing or future international research programmes, such as Future Earth. Such approaches are fostering and calling for the co-design and co-production of knowledge—by means of inter- and transdisciplinary methods—to provide scientifically sound input for a transformation towards sustainability [60]. Here, ES research may play a pivotal role, for example, to determine options and constraints for further land use and biomass production.

Thus, the ongoing debates on the further development of the conceptual framework used in the IPBES process, in particular with respect to the different perceptions of nature as well as appropriate ways of value articulating, can and should be supported by a systematic application of inter- and transdisciplinary methods. There is a vast variety of useful entry points or opportunities to demonstrate the added value of inter- and transdisciplinary research methods and the need for their further development. Many policies and other regulatory frameworks at all levels are dealing either directly (e.g., via research frameworks such as the EU Horizon 2020 or similar) or indirectly with ES research, by making ES-related assessments (e.g., the UK National Ecosystem Assessment, NEA) compulsory or at least part of the process of designing and implementing respective policies.

There are tremendous governance challenges resulting from trade-offs between ecosystem services or land-use conflicts which can be identified and made transparent through ES research. In particular, transdisciplinary research can help create 'sustainable' spaces or fora for communication and negotiation. If land use and biomass production become more important in the future, inter- and transdisciplinary ES research can help to account for and deal with such trade-offs and land-use conflicts by making different normative values and different monetary and non-monetary benefits more explicit, and to address the ESPs as well as the biophysical and societal constraints of land use (e.g., the distributional effects and the power relations behind them).

Finally, one should also reflect on the discrepancy between a concept that inherently hinges on inter- and transdisciplinary research approaches and 'thinking' as well as respective forms of collaboration, and an academic system that is (still) rather cherishing disciplinarily publications and 'rigorous' research.

**Acknowledgments:** Christian Schleyer and Christoph Görg acknowledge funding by the European Union (7th Framework Program; Grant Agreement no: 308428).

**Author Contributions:** Christian Schleyer had the idea and developed the concept of the paper. Further, he was responsible for the introduction and summarised the scientific discourse on the socio-ecological nature of the ES concept. Christoph Görg was responsible for summarising the arguments why the ES concept should be seen as a boundary concept and for establishing the link between the ES concept and the colonisation of nature approach. Alexandra Lux was responsible for the background on transdisciplinary research and Marion Mehring provided the arguments for including it into ES research. All authors contributed to the discussion equally. Christian Schleyer was responsible for the overall writing process of the paper.

**Conflicts of Interest:** The authors declare no conflict of interest. The founding sponsors had no role in the design of the study; in the collection, analyses, or interpretation of data; in the writing of the manuscript, and in the decision to publish the results.

## References

1. Millennium Ecosystem Assessment. *Ecosystems and Human Well-Being. A Framework for Assessment*; Island Press: Washington, DC, USA, 2005.
2. Mehring, M.; Zajonz, U.; Hummel, D. Social-Ecological Dynamics of Ecosystem Services: The Functional Relation Between Supply and Demand—Evidence from Socotra Archipelago, Yemen and the Sahel Region, West Africa. *Sustainability* **2017**, under review.
3. Von Becker, E.; Jahn, T. (Eds.) *Soziale Ökologie: Grundzüge einer Wissenschaft von den Gesellschaftlichen Naturverhältnissen*; Campus: Frankfurt/M., Germany, 2006.
4. Haberl, H.; Fischer-Kowalski, M.; Krausmann, F.; Winiwarter, V. (Eds.) *Social Ecology. Society-Nature Relations across Time and Space*; Springer International: Basel, Switzerland, 2016.
5. Kramm, J.; Pichler, M.; Schaffartzik, A.; Zimmermann, M. Societal relations to nature in times of crisis—Social ecology's contributions to interdisciplinary sustainability studies. *Sustainability* **2017**, under review.
6. Schröter, M.; van der Zanden, E.H.; van Oudenhoven, A.P.E.; Remme, R.P.; Serna-Chavez, H.M.; De Groot, R.S.; Opdam, P. Ecosystem services as a contested concept: A synthesis of critique and counter-arguments. *Conserv. Lett.* **2014**, *7*, 514–523. [CrossRef]
7. Boyd, J.; Banzhaf, S. What are ecosystem services? The need for standardized environment accounting units. *Ecol. Econ.* **2007**, *63*, 616–626. [CrossRef]
8. Fisher, B.; Turner, R.K.; Morling, P. Defining and classifying ecosystem services for decision-making. *Ecol. Econ.* **2009**, *68*, 643–653. [CrossRef]
9. Barkmann, J.; Glenk, K.; Keil, A.; Leemhuis, C.; Dietrich, N.; Gerold, G.; Marggraf, R. Confronting unfamiliarity with ecosystem functions: The case for an ecosystem service approach to environmental valuation with stated preference methods. *Ecol. Econ.* **2008**, *65*, 48–62. [CrossRef]
10. Lamarque, P.; Tappeiner, U.; Turner, C.; Steinbacher, M.; Bardgett, R.D.; Szukics, U.; Schermer, M.; Lavorel, S. Stakeholder perceptions of grassland ecosystem services in relation to knowledge on soil fertility and biodiversity. *Reg. Environ. Chang.* **2011**, *11*, 791–804. [CrossRef]
11. Nahlik, A.M.; Kentula, M.E.; Fennessy, M.S.; Landers, D.H. Where is the consensus? A proposed foundation for moving ecosystem service concepts into practice. *Ecol. Econ.* **2012**, *77*, 27–35. [CrossRef]
12. Potschin, M.; Haines-Young, R. Ecosystem Services: Exploring a geographical perspective. *Prog. Phys. Geog.* **2011**, *35*, 575–594. [CrossRef]
13. Menzel, S.; Teng, J. Ecosystem services as a stakeholder-driven concept for conservation science. *Conserv. Biol.* **2010**, *24*, 907–909. [CrossRef] [PubMed]
14. Chan, K.M.; Guerry, A.D.; Balvanera, P.; Klain, S.; Satterfield, T.; Basurto, X.; Bostrom, A.; Chuenpagdee, R.; Gould, R.; Halpern, B.S.; et al. Where are Cultural and Social in Ecosystem Services? A Framework for Constructive Engagement. *BioScience* **2012**, *62*, 744–756.
15. Bieling, C.; Plieninger, T. Recording manifestations of cultural ecosystem services in the landscape. *Landsc. Res.* **2013**, *38*, 649–667. [CrossRef]
16. Kosoy, N.; Corbera, E. Payments for ecosystem services as commodity fetishism. *Ecol. Econ.* **2010**, *69*, 1228–1236. [CrossRef]

17. McAfee, K.; Shapiro, E. Payments for ecosystem services in Mexico: Nature, neoliberalism, social movements, and the state. *Ann. Assoc. Am. Geogr.* **2010**, *100*, 579–599. [CrossRef]

18. McCauley, D. Selling out on nature. *Nature* **2006**, *443*, 27–28. [CrossRef] [PubMed]

19. Görg, C.; Aicher, C. Ökosystemdienstleistungen—Zwischen Natur und Gesellschaft. Anforderungen an eine inter- und transdisziplinäre Forschung aus Sicht der Sozialwissenschaften. In *Berlin-Brandenburgische Akademie der Wissenschaften, Berichte und Abhandlungen*; De Gruyter-Akademie Verlag: Berlin, Germany, 2014; Volume 16, pp. 35–58.

20. Loft, L.; Lux, A.; Jahn, T. A Social-Ecological Perspective on Ecosystem Services. In *Routledge Handbook of Ecosystem Services*; Potschin, M., Haines-Young, R., Fish, R., Eds.; Routledge: London, UK; New York, NY, USA, 2016; pp. 88–92.

21. Spangenberg, J.H.; Von Haaren, C.; Settele, J. The ecosystem service cascade: Further developing the metaphor. Integrating societal processes to accommodate social processes and planning, and the case of bioenergy. *Ecol. Econ.* **2014**, *104*, 22–32. [CrossRef]

22. Spangenberg, J.H.; Görg, C.; Truong, D.T.; Tekken, V.; Bustamante, J.V.; Settele, J. Provision of ecosystem services is determined by human agency, not ecosystem functions. Four case studies. *Int. J. Biodivers. Sci. Ecosyst. Serv. Manag.* **2014**, *10*, 40–53. [CrossRef]

23. Hausknost, D.; Gaube, V.; Haas, W.; Smetschka, B.; Schmid, M.; Lutz, J.; Singh, S.J. "Society can't move so much as a chair!" Systems, Structures and Actors in Social Ecology. In *Social Ecology: Society-Nature Relations across Time and Space*; Haberl, H., Fischer-Kowalski, M., Krausmann, F., Winiwarter, V., Eds.; Springer: Dordrecht, The Netherlands, 2016; pp. 125–147.

24. Huntsinger, L.; Oviedo, J.L. Ecosystem services are socialecological services in a traditional pastoral system: The case of California's mediterranean rangelands. *Ecol. Soc.* **2014**, *19*, 8. [CrossRef]

25. Plieninger, T.; van der Horst, D.; Schleyer, C.; Bieling, C. Sustaining ecosystem services in cultural landscapes. *Ecol. Soc.* **2014**, *19*, 59. [CrossRef]

26. Comberti, C.; Thornton, T.F.; Wylliede Echeverria, V.; Patterson, T. Ecosystem services or services to ecosystems? Valuing cultivation and reciprocal relationships between humans and ecosystems. *Glob. Environ. Chang.* **2015**, *34*, 247–262. [CrossRef]

27. Palomo, I.; Felipe-Lucia, M.R.; Bennett, E.M.; Martín-López, B.; Pascual, U. Disentangling the Pathways and Effects of Ecosystem Service Co-Production. *Adv. Ecol. Res.* **2016**, *54*, 245–283. [CrossRef]

28. Fischer-Kowalski, M.; Erb, K.-H. Core Concepts and Heuristics. In *Social Ecology: Society-Nature Relations across Time and Space*; Haberl, H., Fischer-Kowalski, M., Krausmann, F., Winiwarter, V., Eds.; Springer: Dordrecht, The Netherlands, 2016; pp. 29–62.

29. Abson, D.J.; von Wehrden, H.; Baumgärtner, S.; Fischer, J.; Hanspach, J.; Härdtle, W.; Heinrichs, H.; Klein, A.M.; Lang, D.J.; Martens, P.; et al. Ecosystem services as a boundary object for sustainability. *Ecol. Econ.* **2014**, *103*, 29–37. [CrossRef]

30. Potschin, M.; Haines-Young, R. Defining and measuring ecosystem services. In *Routledge Handbook of Ecosystem Services*; Potschin, M., Haines-Young, R., Fish, R., Eds.; Routledge: London, UK; New York, NY, USA, 2016; pp. 25–44.

31. Hauck, J.; Görg, C.; Jax, K.; Varjopurob, R.; Ratamäki, O. Benefits and limitations of the ecosystem services concept in environmental policy and decision making—Some stakeholder perspectives. *Environ. Sci. Policy* **2013**, *25*, 13–21. [CrossRef]

32. Jordan, A.J.; Russel, D. Embedding the concept of ecosystem services? The utilisation of ecological knowledge in different policy venues. *Environ. Plan. C Gov. Policy* **2014**, *32*, 192–207. [CrossRef]

33. Jasanoff, S. *States of Knowledge: The Co-Production of Science and Social Order*; Routledge: London, UK; New York, NY, USA, 2004.

34. Kull, C.A.; Arnauld de Sartre, X.; Castro-Larrañaga, M. The political ecology of ecosystem services. *Geoforum* **2015**, *61*, 122–134. [CrossRef]

35. Kolinjivadi, V.; Van Hecken, G.; Rodríguez de Francisco, J.C.; Pelenc, J.; Kosoy, N. As a lock to a key? Why science is more than just an instrument to pay for nature's services. *Curr. Opin. Environ. Sustain.* **2017**, *26–27*, 1–6. [CrossRef]

36. Hausknost, D.; Grima, N.; Singh, S.J. The political dimensions of Payments for Ecosystem Services (PES): Cascade or stairway? *Ecol. Econ.* **2017**, *131*, 109–118. [CrossRef]

37. Tancoigne, E.; Barbier, M.; Cointet, J.-P.; Richard, G. The place of agricultural sciences in the literature on ecosystem services. *Ecosyst. Serv.* **2014**, *10*, 35–48. [CrossRef]

38. Alison, G. Power Ecosystem services and agriculture: Tradeoffs and synergies. *Philos. Trans. R. Soc. B* **2010**, *365*, 2959–2971. [CrossRef]

39. Laurans, Y.; Rankovic, A.; Billé, R.; Pirard, R.; Mermet, L. Use of ecosystem services economic valuation for decision making: Questioning a literature blindspot. *J. Environ. Manag.* **2013**, *119*, 208–219. [CrossRef] [PubMed]

40. Gómez-Baggethun, E.; Ruiz-Perez, M. Economic valuation and the commodification of ecosystem services. *Prog. Phys. Geogr.* **2011**, *35*, 613–628. [CrossRef]

41. Norgaard, R.B. Ecosystem services: From eye-opening metaphor to complexity blinder. *Ecol. Econ.* **2010**, *69*, 1219–1227. [CrossRef]

42. The Economics of Ecosystem Services and Biodiversity (TEEB). *The Economics of Ecosystems and Biodiversity: Ecological and Economic Foundations*; Kumar, P., Ed.; Earthscan: London, UK; Washington, DC, USA, 2010.

43. Díaz, S.; Demissew, S.; Carabias, J.; Joly, C.; Lonsdale, M.; Ash, N.; Larigauderie, A.; Adhikari, J.R.; Arico, S.; Báldi, A.; et al. The IPBES Conceptual Framework—connecting nature and people. *Curr. Opin. Environ. Sustain.* **2015**, *14*, 1–16. [CrossRef]

44. Seppelt, R.; Fath, B.; Burkhard, B.; Fisher, J.L.; Grêt-Regamey, A.; Lautenbach, S.; Pert, P.; Hotes, S.; Spangenberg, J.H.; Verburg, P.H.; et al. Form follows function? Proposing a blueprint for ecosystem service assessments based on reviews and case studies. *Ecol. Indic.* **2012**, *21*, 145–154. [CrossRef]

45. Hauck, J.; Görg, C.; Varjopuro, R.; Ratamäki, O.; Maes, J.; Wittmer, H.; Jax, K. "Maps have an air of authority": Potentials and challenges of ecosystem service maps in decision making at different levels of decision making. *Ecosyst. Serv.* **2013**, *4*, 25–32. [CrossRef]

46. Carmen, E.; Young, J.; Watt, A. *Knowledge Needs for the Operationalization of the Concept of Ecosystem Services (ES) and Natural Capital (NC)*; EU FP7 OpenNESS Project Deliverable 2.3; European Commission FP7: Brussels, Belgium, 2015.

47. Dempsey, J.; Robertson, M.M. Ecosystem Services: Tensions, impurities and pints of engagement within neoliberalism. *Prog. Hum. Geogr.* **2012**, *36*, 758–779. [CrossRef]

48. Barnaud, C.; Antona, M. Deconstructing ecosystem services: Uncertainties and controversies around a socially constructed concept. *Geoforum* **2014**, *56*, 113–123. [CrossRef]

49. Gómez-Baggethun, E.; De Groot, R.; Lomas, P.L.; Montes, C. The history of ecosystem services in economic theory and practice: From early notions to markets and payment schemes. *Ecol. Econ.* **2010**, *69*, 1209–1218. [CrossRef]

50. Haines-Young, R.; Potschin, M. Common International Classification of Ecosystem Services (CICES): Consultation on Version 4, August–December 2012. EEA Framework Contract No. EEA/IEA/09/003, 2013. Available online: www.cices.euorwww.nottingham.ac.uk/cem (accessed on 27 December 2016).

51. Fischer-Kowalski, M.; Haberl, H. Sustainable Development: Socio-Economic Metabolism and Colonization of Nature. *Int. Soc. Sci. J.* **1998**, *158*, 573–587. [CrossRef]

52. Haberl, H.; Erb, K.-H.; Krausmann, F.; Gaube, V.; Bondeau, A.; Plutzar, C.; Gingrich, S.; Lucht, W.; Fischer-Kowalski, M. Quantifying and mapping the human appropriation of net primary production in earth's terrestrial ecosystems. *Proc. Natl. Acad. Sic. USA* **2007**, *104*, 12942–12947. [CrossRef] [PubMed]

53. Fischer-Kowalski, M.; Weisz, H. Society as Hybrid between Material and Symbolic Realms. Toward a Theoretical Framework of Society-Nature Interrelation. *Adv. Hum. Ecol.* **1999**, *8*, 215–251.

54. Cardinale, B.J.; Duffy, J.E.; Gonzalez, A.; Hooper, D.U.; Perrings, C.; Venail, P.; Narwani, A.; Mace, G.M.; Tilman, D.; Wardle, D.A.; et al. Biodiversity loss and its impact on humanity. *Nature* **2012**, *486*, 59–67. [CrossRef] [PubMed]

55. Marx, K. *Capital*; Vintage: New York, NY, USA, 1981; Volume I.

56. Foster, J.B. Marx's Theory of Metabolic Rift: Classical Foundations for Environmental Sociology. *Am. J. Sociol.* **1999**, *105*, 381. [CrossRef]

57. Moore, J.W. Transcending the Metabolic Rift: Towards a Theory of Crises in the Capitalist World-Ecology. *J. Peasant Stud.* **2011**, *38*, 1–46. [CrossRef]

58. Polanyi, K. *The Great Transformation*; Farrar & Rinehart: New York, NY, USA, 1944.

59. Marx, K. *Capital*; Penguin Books: London, UK, 1990 [1867]; Volume I.

60. Görg, C.; Brand, U.; Haberl, H.; Hummel, D.; Jahn, T.; Liehr, S. Challenges for Social-Ecological Transformations. Contributions from Social and Political Ecology. *Sustainability* **2017**, accepted.

61. Jahn, T.; Bergmann, M.; Keil, F. Transdisciplinarity: Between mainstreaming and marginalization. *Ecol. Econ.* **2012**, *79*, 1–10. [CrossRef]

62. Hirsch Hadorn, G.; Bradley, D.; Pohl, C.; Rist, S.; Wiesmann, U. Implications of Transdisciplinarity for Sustainability Research. *Ecol. Econ.* **2006**, *60*, 119–128. [CrossRef]

63. Lang, D.J.; Wiek, A.; Bergmann, M.; Stauffacher, M.; Martens, P.; Moll, P.; Swilling, M.; Thomas, C.J. Transdisciplinary research in sustainability science: Practice, principles, and challenges. *Sustain. Sci.* **2012**, *7*, 25–43. [CrossRef]

64. Mace, G.M.; Cramer, W.; Díaz, S.; Faith, D.P.; Larigauderie, A.; Le Prestre, P.; Palmer, M.; Perrings, C.; Scholes, R.J.; Walpole, M.; et al. Biodiversity targets after 2010. *Curr. Opin. Environ. Sustain.* **2010**, *2*, 3–8. [CrossRef]

65. Carpenter, S.R.; Folke, C.; Norström, A.; Olsson, O.; Schultz, L.; Agarwal, B.; Balvanera, P.; Campbell, B.; Castilla, J.C.; Cramer, W.; et al. Program on ecosystem change and society: An international research strategy for integrated social-ecological systems. *Curr. Opin. Environ. Sustain.* **2012**, *4*, 1–5. [CrossRef]

66. Bergmann, M.; Jahn, T.; Knobloch, T.; Krohn, W.; Pohl, C.; Schramm, E. *Methods for Transdisciplinary Research. A Primer for Practice*; Campus Verlag: Frankfurt/M., Germany; New York, NY, USA, 2012.

67. Daedlow, K.; Podhora, A.; Winkelmann, M.; Kopfmüller, J.; Walz, R.; Helming, K. Socially responsible research processes for sustainability transformation: An integrated assessment framework. *Curr. Opin. Environ. Sustain.* **2016**, *23*, 1–11. [CrossRef]

68. Hummel, D.; Jahn, T.; Keil, F.; Liehr, S.; Stieß, I. Social Ecology as Critical, Transdisciplinary Science—Conceptualizing, Analyzing, and Shaping Societal Relations to Nature. *Sustainability* **2017**, under review.

69. Ostrom, E. A General Framework for Analyzing the Sustainability of Social-Ecological-Systems. *Science* **2009**, *325*, 419–422. [CrossRef] [PubMed]

70. Folke, C. Resilience: The emergence of a perspective for social–ecological systems analyses. *Glob. Environ. Chang.* **2006**, *16*, 253–267. [CrossRef]

71. Haberl, H.; Wiedenhofer, D.; Erb, K.-H.; Görg, C.; Krausmann, F. Interrelations between material stocks, flows and services: A new approach to tackle the decoupling conundrum. *Sustainability* **2017**, under review.

*sustainability*

MDPI

*Article*

# Social-Ecological Dynamics of Ecosystem Services: Livelihoods and the Functional Relation between Ecosystem Service Supply and Demand—Evidence from Socotra Archipelago, Yemen and the Sahel Region, West Africa

**Marion Mehring [1,2,*], Uwe Zajonz [2] and Diana Hummel [1,2]**

[1]   ISOE—Institute for Social-Ecological Research, 60486 Frankfurt am Main, Germany; hummel@isoe.de
[2]   Senckenberg Biodiversity and Climate Research Centre BiK-F, 60325 Frankfurt am Main, Germany; Uwe.Zajonz@senckenberg.de
*   Correspondence: mehring@isoe.de; Tel.: +49-69-707-691939

Received: 27 January 2017; Accepted: 6 June 2017; Published: 26 June 2017

**Abstract:** In aiming to halt global biodiversity loss, it is essential to address underlying societal processes. The concept of ecosystem services claims to bridge between biodiversity and society. At the same time there is a considerable research gap regarding how ecosystem services are provided, and how societal activities and dynamics influence the provision of ecosystem services. Interactions and dependencies between ecosystem services supply and demand come to the fore but context-specific dynamics have largely been neglected. This article is a critical reflection on the current research of ecosystem services supply and demand. We argue that there is a functional relation between the supply and demand for ecosystem services, with the two influencing each other. Scientific interest should focus on both the temporal and spatial dynamics of ecosystem services supply and demand. Presenting two studies from Socotra Archipelago, Yemen and the Sahel regions in Senegal and Mali, West Africa, we illustrate that the society behind the demand for ecosystem services is highly interrelated with ecosystem services supply. We thus advocate the adoption of a social-ecological perspective for current research on ecosystem services supply and demand in order to address these context-specific temporal and spatial dynamics.

**Keywords:** dynamics of use; ecosystem services; social-ecological system; spatial and temporal dynamics; supply and demand

## 1. Introduction

Global biodiversity still continues to decline [1,2]. Biodiversity loss is regarded to be one of the "grand challenges" for humanity, alongside climate change, globalisation, demographic change and food security. The changes in biodiversity alter ecosystem processes and functions, and subsequently influence the ability to provide goods and services to society [3,4].

There has been great progress in recent years in understanding and assessing ecosystem functions; at the same time, the benefits that humans derive from ecosystems are more and more appreciated in the social sciences [5]. The ecosystem services (ES) concept has helped raise awareness about the societal relevance of ecosystems [6]. Furthermore, the concept has proved its purpose to support the maintenance of biodiversity, at least since the publication of the TEEB (The Economics of Ecosystems and Biodiversity) reports [7,8], and the recent constitution of the Intergovernmental Platform on Biodiversity and Ecosystem Services (IPBES). The ES concept promises to bridge between society and nature, namely society's demand for certain services and the supply from ecosystems to satisfy these needs.

At the same time there is a considerable research gap regarding how societal activities and dynamics influence the provision of ES [9,10]. From a system-related perspective, the interlinkages between nature and society come to the fore [11,12] while nature-society interactions, particularly the dynamic aspects of ES, have largely been neglected [13,14].

This article advocates the adoption of a social-ecological perspective on ES in order to better understand the dynamics in supply and demand as well as any interlinkages that exist. We argue that there is functional relation between ES supply and demand, where both nature and society interact and influence each other. Here we present the latest results from two ES studies in the West African Sahel region and Socotra Archipelago, Yemen, respectively, and show that temporal and spatial dynamics in both, nature and society do occur. Finally, we critically reflect on current ES research and ultimately deduce future directions for research.

## 2. Conceptual Reflection: The Social-Ecological Perspective

From a social-ecological perspective, dynamics of supply and demand for ES represent historically and culturally specific societal relations to nature [15]. For the empirical analysis, they can be conceptualised as social-ecological systems (SES) [12] (Figure 1). Societal actors influence the natural system and related ecosystem functions. It is via the latter that nature provides ES, but also ecosystem disservices that harm society. Societal actors directly or indirectly influence the SES via intended management activities or unintended side effects thereof [16]. The four mediating dimensions of knowledge, practices, institutions and technology shape the social-ecological structures and processes as the core part of the SES [16]. Biodiversity plays a critical role at different hierarchical levels of the ecosystem functions and services. It acts as a regulator of the ecosystem processes that underpin ES, and can be a final good in its own right [17,18]. Hence there is an urgent need to consider biodiversity in all its complexity, including its relation to and interaction with the ecosystem and society [9,19–22].

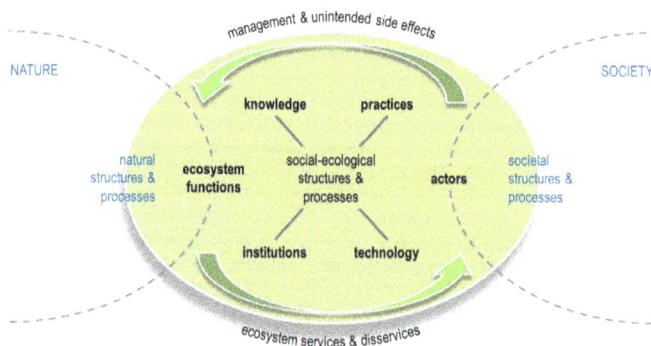

**Figure 1.** The concept of social-ecological systems [22].

The ES concept is generally discussed as being adequately conceived to address the relation between nature and society [6]. Focusing on the interlinkages between society and nature, it inherently captures this social-ecological perspective [11,12,23]. The ES concept has increasingly been included in research objectives, policies and international initiatives such as The EU Biodiversity Strategy to 2020 [24] as well as the IPBES Conceptual Framework [25]. These activities resulted in different research projects to map and assess biodiversity and ES by applying the conceptual dichotomy of ES supply and demand [5]. These authors [5] interpret ES supply as the capacity of a particular area to provide a specific bundle of ES within a given time period. ES demand is understood as the need for specific ES on the part of society, particular stakeholder groups or individuals [26]. This demand is interrelated with other factors, e.g., behavioural norms, consumption patterns, demographic changes, technical regimes or socio-economic conditions [27]. As yet, a variety of different methods from both the natural and social sciences and the humanities have been developed in order to address the respective research questions on the supply [28] and demand [29] for ES.

Shortcomings still exist in the application of the ES concept [23]. The authors argue that the social-ecological nature of the ES concept must be elaborated more carefully. The recent discussion on relational values [30] is a first step in this direction of underlining the social-ecological nature. Besides intrinsic and instrumental values, relational values address those values that pertain to all manner of relationships between nature and people, including relationships that are between people but involve nature [30]. This means e.g., hiking in nature is not only valuable due to a beautiful and remote landscape but also due to being together with good friends for this hiking tour.

However, the dynamic aspects of ES supply and demand have hardly been considered yet [13]. While in some cases ES providing and benefiting areas overlap, in others they do not. In case of spatial disconnection of areas providing and demanding ES, the concept of ES flow is applied. ES flows can thus reflect the spatial relationship between ES supply and ES demand. The flows can work passively through biophysical processes (e.g., climate regulation), actively through societal processes (e.g., transport and trade), or through a combination of both [27]. However, up to now there have been very few attempts to quantify ES flows. In addition, temporal dynamics, such as historic and future land use change, can also influence ES supply and demand [13]. Finally, a recent study [31] on marine and coastal ES explores the social and ecological dynamics of ES, called ecosystem service elasticity. The authors present a conceptual framework that unpacks the chain of causality from ecosystem stocks through flows, goods, value, and shares to contribute to the well-being of different people. However, this framework does not address the functional relation between ES supply and demand, i.e., that nature and society mutually influence each other via spatial and temporal dynamics of ES supply and demand.

Against this background, this article will contribute to the discussion on the dynamics of ES and calls for a social-ecological perspective in current research on ES supply and demand.

## 3. The Social-Ecological Dynamics of Ecosystem Services: Empirical Evidence

Below we present two studies from the West African Sahel and the Socotra Archipelago, Yemen demonstrating the spatial and temporal dynamics of ES supply and demand, respectively. Both studies particularly exemplify the functional relation between ES supply and demand as well as temporal and spatial ES dynamics. However, spatial effects are more pronounced in the Sahel case study whereas temporal effects are more important in the Socotra case study, and hence are illuminated accordingly.

### 3.1. Spatial Dynamics of Ecosystem Services Supply and Demand: The Case of the West African Sahel

The case study of the West African Sahel illustrates the complex interactions between societal factors and ES demand and supply. The SES (Figure 1) investigated here consists of relations between actors such as villagers, farmers and migrants, and different ecosystem functions such as primary production. This relation is mediated by practices (e.g., seasonal migration), knowledge such as local knowledge about soil and plants, technology such as irrigation schemes or use of fertilizers, and institutions such as social networks or migration policies. Important ES are food production, i.e., crop yields, an example for disservices are pests. The actors influence the system for example by agricultural management or unintended side effects thereof [32]. In this way, the SES concept introduced above allows for an analysis of the specific functional relation between ES supply and demand, with a special focus on spatial dynamics.

The majority of the rural population in the West African Sahel depends on subsistence and small-scale farming. The Fulani, Wolof and Dogon are the largest ethnic groups in the Sahel. The Wolof and Dogon are traditionally farmers, while the Fulani are traditionally semi-nomadic pastoralists. Climate change, especially increasing temperatures and rainfall variability, impacts on ecosystems and poses considerable risks to society. Linguère in Senegal and Bandiagara in Mali (Figure 2) are areas in the semi-arid Sahel zone which have always suffered from drought periods and from land degradation. The Bandiagara region in Mali is predominantly inhabited by the Dogon and Linguère in Senegal by the Fulani and Wolof.

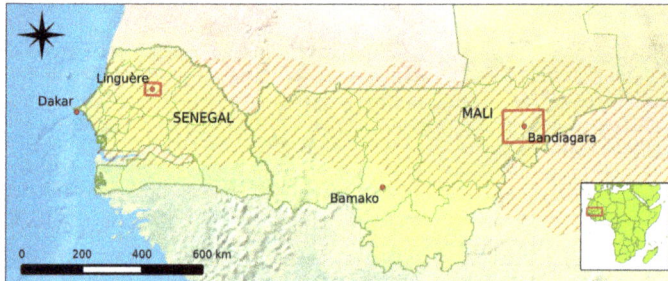

**Figure 2.** Location of study areas in Senegal and Mali (red rectangles). Shaded areas indicate the Sahel's extent, delimited by the 250 mm/a isohyet in the north and 900 mm/a in the south.

Our study followed an inter—and transdisciplinary research approach which integrated natural-scientific and social-scientific knowledge, as well as practical knowledge. The natural-scientific analysis investigated interactions of climate change and land degradation. Changes in temperature, precipitation and vegetation have been analysed by combining global remote sensing techniques with high-resolution images and detailed field work on a local scale [33,34]. The social-empirical research investigated perceptions of environmental changes affecting the local population, as well as their motives for migration and migration patterns. The data collection combined qualitative interviews, focus group discussion and participant observation. In 2012, a survey was conducted with 905 individuals in villages in Bandiagara (324 questionnaires) and Linguère (337 questionnaires) and with migrants from these regions in Bamako (121 questionnaires) and Dakar (123 questionnaires). The survey solely included the Wolof, Dogon and sedentary Fulani [34]. The same questionnaire was used for both study sites. Interviewees and survey participants were selected randomly and differed with respect to age, education, ethnicity, gender, and migration experience. The sampling of the survey participants based on a quota sample and defined three age categories (18 to 30 years, 31 to 50 years, 51 years and older); while the qualitative interviews were also conducted with younger people [32,35]. For the integration of natural-scientific and social-scientific data, Bayesian Belief Networks were used as modelling method [36].

Varying rainfall patterns, but also human activities have contributed to massive changes in flora, fauna and soils. After an extremely dry period from the 1970s to the 1990s, rainfall has been increasing in both regions, but its temporal variability is increasing too. These changes have caused an increase in vegetation in many areas (the so called 'greening phenomenon'). This trend is supported by farmer-managed agroforestry, reforestation programs and nature conservation measures [33,34]. Given the crucial role of agriculture, the demand for ES when it comes to food production and livelihoods is considerable. Local farmers and livestock breeders demand particularly provisioning ES such as crops, wild fruits, fuel wood and fodder, but also supporting services, particularly soil formation, and regulating services such as purification of water. The main economic activities in Bandiagara, Mali are rain-fed agriculture, particularly the cultivation of groundnut, sorghum and millet, as well as vegetable gardening, primarily onions by the Dogon. Linguère, Senegal is a traditionally silvo-pastoral zone. Together with livestock breeding, cropping represents an important occupation for the Wolof and the sedentary Fulani [35]. Shrubs and trees provide considerable ES and thus play a major role in people's everyday life in both study areas. They are a main source of firewood and timber, and are traditionally used for cooking, construction, medicine, and for religious purposes. Leaves and fruits are used for animal feed, and the selling of firewood and charcoal is a common practice which makes a considerable contribution to the income of the local population. In both study sites a survey was conducted with 905 participants in total (Mali: 445; Bandiagara: 324 and Bamako: 121, Senegal: 460; Linguère: 337 and Dakar: 123).

The strong significance of rainfall patterns is very apparent in rain-fed agriculture in the study areas. The majority of the population in Linguère and Bandiagara rely on small-scale farming and livestock breeding. They adapted to the seasonal dynamics in order to satisfy their needs. In this regard, human migration within the countries, especially to urban centres and the capitals, is a significant characteristic of the culture in the region. Few respondents migrated within the departments of Linguere and Bandiagara, while the majority from both study areas moved either to long-distance destinations within the country (i.e., rural departments of Kaolak, Tambacounda and Matam in Senegal or Ségou and Sikasso in Mali) or—in the case of respondents in Mali, to Cote d'Ivoire. One third of respondents chose the capital as destination [35]. The distance between Bamako and the Bandiagara region is at around 670 km, while the distance between Dakar and Linguère is at around 310 km. This means a whole-day journey by bus.

The survey showed that a vast majority in both study regions (87%) have personally experienced migration [32,36]. While in Mali the share of men with migration experience is higher than for women (94% male vs. 70% female), there is no significant difference between the migration experience of men and women in Senegal (80% male vs. 78% female) [35]. Regarding the duration of migration, seasonal (3 to 9 months) and temporary (10 months to 5 years) are the most dominant types. These results highlight the strong interdependence of ES supply and migration [32]. The seasonal variability in the food production system required adaptation strategies from the local people, namely migration to other regions, among other strategies presented below.

Climate change particularly influences crop yields in this region, and thus the ES-dependent food supply [37,38]. Altered rainfall and land degradation directly influence ES supply and compromise the income basis of rural households, because these factors impair both agricultural production and returns [36,39,40]. The dynamics at work in the ES supply trigger a complex compensatory process. More diversified sources of income and multi-local income become relevant for a large proportion of the population as a result. Local people apply different compensatory strategies (Figure 3). Migration is just one coping strategy: increased money transfers from migrants seem to be more important than increasing the number of migrants. The migration of entire families is very rare in both study regions. Usually only few family members migrate to find employment, while another part of the family remains in the village and is still depending on local food supply. The money earned in migration is then targeted at supplementing the agricultural outcome [35]. Other adaptation strategies entail falling back on alternative ES: the selling of livestock represents the most important coping strategy in both areas. The two regions differ with regard to activities such as the selling of wood, or collection and selling of wild fruits, herbs and straw, and the pursuit of gardening activities. In Bandiagara, Mali, in particular, the income from irrigated agriculture alongside small dams, seasonal waterholes or small rivers plays a crucial role. Wood or wild fruits, herbs and straw are mainly traded against crops at the market.

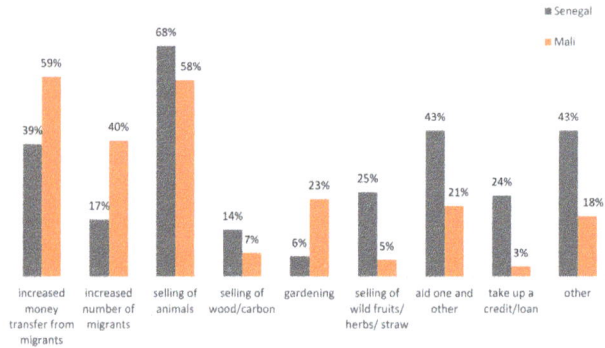

**Figure 3.** Distribution of strategies to compensate for bad harvests (multiple answers possible) in both study regions, Linguère, Senegal and Bandiagara, Mali. The figure illustrates the livelihood activities for those people whose family is involved in an agricultural activity. This is the case for 854 individuals (=100%) (Senegal $n$ = 411; Mali $n$ = 443) [32].

The decrease in crop yields in rural areas makes it necessary to find ways of generating multiple sources of local income in other economic sectors at different places, thus contributing to the spatial as well as temporal alteration of the demand for ES. Seasonal migration (3–9 months) in particular is a traditionally long-established strategy to deal with the seasonality and variability of rainfall, and thus food production. People leave the village during the dry season, when there is little work to do on the fields, and move to the cities to work in the construction or service sector, or, in the case of women, in the domestic sector, for example. They return to the villages for the harvest season [32,36]. This seasonal migration decreases demand in the region of origin. For example, there is less demand for firewood required for cooking and thus shrubs and trees can recover during this period. This temporary shift of residence might also compensate for the alteration of ES supply .Though our study did not quantify the demand for ES during periods of migration and local harvest periods, the results of the social-empirical analysis suggest that people respond to decreasing rainfall and crop yields by extending their duration of migration, either by leaving the village earlier or by staying longer in migration—or by increasing the number of migrants within the household [32,35,36]. The interplay of one ES supply and demand for other ES has an explicitly spatial characteristic here. Besides migration, the demographic dynamics are of great relevance for spatial-temporal dynamics of ES demand and supply: Both regions still experience high population growth with a current annual growth rate of around 3% in Bandiagara and 2.4% in Linguère as a consequence of high fertility rates and a young age structure [40]. Overall, it must be assumed that population growth is associated with an increase in ES demand, notwithstanding the high mobility of the population in both study regions. The level of formal education of the population must be considered here as one of the most important social characteristics. The lower the level of formal education, the more likely the survey participants are to rely on agriculture as their main source of income. Agriculture is an economic activity for 58% of respondents, but only for 18% of participants with a high education level. Participants with a higher level of formal education are more likely to be involved in business, administration, or health, i.e., sectors which are not directly depending on ES. Among the survey participants, the level of education is considerably low: only 24% have obtained formal education, and the level of education is significantly lower among women than men [41]. Furthermore, people fall back on alternative ES within the given SES, such as the collection of wild fruits and wood in order to compensate for deteriorating environmental conditions (Figure 3). These strategies might also augment the pressure on ES. For example, the most dominant adaptation strategy, i.e., the selling of animals, can increase the demand for ES in other target areas where the animals move to, as animal husbandry poses high demands on ecosystem services particularly fodder and water.

Summing up, the study from West African Sahel illustrates the strong dependence on crop yields for agricultural production and food supply. The seasonality of this ES supply is crucial. Overall, the complex dynamics in the ES supply trigger complex compensatory processes which in turn have impacts on both ES supply and demand. Migration, particularly seasonal migration, is one of the most important drivers with great impacts on spatial and temporal dynamics, but must be considered in conjunction with social differences such as age, gender, and education level.

*3.2. Temporal Dynamics of Ecosystem Service Supply and Demand: The Case of Socotra Archipelago, Yemen*

The case study of the Socotra Archipelago is a fishery-based SES representing complex social-ecological processes. In particular, the interaction between local fisher communities (actors) and ecosystem functions such as marine primary productivity is investigated. The seasonal variation in ES supply, namely fish biomass, and its impact on societal structures and processes (i.e., adaptation of local livelihood strategies) is examined. This interrelation is considered as being shaped by the mediating dimensions [16] of practices (fishing customs of local fishermen), technology for fishing, knowledge about the local monsoon patterns and institutions such as local associations or policies. In this sense, applying the SES concept as described above (see Figure 1) helps analyzing the respective functional relations between ES supply and demand with a particular focus on temporal dynamics at this case study site.

The Socotra Archipelago, part of Yemen, lies in the northern Indian Ocean, at the Horn of Africa (Figure 4). It is globally recognised for its outstanding biodiversity and endemism, the reason why the entire island group was designated a UNESCO World Heritage Site in 2008 [42,43]. The archipelago is characterised by a unique cultural heritage: the Socotri people speak a unique non-written pre-Islamic language of ancient origin, and their culture host a wealth of traditional knowledge on the sustainable use of natural resources and biodiversity [44–46]. As for Yemen as a whole the human island population of some 60,000 ranks very low on the Human Development Index (HDI) (rank 168 for Yemen in 2015, HDI 0.482 [47]).

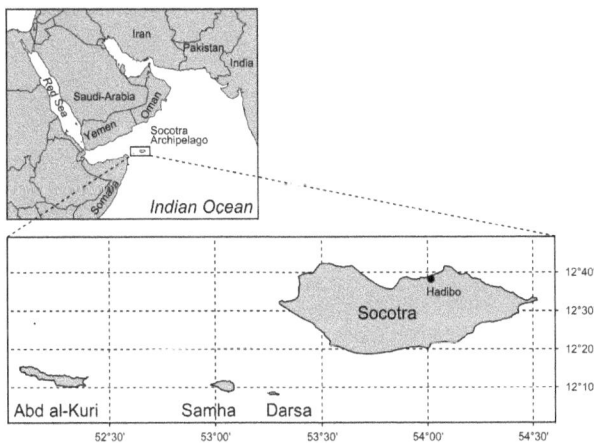

**Figure 4.** Overview map of the Socotra Archipelago, showing its geographic position in the Indian Ocean.

The research program started in 2007 and was conducted in four coastal pilot areas at the north Coast of Socotra Island, including six study villages, and fourteen permanent subtidal monitoring sites in close proximity to these villages. In particular the research focused on marine and coastal ecosystem services being critical to a changing subtropical social-ecological island system, environmental variables determining key ES as well as the socio-economic context of the local fisher communities. Biological and environmental data were generated between 2009 and 2013 primarily by subtidal visual fish biomass estimates, in-situ recording of environmental and ES data, analyses of ocean-color based environmental proxies (i.e., Chlorophyll a), carbon-based productivity models and sea surface temperature. A pilot study on ES supply and demand was completed in 2011. In 2013 empirical social research addressing the vulnerability of local livelihoods was conducted. In particular a household survey in the six study villages (45 questionnaires) was accomplished. Furthermore, focus groups discussions in each village were organized. Two types of groups were distinguished, fishermen from the village (5 to 7 participants per group) and women groups (5 to 15 participants per group). Finally, key informant interviews with 7 experts from 4 different institutions (administration, tourism, protected area team, and fishery organizations) were conducted. The variation and dynamics of ES demand were captured by respective questions during the interviews and from anecdotal reports by the communities, based on existing knowledge [44–46]. The methods and further results are primarily documented in [48–51].

The islands are subject to the alternating monsoon seasons in the northern Indian Ocean: the 'weak' and wet winter or north-east monsoon (November–February) and the 'strong' and dry summer or south-west monsoon (May–September). This seasonally reversing monsoon system creates particular oceanographic conditions around these islands during the summer monsoon, which result in seasonally very high marine primary productivity levels, providing the basis for exceptionally high productivity of fish (Figure 5, representing the related key ES) and other marine biomass [44,51].

Socotra ranks among the most productive inshore marine areas in the Indo-West Pacific. Fish biomass and community pattern are highly seasonal, linked to the monsoon dynamics and the associated local upwelling systems [51,52]. The seasonally varying marine productivity pattern [49] accordingly results in a strong temporal variability of key ES supply, namely fish biomass (Figure 5) [48].

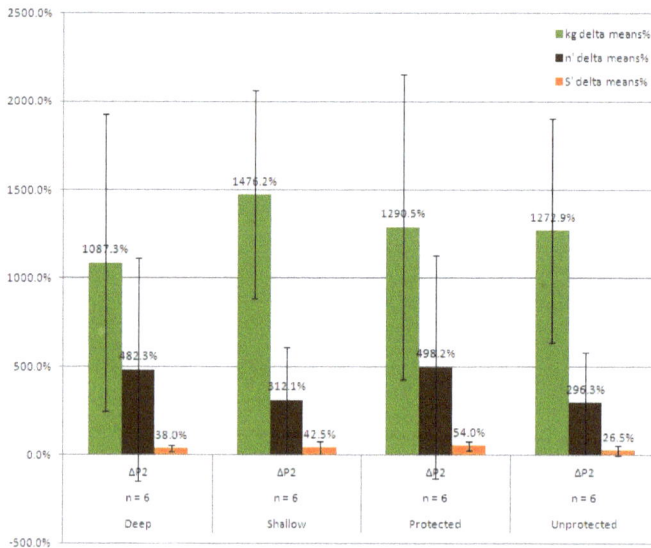

**Figure 5.** Comparison of the mean difference (%) at six study sites of fish biomass (green), fish individuals (brown) and fish diversity (species) (orange) between the post-monsoon and the pre-monsoon seasons on Socotra Archipelago, according to depth (deep versus shallow) and protection status (protected versus unprotected) [53].

The inhabitants of the coastal plains rely primarily on rural fisheries, while the population of the island interior consists of semi-nomadic pastoralists whose main source of livelihood is livestock herding. Both population groups complement their livelihoods with a variety of additional services provided by the specific ecosystems they dwell on, and by traditionally exchanging (trading) benefits among themselves and with the other group, respectively, especially at times of seasonal supply shortages [45,46]. The income of the coastal dwellers, mainly fishermen, is relatively low and highly dependent on fishing during the non-monsoon periods, which contributes more than half of the annual income, including the direct consumption (Figure 6a). Moreover, if compared to the actual effort (time and energy) allocated to the different livelihood activities (Figure 6b), the revenues from fishing in the non-monsoon periods make it the economically most viable activity compared to all other sources of livelihood, which contribute relatively little to the total income. Therefore, the total income of a fishing family on Socotra Archipelago is highly dependent on a temporal (seasonal) ES supply pattern. As with marine productivity patterns also terrestrial productivity is partly climate-dependent, with fodder for livestock becoming increasingly scarce in the interior during the dry period (before the onset of the monsoon).

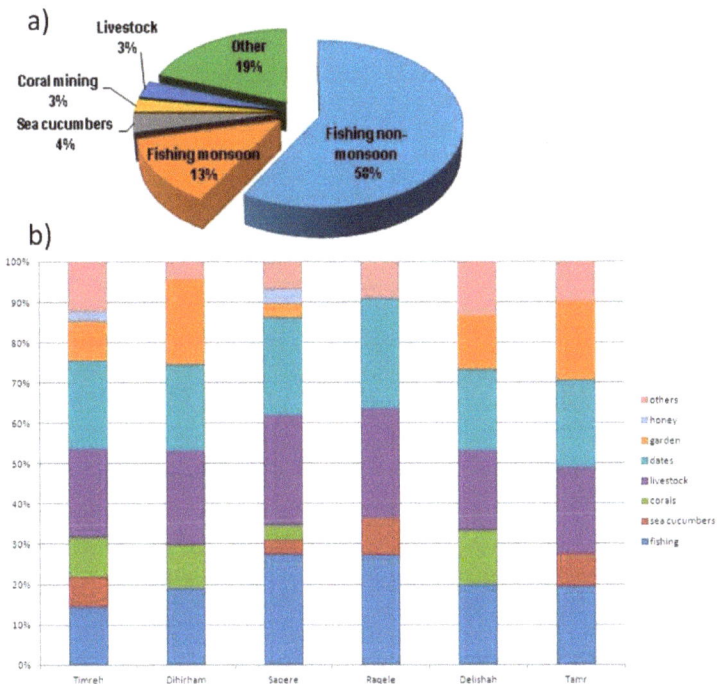

Figure 6. ES benefits to the small-scale fishing (household) economies, (a) showing the diversity and relative share of economic income sources across the year (differentiating between the contribution of fishing during the monsoon and non-monsoon periods, and (b) comparing the relative effort (time and energy) expended on different livelihood activities throughout the year; based on assessments in six coastal villages [54].

Communities traditionally offset the resulting shortages of fishery products for subsistence and marketing during the end of the pre-summer monsoon period (when biomass is fished down) and during the summer monsoon (when fishing is not possible due to the harsh sea conditions) by engaging in an array of alternative but economically less productive livelihood activities [45]. People fall back on alternative marine ES (e.g., coral mining) or even on alternative ES in different ecosystems (e.g., livestock, horticulture). Fishing families relocate their livestock to the interior during the wet (green) periods, thus providing meat from the interior to the coastal population during the summer monsoon when fishing ceases. Likewise, pastoralists from the island interior profit from seasonal migration to the coast with their livestock during the dry seasons when fodder for livestock becomes scarce [44,55]. This intricate web of responses to temporal ES supply patterns is vested in cultural norms and enforced by traditional leaders such as the clan elders and tribal sheikhs, whereby migration patterns and the regular exchange between fishers and pastoralists are also significant for socialization processes among the people as important element of the social system and essential for their identity. Environmental changes such as climate change, which potentially impact on temporal replenishment processes, are bound to result in a pronounced socio-economic vulnerability of both the fishing and the pastoralist population, not least because of limited economic alternatives and socio-cultural adaptation capacities [50,56]. The latter relates to the ability or potential to adjust to changing conditions e.g., modify practices by learning or changing to different technologies [56].

The study from Socotra, Yemen, presented here, illustrates that the users behind the demand for ES are highly dependent on one key ES (i.e., the fish biomass relevant for food production and household economies). The provision of this key ES is subject to strong temporal variation, being significantly correlated to the monsoon patterns, and consequently to the fishing behavior of the local communities. This temporal dynamic in the ES supply triggers complex compensating processes in society, including falling back on alternative ES or alternative ecosystems.

## 4. Discussion

Both empirical studies underscore the existence of relevant and complex spatial and temporal dynamics in both ES supply and ES demand. The study from the West African Sahel illustrates in particular the spatial dynamics. People in this region adapted to the seasonality in the agricultural systems by means of seasonal and temporal migration to other regions, particularly urban centres. These migration activities are deeply rooted in culture and are part of the local livelihood strategies. Climate change is now exerting additional pressure in terms of altered rainfall patterns and land degradation. Besides migration, the people tend to compensate by falling back on alternative ES. The decrease in ES supply thus intensifies societal processes such as human migration, a local livelihood strategy and a significant characteristic of the culture in the region, but also influences other compensation strategies for crop failure. The study from Socotra Archipelago illustrates in particular the role of temporal dynamics in ES supply, in this case fish biomass related to the Monsoon. The population concerned has to cope with the strong seasonal variability of marine ES, i.e., by adjusting consumption and cooperative socio-economic activities. People from the coast are provided with meat from the interior during the summer monsoon, when fishing ceases.

Furthermore, the results confirm a previous study [29], who conclude that temporal and spatial dynamics have not yet been sufficiently integrated into studies of ES supply and demand. The concept of ES flows seeks to bridge the gap between spatially separated 'provider' areas and 'recipient' areas [27]. Problems emerge when services and goods are part of a complex supply chain and it becomes difficult to locate the demand [57]. However, the temporal dynamics in both ES supply and ES demand, as exemplified at our study sites (i.e., seasonality in fish biomass and temporal human migration) are not yet represented in the concept of ES flow.

From the social-ecological perspective described above, which addresses the social-ecological structures and processes including spatial and temporal dynamics in nature and society, we argue that this ES demand and supply relation is even more complex. Both studies provide evidence for the diversification of livelihood strategies due to varying ES supply (reduced ES supply: crops—during the dry season and fish biomass—during the monsoon) (Figure 7). As result of food shortages, remaining ES demand leads to the diversification of livelihood strategies including migration or the practice of other (alternative) livelihood activities. The need for livelihood stability (i.e., the need for stable income for survival and other needs all year round) triggers a shift to a second or third-desired income source in case of the preferred income source declines. The West African study affirms the occurrence of societal processes such as human migration with both a temporal and spatial dimension such as seasonal migration and rural-urban migration, respectively as livelihood strategy (Figure 7). Whereas the Socotra study supports the relevance of falling back on alternative ES, such as coral mining, horticulture or livestock breading as alternative livelihood strategies. Both studies highlight the dependency on ES supply throughout the year. As a result it is important to focus ES analysis on ES bundles (i.e., sets of ecosystem services that repeatedly appear together across space and time [58]), rather than on isolated ES as well as to expand ES analysis to livelihood analysis including all compensation strategies. These results are in line with another study [31], that argues that the relationships between ES and human well-being are complex interrelations between nature and society.

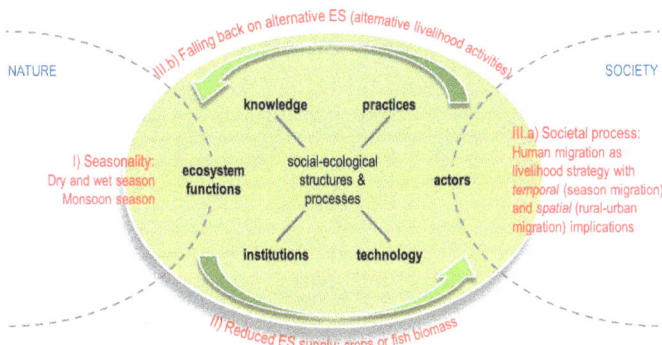

**Figure 7.** Diversification of livelihood strategies: The concept of social-ecological systems applied to the case study sites illustrating the diversification of livelihood strategies (III.a and III.b) as a result of reduced ES supply (II) due to seasonal variation (I).

Our article particularly highlights the functional relation between ES supply and demand. Both case studies showcase that there is seasonality in ES supply, i.e., fish biomass in the Socotra study and food supply in the West African study and this temporal dynamic determines complex compensation strategies in society. As the people are highly dependent on the supply of food, they fall back either on other marine and even terrestrial ES (Socotra study) or migrate to other regions (West African study). The local people adapted their livelihood strategies to the seasonal variations in ES supply. ES supply and demand are interrelated, with temporal (seasonal ES supply) and spatial dynamics (human migration) influencing each other. In literature, ES demand is the "need for specific ES by society, particular stakeholder groups or individuals" [26]. The results from the West African Sahel indicate that by following this understanding of ES demand, one fails to address the functional chain of the ES supply and demand relation. This empirical study clearly demonstrates that variability in the seasonality of ES supply (crop yields) directly influences societal processes such as human migration, a historically rooted adaptation strategy. ES demand is not a static term and instead stands in a functional relation to ES supply. This result goes beyond the current debate, i.e., [29] that differing conceptions of demand (such as desired goods or utilised goods) alter our understanding of ES. We argue that ES demand has to be considered in a functional relation to ES supply with ensuing spatial and temporal implications.

Finally, referring to the current research on mapping and assessing ES, our results underline the findings from other authors [28] claiming that there is an urgent need to develop methods and concepts for deepening our understanding of the social-ecological processes behind the supply and demand for ES in order to improve our ability to map ES for decision making.

## 5. Conclusions

The present article refers to the current debate on ES supply and demand against the backdrop of two empirical studies. Addressing the current debate from a social-ecological perspective we conclude that

- ES supply and demand stand in a functional rather than a static and linear relation to each other: nature and society exert mutual influence. Changes in ES supply can impact on ES demand and vice versa.
- there are interdependent temporal and spatial dynamics in both ES supply and ES demand.

To sum up, this is what we call social-ecological dynamics of ES. We thus argue to (1) not analyse isolated ES, but rather look at all relevant ES bundles, and (2) expand ES analysis to livelihood analysis including all compensation strategies.

From a policy perspective our study highlights the importance of acknowledging these dynamics of ES supply and demand. We claim that this aspect has to be considered in biodiversity policies and deliberation processes such as the assessments of IPBES. Without considering these dynamics dependency of societal groups on respective ES, and interrelation of altered ES supply triggering societal process, cannot be integrated into decision making processes.

These conclusions point to novel research directions. Particular attention should be given to the social-ecological dynamics of ES. From a conceptual perspective, these dynamics have to be integrated into current ES frameworks. The temporal and spatial dynamics of both ES supply and ES demand have to be considered, particularly their interdependency. Further empirical studies are needed in order to be able to draw general conclusions. New and problem-oriented research questions arise about how to compensate for variations in ES supply, and about the consequences of falling back on alternative ES and/or alternative ecosystems, such as conflicts of use, access to ES or regulation of ES.

**Acknowledgments:** We thank our colleagues from ISOE—Institute of Social-Ecological Research, Frankfurt am Main/Germany: Thomas Jahn, Engelbert Schramm, Stefan Liehr, Alexandra Lux, Robert Lütkemeier and Lukas Drees for the joint conceptual work on social-ecological systems. The former students Matthias Goerres, Hannes Pulch, Moteah Sheikh Aideed and Marie Martin prepared Bachelor and Master Theses within the framework of the Socotra study programme, contributing data and insights, and are cordially acknowledged. The Socotra study was partly funded by the LOEWE programme of the State of Hesse, and its execution kindly supported by the Environment Protection Authority Socotra. The preparation of this article benefitted from the ongoing UNEP-GEF Socotra Project (GEF grant #5347). The study in West Africa was conducted as part of the micle project "Climate Change, Changes to the Environment and Migration in Sahel—Social-ecological conditions of population movements in Mali and Senegal" and was funded by the German Federal Ministry for Education and Research (BMBF) (grant numbers 01UV1007A/B). We further thank Lukas Drees and Engelbert Schramm from ISOE for comments on the manuscript and Nicolai Mehlhaus for support in preparing the manuscript.

**Author Contributions:** M.M. had the idea and conceived the concept of the article; U.Z. was responsible for the study on the Socotra Archipelago, wrote the chapter on temporal dynamics, and contributed to the discussion; D.H. was responsible for the study from West Africa, wrote the chapter on spatial dynamics, and contributed to the discussion; M.M. was responsible for the overall writing process.

**Conflicts of Interest:** The authors declare no conflict of interest.

## References

1.  Secretariat of the Convention on Biological Diversity. *Global Biodiversity Outlook 4: A Mid-Term Assessment of Progress towards the Implementation of the Strategic Plan for Biodiversity 2011–2020*; Secretariat for the Convention on Biological Diversity: Montreal, QC, Canada, 2014.
2.  Tittensor, D.P.; Walpole, M.; Hill, S.L.L.; Boyce, D.G.; Britten, G.L.; Burgess, N.D.; Butchart, S.H.M.; Leadley, P.W.; Regan, E.C.; Alkemade, R.; et al. A mid-term analysis of progress toward international biodiversity targets. *Science* **2014**, *346*, 241–244. [CrossRef] [PubMed]
3.  Chapin, F.S.C., III; Zavaleta, E.S.; Eviner, V.T.; Naylor, R.L.; Vitousek, P.M.; Reynolds, H.L.; Hooper, D.U.; Lavorel, S.; Sala, O.E.; Hobbie, S.E.; et al. Consequences of changing biodiversity. *Nature* **2000**, *405*, 234–242. [CrossRef] [PubMed]

4.  Cardinale, B.J.; Duffy, J.E.; Gonzalez, A.; Hooper, D.U.; Perrings, C.; Venail, P.; Narwani, A.; Mace, G.M.; Tilman, D.; Wardle, D.A.; et al. Biodiversity loss and its impact on humanity. *Nature* **2012**, *486*, 59–67. [CrossRef] [PubMed]

5.  Burkhard, B.; Kroll, F.; Nedkov, S.; Müller, F. Mapping ecosystem service supply, demand and budgets. *Ecol. Indic.* **2012**, *21*, 17–29. [CrossRef]

6.  Schröter, M.; Barton, D.N.; Remme, R.P.; Hein, L. Accounting for capacity and flow of ecosystem services: A conceptual model and a case study for Telemark, Norway. *Ecol. Indic.* **2014**, *36*, 539–551. [CrossRef]

7.  TEEB. *The Economics of Ecosystems & Biodiversity: An Interim Report*; E ISB N-1 3 978-9 2-79-0; European Commission and the German Ministry for the Environment: Wesseling, Germany, 2008.

8.  Ten Brink, P. (Ed.) *The Economics of Ecosystems and Biodiversity in National and International Policy Making*; Earthscan: London, UK, 2011.

9.  Carpenter, S.R.; Mooney, H.A.; Agard, J.; Capistrano, D.; Defries, R.S.; Diaz, S.; Dietz, T.; Duraiappah, A.K.; Oteng-Yeboah, A.; Pereira, H.M.; et al. Science for managing ecosystem services: Beyond the Millennium Ecosystem Assessment. *Proc. Natl. Acad. Sci. USA* **2009**, *106*, 1305–1312. [CrossRef] [PubMed]

10. Stoll, S.; Frenzel, M.; Burkhard, B.; Adamescu, M.; Augustaitis, A.; Baeßler, C.; Bonet, F.J.; Carranza, M.L.; Cazacu, C.; Cosor, G.L.; et al. Assessment of ecosystem integrity and service gradients across Europe using the LTER Europe network. *Ecol. Model.* **2015**, *295*, 75–87. [CrossRef]

11. Görg, C.; Aicher, C. Ökosystemdienstleistungen-zwischen Natur und Gesellschaft. Anforderungen an eine inter-und transdisziplinäre Forschung aus Sicht der Sozialwissenschaften. In *Berichte und Abhandlungen: Band 16*; Berlin-Brandenburgische Akademie der Wissenschaften (Hg.), Ed.; De Gruyter Akademie Forschung: Berlin, Germany, 2014.

12. Loft, L.; Lux, A.; Jahn, T. A Social-Ecological Perspective on Ecosystem Services. In *Handbook of Ecosystem Services*; Potschin, M., Haines-Young, R., Fish, R., Turner, R., Eds.; Routledge: Abingdon, UK, 2016; pp. 88–94.

13. Stürck, J.; Schulp, C.J.; Verburg, P.H. Spatio-temporal dynamics of regulating ecosystem services in Europe—The role of past and future land use change. *Appl. Geogr.* **2015**, *63*, 121–135. [CrossRef]

14. Burkhard, B.; Petrosillo, I.; Costanza, R. Ecosystem services—Bridging ecology, economy and social sciences. *Ecol. Complexity* **2010**, *7*, 257–259. [CrossRef]

15. Hummel, D.; Jahn, T.; Keil, F.; Liehr, S.; Stieß, I. Social Ecology as Critical, Transdisciplinary Science—Conceptualizing, Analyzing, and Shaping Societal Relations to Nature. *Sustainability* **2017**. accepted.

16. Liehr, S.; Röhrig, J.; Mehring, M.; Kluge, T. How the Social-Ecological Systems Concept Can Guide Transdisciplinary Research and Implementation: Addressing Water Challenges in Central Northern Namibia. *Sustainability* **2017**. accepted.

17. Mace, G.M.; Norris, K.; Fitter, A.H. Biodiversity and ecosystem services: A multilayered relationship. *Trends Ecol. Evol.* **2012**, *27*, 19–26. [CrossRef] [PubMed]

18. Balvanera, P.; Siddique, I.; Dee, L.; Paquette, A.; Isbell, F.; Gonzalez, A.; Byrnes, J.; O'Connor, M.I.; Hungate, B.A.; Griffin, J.N. Linking Biodiversity and Ecosystem Services: Current Uncertainties and the Necessary Next Steps. *BioScience* **2014**, *64*, 49–57. [CrossRef]

19. Martin-López, B.; Montes, C. Restoring the human capacity for conserving biodiversity: A social–ecological approach. *Sustain. Sci.* **2015**, *10*, 699–706. [CrossRef]

20. Ban, N.C.; Mills, M.; Tam, J.; Hicks, C.C.; Klain, S.; Stoeckl, N.; Bottrill, M.C.; Levine, J.; Pressey, R.L.; Satterfield, T.; et al. A social–ecological approach to conservation planning: Embedding social considerations. *Front. Ecol. Environ.* **2013**, *11*, 194–202. [CrossRef]

21. Fischer, J.; Gardner, T.A.; Bennett, E.M.; Balvanera, P.; Biggs, R.; Carpenter, S.; Daw, T.; Folke, C.; Hill, R.; Hughes, T.P.; et al. Advancing sustainability through mainstreaming a social–ecological systems perspective. *Curr. Opin. Environ. Sustain.* **2015**, *14*, 144–149. [CrossRef]

22. Mehring, M.; Bernard, B.; Hummel, D.; Liehr, S.; Lux, A. Halting biodiversity loss: How social–ecological biodiversity research makes a difference. *Int. J. Biodivers. Sci. Ecosyst. Serv. Manag.* **2017**, *13*, 172–180. [CrossRef]

23. Schleyer, C.; Lux, A.; Mehring, M.; Görg, C. Ecosystem Services as a boundary concept: Social-ecological perspective on a contested concept and the need for inter- and transdisciplinary collaboration. *Sustainability* **2017**, submitted.

24. European Commission. *The EU Biodiversity Strategy to 2020*; Publications Offfice of the European Union: Luxembourg, 2011.

25. Díaz, S.; Demissew, S.; Carabias, J.; Joly, C.; Lonsdale, M.; Ash, N.; Larigauderie, A.; Adhikari, J.R.; Arico, S.; Báldi, A.; et al. The IPBES Conceptual Framework—Connecting nature and people. *Curr. Opin. Environ. Sustain.* **2015**, *14*, 1–16. [CrossRef]

26. Albert, C.; Bonn, A.; Burkhard, B.; Daube, S.; Dietrich, K.; Engels, B.; Frommer, J.; Götzl, M.; Grêt-Regamey, A.; Job-Hoben, B.; et al. Towards a national set of ecosystem service indicators: Insights from Germany. *Ecol. Indic.* **2015**, *61*, 38–48. [CrossRef]

27. Villamagna, A.M.; Angermeier, P.L.; Bennett, E.M. Capacity, pressure, demand, and flow: A conceptual framework for analyzing ecosystem service provision and delivery. *Ecol. Complex.* **2013**, *15*, 114–121. [CrossRef]

28. Martínez-Harms, M.J.; Balvanera, P. Methods for mapping ecosystem service supply: A review. *Int. J. Biodivers. Sci. Ecosyst. Serv. Manag.* **2012**, *8*, 17–25. [CrossRef]

29. Wolff, S.; Schulp, C.; Verburg, P.H. Mapping ecosystem services demand: A review of current research and future perspectives. *Ecol. Indic.* **2015**, *55*, 159–171. [CrossRef]

30. Chan, K.M.A.; Balvanera, P.; Benessaiah, K.; Chapman, M.; Diaz, S.; Gomez-Baggethun, E.; Gould, R.; Hannahs, N.; Jax, K.; Klain, S.; et al. Opinion: Why protect nature? Rethinking values and the environment. *Proc. Natl. Acad. Sci. USA* **2016**, *113*, 1462–1465. [CrossRef] [PubMed]

31. Daw, T.M.; Hicks, C.C.; Brown, K.; Chaigneau, T.; Januchowski-Hartley, F.A.; Cheung, W.W.L.; Rosendo, S.; Crona, B.; Coulthard, S.; Sandbrook, C.; et al. Elasticity in ecosystem services: Exploring the variable relationship between ecosystems and human well-being. *Ecol. Soc.* **2016**, *21*. [CrossRef]

32. Hummel, D. Climate change, land degradation and migration in Mali and Senegal—Some policy implications. *Migr. Dev.* **2016**, *5*, 211–233. [CrossRef]

33. Brandt, M.; Romankiewicz, C.; Spiekermann, R.; Samimi, C. Environmental change in time series—An interdisciplinary study in the Sahel of Mali and Senegal. *J. Arid Environ.* **2014**, *105*, 52–63. [CrossRef]

34. Brandt, M.; Verger, A.; Diouf, A.; Baret, F.; Samimi, C. Local Vegetation Trends in the Sahel of Mali and Senegal Using Long Time Series FAPAR Satellite Products and Field Measurement (1982–2010). *Remote Sens.* **2014**, *6*, 2408–2434. [CrossRef]

35. Van der Land, V. The Environment-Migration Nexus Reconsidered: Why Capabilities and Aspirations Matter. Ph.D. Thesis, Universität Goethe, Frankfurt am Main, Germany, 2015.

36. Liehr, S.; Drees, L.; Hummel, D. Migration as Societal Response to Climate Change and Land Degradation in Mali and Senegal. In *Adaptation to Climate Change and Variability in Rural West Africa*; Yaro, J.A., Hesselberg, J., Eds.; Springer: Heidelberg, Germany, 2016; pp. 147–169.

37. Roudier, P.; Sultan, B.; Quirion, P.; Berg, A. The impact of future climate change on West African crop yields: What does the recent literature say? *Glob. Environ. Chang.* **2011**, *21*, 1073–1083. [CrossRef]

38. Sissoko, K.; van Keulen, H.; Verhagen, J.; Tekken, V.; Battaglini, A. Agriculture, livelihoods and climate change in the West African Sahel. *Reg. Environ. Chang.* **2011**, *11*, 119–125. [CrossRef]

39. Pardoe, J.; Kloos, J.; Assogba, N.P. Seasonal Variability: Impacts, Adaptations and the Sustainability Challenge. In *Adaptation to Climate Change and Variability in Rural West Africa*; Yaro, J.A., Hesselberg, J., Eds.; Springer: Heidelberg, Germany, 2016; pp. 41–57.

40. KC, S.; Lutz, W. *Demographic, Migration and Human Capital Scenarios for Mali and Senegal with Special Emphasis on Bandiagara and Linguere Districts: Final Report of the Demographic Analysis in the Micle-Project*; ISOE—Institute for Social-Ecological Research: Frankfurt am Main, Germany, 2014.

41. Van der Land, V.; Hummel, D. Vulnerability and the Role of Education in Environmentally Induced Migration in Mali and Senegal. *Ecol. Soc.* **2013**, *18*, 414. [CrossRef]

42. Scholte, P.; Al-Okaishi, A.; Suleyman, A.S. When conservation precedes development: A case study of the opening up of the Socotra archipelago, Yemen. *Oryx* **2011**, *45*, 401–410. [CrossRef]

43. Van Damme, K.; Banfield, L. Past and present human impacts on the biodiversity of Socotra Island (Yemen): Implications for future conservation. *Zool. Middle East* **2013**, *54*, 31–88. [CrossRef]

44. Cheung, C.; DeVantier, L. *Socotra: A Natural History of the Islands and Their People*; Odyssey Books & Guides: Hong Kong, China, 2006.

45. Morris, M. *Manual of Traditional Land Use in the Soqotra Archipelago: GEF YEM/96/G32*; Royal Botanic Garden Edinburgh: Edinburgh, UK, 2002.

46. Miller, A.G.; Morris, M. *Ethnoflora of the Soqotra Archipelago*; Royal Botanic Garden Edinburgh: Edinburgh, UK, 2004.

47. UNDP. *Human Development Report 2016*; United Nations: New York, NY, USA, 2017.

48. Goerres, M. Coastal and Terrestrial Ecosystem Services, Uses and Users: A Case Study From Socotra Island, Yemen. Bachelor's Thesis, Johannes Gutenberg-Universität Mainz, Mainz, Germany, March 2011.

49. Aideed, M.S. Spatial and Temporal Dynamics of Fish Biomass Productivity at Socotra Island and the Impact of Small-Scale Fisheries on Fish Community Structure. Master's Thesis, Hadrahmout University, Mukallah, Yemen, December 2013.

50. Martin, M. Vulnerability of Fishery-Based Social-Ecological Systems to Climate Change. A Case Study on the Small-Scale Fishery of Soqotra Island, Yemen. Master's Thesis, Goethe-University, Frankfurt am Main, Germany, November 2013.

51. Zajonz, U.; Lavergne, E.; Klaus, R.; Krupp, F.; Aideed, M.S.; Saeed, F.N. The coastal fishes and fisheries of the Socotra Archipelago, Yemen. *Mar. Pollut. Bull.* **2016**, *105*, 660–675. [CrossRef] [PubMed]

52. Lavergne, E.; Zajonz, U.; Sellin, L. Length-weight relationship and seasonal effects of the Summer Monsoon on condition factor of Terapon jarbua (Forsskål, 1775) from the wider Gulf of Aden including Socotra Island. *J. Appl. Ichthyol.* **2013**, *29*, 274–277. [CrossRef]

53. Zajonz, U.; Sheikh Aideed, M.; Martin, M.; Naseeb Saeed, F.; Lavergne, E. Unpublished data.

54. Martin, M.; Zajonz, U.; Hummel, D.; Sheikh Aideed, M.; Naseeb Saeed, F. Unpublished data.

55. Van Rampelbergh, M.; Fleitmann, D.; Verheyden, S.; Cheng, H.; Edwards, L.; de Geest, P.; de Vleeschouwer, D.; Burns, S.J.; Matter, A.; Claeys, P.; et al. Mid—To late Holocene Indian Ocean Monsoon variability recorded in four speleothems from Socotra Island, Yemen. *Quat. Sci. Rev.* **2013**, *65*, 129–142. [CrossRef]

56. Allison, E.H.; Perry, A.L.; Badjeck, M.-C.; Neil Adger, W.; Brown, K.; Conway, D.; Halls, A.S.; Pilling, G.M.; Reynolds, J.D.; Andrew, N.L.; et al. Vulnerability of national economies to the impacts of climate change on fisheries. *Fish Fish.* **2009**, *10*, 173–196. [CrossRef]

57. Burkhard, B.; Kandziora, M.; Hou, Y.; Müller, F. Ecosystem Service Potentials, Flows and Demands—Concepts for Spatial Localisation, Indication and Quantification. *Landsc. Oniline* **2014**, *34*, 1–32. [CrossRef]

58. Raudsepp-Hearne, C.; Peterson, G.D.; Bennett, E.M. Ecosystem service bundles for analyzing tradeoffs in diverse landscapes. *Proc. Natl. Acad. Sci USA* **2010**, *107*, 5242–5247. [CrossRef] [PubMed]

![sustainability logo] *sustainability*

MDPI

*Article*

# More Energy and Less Work, but New Crises: How the Societal Metabolism-Labour Nexus Changes from Agrarian to Industrial Societies

**Willi Haas [1],\* and Hailemariam Birke Andarge [2]**

[1]     Institute of Social Ecology, Alpen Adria Universitaet, 1070 Vienna, Austria
[2]     Department of Geography and Environmental Studies, Debre Berhan University,
        445 Debre Berhan, Ethiopia; hailsha7@gmail.com
\*      Correspondence: willi.haas@aau.at; Tel.: +43-1-522-4000-422

Received: 30 January 2017; Accepted: 4 June 2017; Published: 26 June 2017

**Abstract:** The scientific finding that humanity is overburdening nature and thus risks further ecological crises is almost uncontroversial. Main reason for the crises is the drastic increase in the societal metabolism, which is accomplished through labour. In this article, we examine the societal metabolism-labour nexus in two energy regimes: a valley in the Ethiopian highlands, typical of an agrarian society, and a village in Austria, typical of an industrial society. In the Ethiopian village, the supply of food demands almost the entire labour force, thus limiting the capacity to facilitate material flows beyond food provision. In the Austrian village, fewer working hours, lower workloads but 50 times higher useful energy allow to accumulate stocks like buildings 70 times higher than the Ethiopian case. With fossil energy, industrial societies decisively expand their energy supply and reduce labour hours at the cost of high carbon emissions, which are almost non-existent in the Ethiopian case. To overcome the resulting ecological crises, there is a call to drastically reduce fossil fuel consumption. Such an abandonment of fossil fuels might have as far reaching consequences for the societal metabolism-labour nexus and consequently human labour as the introduction of fossil fuels has had.

**Keywords:** labour; sociometabolic transition; energy regimes; primary energy; exergy

---

## 1. Introduction: Human Labour and the Sociometabolic Transition

Global ecological crises such as climate change or the progressive loss of species have already crossed the planetary boundaries for a safe operating space for humanity [1]. Scientific analyses even go so far as to characterize mankind as a geological factor in the newly-defined Anthropocene [2]. In this epoch, humans have become the dominant influence on biological, geological and atmospheric processes of the earth system. The insight that this increasing human dominance of the earth system is due to the rapidly growing societal metabolism, global society's material and energetic exchange with nature, becomes more and more accepted [3]. Main driver of this development are especially industrialized and emerging economies.

The growing social metabolism has become possible due to an energy transition, namely a shift away from the biomass-based energy regime of agrarian societies to fossil fuel-based industrial societies [4,5]. In this article, we want to show that such energy transitions go hand in hand with a transition of human labour itself. This is important, since intensifying global ecological crises, sooner or later, require a new transition away from the use of fossil fuels which is expected to have as many and equally far-reaching implications for human labour as the previous transition to a fossil-fuel based industrial society had [6]. Thus, it seems to be rewarding to better understand the interplay between material use, energy consumption, and labour in the context of the different energy regimes.

To investigate what we call the societal metabolism-labour nexus, we make use of a sociometabolic perspective. To this end, in a first step, we develop a simple analytical scheme in the following section (Section 2). In a second step, before presenting two case studies, we briefly introduce methods and describe surveys as well as data generation for the cases (Section 3). In a third step, we apply the scheme to two local cases, one as illustration for an agrarian (Section 4) and one as illustration for a rural village in an industrial society (Section 4). The first local case is in Ethiopia, in a valley in the highlands. The local agrarian community is based on biomass as the primary energy source and uses fossil energy carriers only to a minor extent. The second case is an Austrian village which is part of an industrial society based on the use of fossil energy. After a detailed presentation of the two cases, in a forth step, we contextualise the results on the societal metabolism-labour nexus with average data taken from literature and compare the two cases with each other for better understanding the societal metabolism-labour nexus (Section 5).

## 2. The Social Metabolism-Labour-Nexus

Society changes nature both in an intentional and an unintentional way, and the changes in turn affect society. These society-nature interactions are analysed with the concepts of social metabolism, society's material and energetic exchange with nature, and the colonization of nature [7–11]. Based on the idea of the metabolism developed by Marx [12], the concept was further developed analogously to the economic perspective in the national accounts. It is not simply humans, taken on their own, who are dependent and have an impact on nature, but the societal mode of production and consumption which we call sociometabolic regime. Key to metabolic regimes is the energy availability quantitatively across space which on the one hand largely determines the capacity of humans to change nature and to remove, transport and process resources and on the other hand makes a big difference for the societal conditions e.g., how much time remains for other activities than food provisioning. In the history of mankind, several sociometabolic regimes can be distinguished with distinctive transitions between them, often referred to as revolutions. The Neolithic revolution marks the transition from the regime of hunting and gathering to the agrarian society, and the industrial revolution the transition from the agrarian to the industrial regime. Apart from very few exceptions (e.g., small island states where wind as drive energy for sailing boats used for trading and fishing could make up a significant share in the societal energy supply), biomass is by far the most important source of energy until the industrial revolution. As a rule, biomass provides about 99% of the primary energy in preindustrial societies. Food and feed for humans and animals comprises by far the largest share of this energy input. By using biomass, humans make use of solar energy flows through land management. The German environmental historian Rolf Peter Sieferle [4] called this the controlled solar energy system of agricultural societies. Work of a population on a given territory is invested to transform ecosystems and to increase the amount of usable biomass that can be harvested per unit of area. A minimum prerequisite for the long-term functioning of such an energy system is that the harvested biomass at least provides the necessary energy for human and animal work in the form of food and feed. The higher the surplus, the more complex societal structures are possible. However, in agrarian societies, this surplus is quite limited and is around ten % in average years [6] thus limiting the number of people who are not tied to work for society's food provision.

The still ongoing industrialization of agriculture since the 16th century [3] has led to a massive transformation of the agricultural landscape, which has to be made machine-friendly. Large fields have been created, from which all the species-rich field margins with ditches and shrubs, and land morphological obstacles have been removed. Working with heavy equipment promoted soil compaction and erosion. Large-scale monocultures require a high use of agrochemicals [13–15] Consequently, the role of agriculture within society changed dramatically during the transition from an agrarian to an industrial society. While agriculture in agrarian societies is mainly an energy supplier by providing food and animal labour, industrialized agriculture consumes more energy than is ultimately obtained from it [16–18]. This is made possible by fossil fuels, which can be mined in large quantities of deposits as required. Further, fossil fuels have very favourable features like the combination of easy portability with trucks and pipelines, and a high energy density which is about three times as high as the one of biomass. Thus, access to fossil fuels enables high energy availability almost anytime and anywhere.

Today, half of the world's population still lives under more or less agrarian conditions [3], and what is usually called development or progress can be understood as a transition from the agrarian to an industrial regime [19]. To discuss the interplay between work, energy and material more specifically in the two different regimes, we introduce two empirical cases: an agrarian community in a valley of the Ethiopian highlands and an industrialized Austrian rural village. These two examples serve as illustrative cases for the two regimes. In agrarian societies, there are notable differences between local cases with variations in soil fertility, climatic conditions, agricultural practices, favoured plants, importance of livestock, population density and others. The same can be stated for rural cases in industrial societies. However, despite these remarkable differences in their metabolic rates, this is material and energy use per capita, variations within in one energy regime are not as significant as the differences from one to another energy regime [3,4,20,21]. Further, despite the differences between cases of one and the same energy regime, they display noteworthy communalities how they change in the transition like the building up of growing stocks, the increased use of technology hand in hand with less and less man power needed to supply societies with the necessary food, feed and technical energy. Therefore, our analysis is less concerned about differences within one energy regime, but focusses on the more fundamental differences in the relation between work, energy and material in two cases illustrative for two different energy regimes.

To investigate these differences, we use a simple analytical scheme consisting of the following dimensions [6,22]:

(A)  Societal stocks: human population, domesticated animals, colonized land and man-made artefacts.
(B)  Sociometabolic material flows like extraction, trade flows and material consumption.
(C)  Primary energy and useful energy related to human labour, animal labour and technology.

Based on this scheme and the methods described in the following section, we discuss the societal metabolism-labour nexus in two cases, one in an agrarian and one in an industrial society.

## 3. Methods Used and Data Sources

The two case studies follow standard Material Flow Accounting principles. Therefore, we present here only a summary and provide an overview of the references used. In this description, we pay special attention to issues especially relevant for the cases or where the cases slightly deviate from the standards. Further, we describe the surveys performed to generate these data and give references for more detailed information on the case studies. Further, we provide some background on the methodology for calculating useful energy and provide the references, where these method is discussed in more detail. Finally, we briefly describe how carbon emissions are estimated for the two cases.

### 3.1. Material Flow Accounting (MFA)

The base of the analysis relies on the conceptual framework and the system boundaries applied in economy-wide material flow accounting at national level (MFA) [9,11,23–25]. The guide and further references consistently provide detailed information relevant for accounting.

3.1.1. Methodological and Conceptual Considerations

Societal metabolism is based on the premise that any social system reproduces itself biophysically through a constant flow of materials and energy with its natural environment as well as with other social systems. Thus, size and composition of flows are intricately linked to the biophysical stocks of the social system; therefore, we are especially interested in stocks. To measure the biophysical flow characteristic, we use domestic material consumption as a key indicator, which is material extraction plus imports minus exports within a year. This indicator includes not just the consumption of people, but also the feed for livestock and all materials to maintain or expand societal stocks as well as all losses occurring during the processing of materials. Further, flows in the two cases are broken down into main material groups in accordance with material flow accounting conventions (cf. [25]). We usually distinguish into biomass (like cereals, grass, wood, etc.), mineral materials (non-metallic minerals like sand, gravel, stones, cement etc. mainly for construction and metals) and fossil energy carriers (like oil, gas, coal). Practically, flows are measured in tonnes for a specific year and all flows like biomass are reported in fresh weight. Standard moisture contents are provided in the guide. Biggest uncertainties exist for construction minerals since these materials are consumed in large quantities but in irregular intervals. Thus, accounting annual flows in small local communities via surveys or observations might provide an accurate but misleading result, especially if main construction work was performed just in the studied year or, in contrast, in many years but not in the studied year. In our cases, we provide an average picture and therefore asked people about stock changes in the previous years and estimated an annual average of additions to stocks, which still has higher uncertainties than other flow data.

In social metabolism research boundaries are not territorial ones but functional ones. While this issue is of minor relevance for the economy-wide material flow accounting, this becomes quite a tricky issue at the local level. To account for this, further standards were applied as discussed in the local studies manual [26]. In order to follow a functional understanding of system boundaries in both cases, we first defined who belongs to the community or village in accordance to political or economic membership as well as simply by factual presence. Thus, people, a stock of the social system, is defined. If the boundaries are functional and not territorial, it needs to be clarified which land, another societal stock, belongs to the social system. The land under consideration is the land that is under legitimate control of the members of the community. In our two specific cases, land was involved which was positioned so to say outside the territory but under legitimate control of members of the community. In Bahskurit outside the territory means outside the valley respectively the watershed and in Theyern outside means outside the political boundaries of the village. Thus, flows related to this land under legitimate control of the Bashkurit Valley were accounted for. Consequently, land and its related flows under legitimate control of persons not belonging to the respective local society but within the political borders of Theyern or the geographic boundary of Bashkurit Valley was not considered. A social system's livestock is determined by belonging to people of the social system or when their reproduction like feeding, health care and breeding is controlled by local people. Analogous, whether artefacts belong to a social system or not is defined via belonging to or maintenance by a member of the social system.

These are in brief the principles applied in these studies to identify societal stocks and flows for the two cases. All these decisions to identify a social system's stock is discussed in more detail in manuals for the national as well as the local level, also including practical conventions such as definitions where the boundary between society and nature is drawn [23–26].

### 3.1.2. Surveys and Data

While economy-wide material flow accounting at the country level can rely mainly on statistical data, the local level mostly can't use any statistical information since they are not accurate enough or simply are not available at all. Therefore, in both local case studies presented here a survey for generating primary data was performed. In Bashkurit Valley the survey was done during a one year field trip. Based on the data requirements for the analysis of land, material, energy and time use, the researcher and his research assistant applied different survey methods. Based on existing maps from previous studies an inspection of the land was performed to establish the different types of land use. This information was later verified in group discussions. Number of livestock was established during a household survey and cattle numbers were verified by counting animals in stables. Extraction of harvest and fire wood was estimated by weighing the bundles typically put together by inhabitants.

In combination with results from household surveys about the number of bundles produced during a year, harvest and fire wood amounts were calculated. This was cross-checked with yield factors from the local university in Gondar. Grazing amounts were estimated by feed requirements per head of the different types of domestic animals minus the fodder provided by the farmers. Information on the later was gathered during the household survey. Time use data were obtained by a mix of in-depth participatory observations of a variety of different persons (representing different age groups, gender, family sizes and wealth), household surveys and group discussions. The data on the household possessions was gained during the household survey by weighing the typical artefacts in some households and by counting number of artefacts in all surveyed households. Weight of buildings was calculated by estimating the weight of each material component (wood, grass, metal, stone, clay, bricks and cement) used for two different types of houses (rectangular tin-roof houses and circular thatched-roof houses). Tin-roof houses are mainly used as residential houses, while thatched-roof houses are mainly used as kitchen buildings and stables. Detailed information is documented in Andarge [27].

In Theyern data were mainly gained by household surveys and methods and data described in detail in Haas and Krausmann [28]. About 80% of the households were interviewed and almost all of them were very collaborative. Each interview was performed by two interviewers who were part of a research team with very good knowledge on the method of material flow accounting. Since an official map with demarcation lines was prepared in the year before the interview, it was used to discuss with each household which land is under their control and what is the actual land use (in case of fields farmed crops were inquired). Several data on the buildings like square meters, floors, roof type, key materials used and date of construction were collected. This information was cross-checked with aerial photographs. Weight of buildings was estimated by using a factor related to the habitable area of the buildings [29]. Data on vehicles like type and brand of vehicle, mileage per year, fuel consumption per 100km was collected. With these figures and standard factors for vehicle weights the overall stocks of vehicles as well as the annual fuel consumption was estimated. Estimating the household possession was a difficult task, since a vast number of artefacts were present in the houses. Thus, possessions were established by a check list containing different product groups, with heavier products being listed in more detail. Electricity bills were analysed together with the interviewees to obtain the kilowatt hours per year. Harvest data from agricultural fields as well as the local forests were well known by farmers and thus data collection was simple. Cross-checking with typical yield data confirmed the information. Local consumption of locally produced food was surveyed as well. Food requirements for the local population was estimated on the base of average nutrition data including average food waste losses for Austria. Thus, imports of food were estimated by deducting the consumed food locally produced from the food requirements, since this can be assumed to be purchased elsewhere. Regarding time use only time spent for economic activities was inquired (employment, farming, assistance in farming activities).

In principle time use was surveyed on an annual base, this means that weekly and seasonal differences including holidays (in the case of Austria) were specifically asked for to estimate the annual time spent for labour. The daily time use for labour was then the division of the annual time spent by 365.

### 3.2. Primary and Useful Energy

In principle, the method applied here follows the detailed description presented in a two-part article by Haberl [20,30]. In addition to materials, the biophysical reproduction of social systems (and thus it's societal stocks) requires a continuous supply with energy, which can be provided by materials in the form of material energy carriers like biomass or fossil fuels or by electricity, which flows without material. In our study, we distinguish between primary energy and useful energy.

Primary energy is defined as the energy in the form in which it is extracted from the natural environment, for example, extracted energy-rich materials (biomass, fossil fuels), harnessed flows of mechanical energy (hydropower, wind power, etc.), nuclear energy transformed to heat, or radiant solar energy used to produce heat or electricity [30]. To highlight two case-specific examples: primary energy is the energetic value of fire wood for cooking/heating or of cereals for human nutrition at the point the respective energy carrying materials are collected from forests or harvested from fields. When we account for primary energy in our two cases, we always consider the energy content at the point of extraction, even if it is imported. For example, in the case of local electricity consumption, we consider the related primary energy harnessed in a hydropower plant or the oil extracted from nature which was used to produce electricity in a thermal power plant.

Useful energy ("exergy") is the energy that is ultimately used or available to society (see Figure 1). This refers to the actually useful work such as moving a pile of sand from one place to another or the physical drive power pulling a plough for a certain distance, or the thermal energy that heats the food in the cooking pot. Societies dispose over useful energy delivered by humans, domesticated animals and technology. These different types of work taken together, we call societal useful energy. Of special interest is how much useful energy is invested by humans in the form of human labour. Useful work including human and animal labour as well as work performed by technologies can be quantitatively recorded in joule [31,32]. In addition, human labour spent in the economic system like for agriculture or wage labour can be expressed in average hours worked per person and day [26]. In practice, all hours annually worked in a local community are surveyed and divided by total population and 365 days. The useful energy provided by humans is then the product of hours worked and the average work load per hour (J/hour) [20].

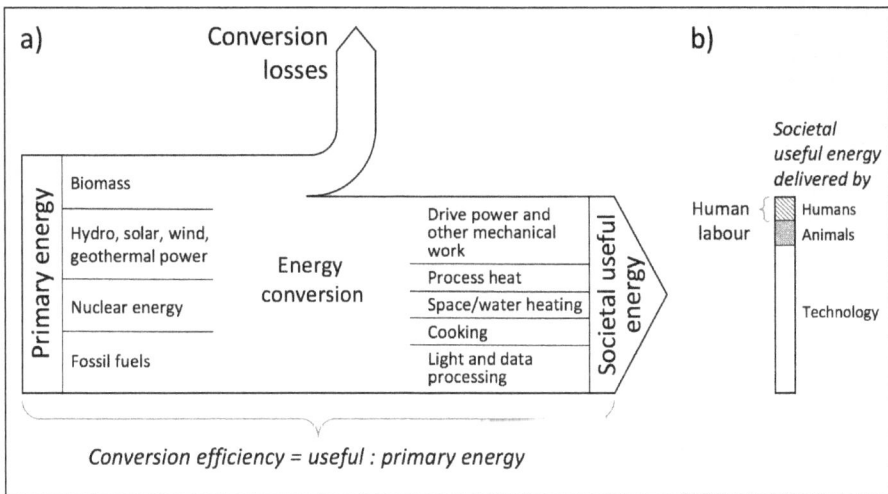

**Figure 1.** (**a**) Conversion from primary energy as extracted from nature to societal useful energy at point of societal use by type of useful energy delivered and conversion efficiency as output to input ratio [20,30]; (**b**) Alternative view of societal useful energy by the three converters humans, animals and technology to deliver useful energy and indication of human labour as sub-category of societal useful energy (schematic sketch presents conceptual relations but no real flow proportions).

Technically, useful energy is the remaining energy after all conversion losses are deducted from the primary energy and the conversion efficiency is the ratio between useful to primary energy.

In the case of human labour in Bashkurit Valley (agrarian society) the assumptions are that humans can deliver up to 100 Watt per hour in continuous work [33]. This is, however, a maximum. A continuous delivery of 40–50 Watt useful power per hour may be an appropriate guess; based on the 5 h of work per day in Bashkurit Valley this equals about 300 MJ per capita and year. In Theyern (industrialized society), it is estimated that the continuous delivery for farming activities is about 50% of the one in agricultural societies and for desk work about 30% of agrarian farming; based on 2.5 h of work per day in average in Theyern, this equals about 50 MJ per capita and year. In a similar way, the useful energy for animals in Bahskurit was calculated according to Haberl [20]. In the case of Theyern the animal labour of horses was calculated by assuming 250 Watt of useful energy per hour of horseback riding [34]. With one hour of daily riding for both horses this adds up to 5 MJ per person and year. Useful energy delivered by vehicle motors was calculated assuming an efficiency of 15% (from final energy to useful energy) [35] and the efficiency from primary to final energy was assumed with 65% [20]; the efficiency of cooking was assumed to be 6% for Bashkurit valley [36]. In view of the three-stone cooking technology used, this assumption seems to be realistic. If the cooking takes place in the hut, the fire for cooking also heats the hut. This could increase the efficiency of the energy use to as much as 80% albeit only in cold seasons. In this calculation, this was neglected, also because the warmth rapidly escapes in the draughty huts. Due to the indoor pollution, huts need to be draughty to limit harmful health effects. Main reason for not considering the waste heat of cooking has heating is that the heat is mainly required in the residential houses, while the warmth is generated in the kitchen huts. The cooking efficiency for Theyern was estimated to be 30% (according to Haberl). Space heating in Theyern was calculated according to assumptions made by Haberl but adjusted to the local energy mix (biomass, oil, electricity, solar energy for hot water). In a similar way, the conversion from petroleum to lighting in the Bashkurit Valley is based on Haberl's assumptions but adjusted to petroleum as the only source.

### 3.3. Carbon Emissions

For estimating the carbon emissions a simple calculator was used [37]. The country-specific emissions factors applied here are 2.8 kg $CO_2$/litre gasoline, 3.2 kg $CO_2$/litre diesel or heating oil and 0.184 kg $CO_2$/kWh electricity for Theyern. Petroleum used in the Bashkurit Valley was calculated with 2.8 kg $CO_2$/litre. Biomass was assumed to be carbon neutral in both cases.

## 4. No Time: Insights into an Agrarian Society of the Ethiopian Bashkurit Valley

The first case is located in Ethiopia, a country that has seen fast growth in recent years, but GDP per capita is with 500 US$ (in current prices for 2012) [38] one of the lowest in the world. The agricultural sector accounts for about 47% of national GDP and 85% of employment [39]. Agriculture is mainly rain-fed, dominated by smallholders (Lavers 2012) and shows due to the very low use of fossil energy and machinery typical patterns of an agrarian society [40].

The case study area itself is a valley called Bashkurit and located in the northwest of Ethiopia. It lies north of Gondar (population approximately 200,000), the fourth largest city in Ethiopia (see Figure 2). The closest road to Gondar is a few kilometres away and can only be reached on foot paths. The landscape is about 2300 m above sea level, is slightly hilly and barren in all seasons except for summer (June to September). The inhabitants feed themselves by means of a sophisticated system of field and livestock farming and use fossil fuels for lighting only. As in many agrarian societies, rain is of existential importance in the Bashkurit Valley as it determines whether the next harvest is sufficient for local food supplies and whether—in the best case—a surplus can be sold in the market. In Bashkurit, the precipitation is a yearly average of 1200 mm per year, much more than in the low-precipitation north-east of Austria (450 mm) and roughly at the level of the Austrian average (1100 mm). However, rain falls more concentrated in a shorter time, namely 80% of the rainfall in four summer months (June to September). In addition, rainfall is subject to strong annual fluctuations, and both severe rainfall and droughts occur frequently. The daily temperature maximum ranges from seven degrees Celsius in November to 32 degrees Celsius in April. In accordance with this climate, the Bashkurit Valley is harvested once per year.

We prepared a complex material and energy flow calculation for the valley. For a year, the people in the valley witnessed and participated in weighing, measuring, observing, counting and recording for a detailed stock taking exercise of the material stocks, material and energy flows as well as time use [27]. All data refer to the year 2012.

**Figure 2.** The Bashkurit Valley in summer (Photo: Hailemariam B. Andarge).

*4.1. Societal Stocks: People, Livestocks, Land and Artifacts*

In 2012, 2200 people lived in six villages of two to 13-person families (an average of seven persons live in about 310 households) in the valley. The annual population growth in the valley is between two and three %, the population has doubled in the last 25 years. Life expectancy at birth in Ethiopia is despite improvements with 61 years still low [41]. A national strategy from 1993 aims at a birth rate reduction: for women, instead of 7.7, only 4.0 children are to be born [27]. Because of this strategy, the Bashkurit Valley has recently seen a slight decline in the birth rate, in effect growth continues albeit slower. In the 1700 hectares of the valley, about one third of the area is forest and scrubland, about a third are fields, a quarter are hedges, and the remaining roughly eight % are rocky or used for paths and buildings [27]. The 2200 people in 2012 held more than 400 oxen, 900 cows, 300 donkeys, 1500 sheep and goats as well as 1000 poultry animals. The families lived in over 500 houses (some families had more than one house), 80% of which are covered with corrugated iron. The remaining houses have grass roofs. 2.4 tonnes of material per person are used for houses, about half of the material is biomass and the other half consists of minerals such as stones, clay and iron (for the corrugated iron roofs). By way of comparison, for Europe, the values are between 130 and 325 tonnes per capita [42]. Household and personal possessions add up to about 130 kg per person in Bashkurit. Due to the absence of a power supply network, there are neither electric household appliances nor electronic devices such as mobile phones or computers, except for a few radios. Much of the household items consist of kitchen utensils, furniture and agricultural tools such as pickaxes and ploughs. Some of the households also have petroleum lamps. Aside from clothing, personal possession is minimal and consists mainly of school supplies for children. For adults, possession of the Bible is of high personal importance.

*4.2. Metabolic Flows of a Valley in an Agrarian Society*

In the Bashkurit valley, approximately 7000 tonnes of material or 3.2 tonnes per person are extracted annually from the local environment (Table 1). These consist almost entirely of biomass, roughly 40% of harvested field crops (three-fourths are millet, wheat, corn, teff and barley), 40% of biomass grazed by animals and 20% of collected wood, mainly used as firewood. In addition, very small amounts of honey are taken from nature. The only minerals extracted locally are loam and stones for the construction of huts (less than 1% of the extraction).

The local community imports 165 kg of material per person and year. The bulk of this, 80%, is biomass in the form of seed, feed and food. Food mainly consists of sugar and cooking oil. Other imports are mineral materials such as salt, iron, cement and fertilizer. About nine kilograms of fertilizer per person per year are imported, which amounts to 35 kg per hectare of farmed land (in industrial agriculture this is 100 to 200 kg per hectare). Twelve kilograms of artifacts (kitchen utensils, tools and clothes) and about three kg of fossil fuels are imported per person and year.

Exports amount to about 284 kg per person, accounting for just under 10% of the local extraction. Exports are about 40% unprocessed wood and grain. Furthermore, straw and dried cow dung are exported as fuel. Quantitatively insignificant, but of economic importance, is the export of honey and mud furnaces. For the latter, clay is removed locally and mixed with imported cement and processed into small cooking stoves. Most of the exported goods are brought to nearby markets. Although the income is low, it is necessary to import goods or to finance health care and school education.

The study of the per capita food supply shows an intake by humans of 2127 kcal/cap/day [27]. An estimate of the energy requirements based on FAO data of kcal/kg body mass for different age groups [43] and the body mass per age group in the Bashkurit valley shows that 2250 kcal/cap/day are required. Thus, there is slightly too little food available compared to the requirements which constitutes a situation of food scarcity.

**Table 1.** Per capita material flows and consumption (extraction plus imports minus exports) for the Bashkurit valley in kg fresh weight in 2012 [27].

| Category (kg per Person and Year) | Extraction | Imports | Exports | Consumption |
|---|---|---|---|---|
| Biomass | 3190 | 135 | 280 | 3045 |
| Mineral materials | 2 | 15 | 4 | 13 |
| Fossil energy carriers | 0 | 3 | 0 | 3 |
| Artefacts | 0 | 12 | 0 | 12 |
| Total | 3192 | 165 | 284 | 3073 |

*4.3. Primary Energy, Societal Useful Energy and Human Labour*

Day-to-day life in the Bashkurit Valley is extremely labour-intensive, as is clear from a statement from Wagaw, a local farmer: "We daily struggle with nature to fill our stomachs and the stomachs of our cattle. Even in the heat of the sunshine, we do not interrupt our work, but protect our heads with caps" ([27], p. 98).

When people are asked about the cultural significance of work, one quickly finds that work is closely linked to food procurement. This is illustrated by the frequently used Amharic saying "Yalsera aybla", which means "no work, no food". Subjectively, work creates two opposing meanings for the inhabitants, depending on the harvest. If there is enough food, the residents are proud of their work. If the food is not sufficient, they feel it as a burden.

If we examine human labour quantitatively, we base this analysis on functional time use studies [44]. These studies treat human time as a key resource of social systems. A social system's available time per average day is characterized by 24 h per day and individual multiplied by population size. This time is spent for four different functional subsystems: the person system (PS), the household system (HS), the economic system (ES) and the community system (CS). According to literature, the time invested in the economic system is what we refer to as labour time [26].

Table 2 provides some impressions of daily routines and ascribes the activities to the four functional subsystems.

**Table 2.** Daily routines in a chronological order from getting up to going to bed—Due to the low level of activity of the elderly they are not considered in this qualitative description (nevertheless their time use is included in the overall time use analysis).

| Hour | Daily Activity |
|------|----------------|
| 5:30 | Most people get up around 5:30 a.m. **Women** often rise earlier than men. After completing their personal hygiene (PS), women walk to fetch water (HS), visit the nearby church to pray (CS), prepare breakfast, take care of children, grind grains using stones, and do other domestic chores (HS). **Girls** help the women (HS), visit a nearby church to say their prayers (CS), clean the cattle pen, prepare dung cake, scare away wild animals from fields or travel to the market to sell firewood and dung cake (ES). **Men** feed cattle, maintain farm tools (ES), or look after children (HS). The **boys** fetch water, mind the younger children (HS), protect crops from wild animals, travel to the market and feed the cattle at home (ES). These activities usually extend to breakfast time. |
| 7:30 | Except on fasting days, breakfast is taken from 7:30 to 8.00 a.m. After breakfast, **women** milk cows, clean cattle pen, prepare dung cakes, herd the cattle, walk to the market, work in farm fields, perform wage work or handcraft (ES), participate in public meetings, engage in communal works (CS), and prepare lunch (HS). **Men** work in the fields, herd cattle, undertake wage work or handcrafting, go to the market (ES), and participate in public meetings and communal works (CS). School is open in two shifts, one in the morning and the other in the afternoon. Some **children** attend school after breakfast (PS), while others work with parents (ES). |
| 12:30 | Lunch is usually taken from 12:30 to 1:30 p.m. |
| 1:30 | After lunch, the **women** engage in herding, work in the fields, milk cows, perform handicraft (ES), collect firewood, fetch water, go to grinding mills, care for children, prepare dinner, and other domestic chores (HS), visit relatives and attend festivals (CS). **Men** work in the fields, go to the market, herd, undertake maintenance work, perform wage work (ES), do communal work, attend public meetings, visit relatives, attend festivals (CS) and have an infrequent leisure hour (PS). The time after lunch is school time (PS) for some **children**, while others mind the younger children, go to a grinding mill, collect firewood, perform domestic chores (HS), work in the fields, herd cattle, go to the market (ES), and perform communal work, visit relatives, play games, and sometimes attend festivities (CS). |
| 6:00 | Most people return from outdoor work around 6:00 p.m. Darkness sets in, lamps are lit. **Women** prepare dinner, milk cows, and breastfeed children (HS), while the **girls** study (PS) and some of them help their mothers (HS). **Men** carry out maintenance work (ES), while **boys** study (PS) and feed cattle in the compound (ES). |
| 8:00 | Dinner time is usually from 8:00 to 9:00 p.m. (earlier for younger children). After that, **married couples** may take a rest and discuss the day (PS) and prepare for the next day (HS, ES). **Women** grind grains using stone grinds, care for children (HS) and make baskets (ES). Some **children** study, while others take a rest and have a conversation (PS). With these activities, people usually conclude their day. |
| 10:00 | People go to sleep between 10:00 and 11:30 p.m. |

In Bashkurit the annual average human labour time in the economic system of all persons, whether old or young, is about five hours per day and person, mainly in agriculture (including market walks, making baskets, and other smaller wage labour). In addition to the work in the economic system, two and a half hours are spent for household chores and a little more than two hours are spent for community affairs (work for shared infrastructure or care for neighbours, church visits and meetings). This leaves about 14 h for sleeping, eating, body hygiene and leisure. At the age of six years, children work four hours a day on average. The greatest time demand for work, be it in the economic or in the household system, is on 25 to 50-year-old adults with over ten hours a day, with women more burdened than men. Especially the age group of 25 to 40-year-old women has less than ten hours for sleeping, eating, body hygiene and leisure (PS), men of the same age eleven hours. Compared to other agrarian societies, this is an extremely low level of time for personal reproduction (PS). In Nalang (Laos), a subsistence economy dominated by rain-fed rice farming, and in Campo Bello (Bolivia), a swidden agriculture with fishing, hunting, gathering and raising poultry, the work load in the economic system (ES) is between 2 and 3 h (in Bashkurit five hours) and the time for personal recreation is 18 h (in Bashkurit 14 h) [27,45]. In addition to human labour, oxen and cows provide work for the farmers. The domesticated animals are used for ploughing, sowing, harvesting, threshing and transporting.

In the Bashkurit Valley, people provide around 650 Gigajoule of useful energy per year in the economic system (300 Megajoule per person per year; calculation is based on the hours worked and assumptions on the physical work delivered per hour in accordance to [20]), the energy provided by animals is about 300 Gigajoule (see Figure 3). To make it comparable on the level of communities, we express animal labour per human capita, this is 120 Megajoule per person per year. A further considerable amount of useful energy is consumed for cooking where 1200 Gigajoule are used (about 550 Megajoule per person and year). Apart from battery powered radios and petroleum lamps for lighting, there are neither fossil-fuelled nor electrically operated devices, which provide significant work for the reproduction of the local community. The total useful energy thus amounts to about 2200 Gigajoules (1000 Megajoule per person and year). To provide this useful energy, approximately 100,000 Gigajoules of primary energy per year (or 44,200 Megajoule (MJ) per person and year) must be expended (see Table 3).

**Figure 3.** Ploughing as a combination of human and animal work (Photo: Hailemariam B. Andarge).

**Table 3.** Per capita primary energy (including imports and exclusive exports) and useful energy for the Bashkurit Valley in kg fresh weight in 2012 [27].

| Primary Energy (MJ Per Person and Year) | | Useful Energy (MJ Per Person and Year) | | Conversion Efficiency |
|---|---|---|---|---|
| Food | 5000 | Human labour | 300 | 0.0600 |
| Fodder | 30,000 | Animal labour | 120 | 0.0040 |
| Fire wood and cow dung | 9000 | Heat for cooking [1] | 550 | 0.0610 |
| Petroleum | 200 | Lighting | 30 | 0.1500 |
| Total | 44,200 | | 1000 | 0.0220 |

[1] In contrast to the numbers shown in the source, a change was made here. Instead of using 25% of the primary energy contained in the firewood as useful energy for cooking, 6% were assumed based on [36].

Looking at the conversion efficiency indicates that humans next to stoves have the highest conversion efficiency. Despite this relatively high conversion efficiency of stoves, compared to other technologies it is very low in the Baskurit valley, since cooking can be done with three times higher conversion efficiencies [46]. Animals consume the highest amount of energy with a very limited return, thus having a very low conversion efficiency. It should be noted that human food does not only serve to allow people to work. To feed people is a value in itself. Likewise, the animals provide not only mechanical work but also food (albeit to a very low level) and cow dung (as fertilizer and fuel for cooking). In addition, the number of cattle is very important for the owner's status, a service not reflected in the conversion efficiency since it cannot be expressed in energy terms.

*4.4. An Intermediate Resume: Built-In Dynamics of the Societal Metabolism-Labour Nexus*

Agrarian societies have the potential for ecological sustainability in energy terms because they consume nearly only locally available renewable resources. At the same time, these limited resources hinder them from organising material flows much beyond food provision. However, and this is well visible in the Bashkurit Valley, agrarian societies are in danger of initiating a downward spiral [47]. First, the scarcity of useful energy for food production must be overcome. Since three quarters of the necessary energy for food provision is provided by human labour, the local population hopes that more offspring will lead to a rise in the workforce, thus reducing the workload per person. Second, more people also require more food. Since the amount of food available in agrarian societies is essentially dependent on the area, which usually remains constant, this is likely to lead to soil degradation. Third, in turn, this results in reduced soil fertility, lower yields and thus reduced primary energy in terms of food. The alleged solution of the problem thus exacerbates the original energy shortage and calls into question the viability of survival, since the daily work already affects the population to its physical limits. This labour burden as well as the food insecure situation contributes to a low number of old people. Only eight % of the population in the Bashkurit Valley is older than 50 years (In Ethiopia, the life expectancy at birth in 2012 is 61 years [41]).

Looking at Bahskurit from a biophysical system perspective, reduced animal husbandry could release labour time, reduce food scarcity and mitigate environmental pressures. Currently, there is a cow for each person older than 15 years. Cattle do little work and animal products contribute very little to the human diet (less than two % of the calorie requirement). At the same time, however, cows devour 85% of the primary energy in the form of feed and demand a significant portion of the populations labour time. A reduction of the number of livestock, together with a halt to population growth, would therefore be a biophysically option to overcome food scarcity. However, the high symbolic importance of having a high number of children and the status related to a large cattle herd size counters such a change. Alternatively, an industrialization of agriculture is often supposed to solve the problem of energy shortages and soil degradation. However, this is not feasible without external resources such as financial development aid and is associated with a series of follow-up problems as known from industrial societies (see Section 6).

## 5. Energy in Abundance: A village in Industrialized Lower Austria

The second case is located in Austria. In contrast to Ethiopia Austria's GDP per capita is with 24,380 US$ (in current prices for 2001, the year of study) relatively high [38]. The agricultural sector accounts for about 1.7% of national GDP [38] and 6% of employment [48]. Agriculture in Austria is characterized by small-scale farm operations and comparatively low-technical input as compared to other Central European land use systems. However, compared to Ethiopia it can be described as highly intensified with high input of fossil energy and machinery.

The case study itself takes us to a comparatively small village of about 60 people called Theyern. It is located in Lower Austria, ten kilometres from the next major city Krems (population approximately 20,000) and 15 km from the Lower Austrian state capital St. Pölten. The distance to Vienna, the capital of Austria, is around 50 km. The place itself is located within a landscape of gently rolling hills on a plateau on almost 400 m altitude. The basic structure of the village dates to a 13th century clearing island, which can still be seen on today's aerial images (see Figure 4). In the middle of the clearing is a compact settlement structure, surrounded by fields and finally surrounded by forest. The residential area covers 76 hectares and is dominated by orchards. This case study was carried out by means of qualitative and quantitative data collection [28]. The research team consulted nearly all households, interviewed several representatives of the local council and the mayor of the municipality. All data refer to the year 2001.

**Figure 4.** Aerial photograph of Theyern (Source: Google Earth).

*5.1. Societal Stocks: People, Animals, Land and Artifacts*

Of the 63 villagers at the time of the survey, around 13% are full-time farmers and 6% are part-time farmers. As their main source of income, the part-time farmers work in a beverage factory some ten kilometres from the village. A further 15% have full-time jobs as workers, in technical professions, in administration and as teachers, mainly in Krems and St. Pölten. Altogether, about a third of the inhabitants are employed. A quarter of the inhabitants are children or in training and another quarter pensioners. A little over 10% of the people are mainly concerned with household chores. This is a very typical profile for an industrialized society (the Austrian average for 2001 is 8% [49]). A historic reconstruction of the village shows that around 1830, about 100 people lived there. For them, work in agriculture started in early childhood and continued all through life. Since 1830, the population size has dropped (due mainly to outmigration), but is now stable again (the number of inhabitants in 2016 [50] was still the same as in 2001 [51], indicating a population growth of zero %). The high level of mobility and the proximity to urban centres combined with a high quality of life in the countryside lead to the return of many inhabitants after their secondary, tertiary education or further trainings.

The cattle population is modest compared to the Bashkurit Valley: two cows, three horses, 25 pigs and 161 chickens in the entire village (0.1 cattle per resident, in Bashkurit Valley this is 1 cattle per person aged 15 or older). In contrast, possessions in Theyern are difficult to grasp due to the wide range and large number of artefacts. There are 55 buildings (17,000 tonnes of material in stock), 40 cars, six motorcycles and 35 tractors/combine harvesters/trucks (all vehicles together are 86 tonnes) and 286 larger household appliances (e.g., washing machines) (eight tonnes). There are an estimated 220 tonnes per capita of stock materials in buildings (compared to 2.4 tonnes per capita in the Bashkurit Valley) and about 1.5 tonnes per capita of vehicles.

*5.2. Metabolic Flows of a Rural Village in an Industrial Society*

An analysis of the material flows shows that 12 tonnes of biomass per person (Table 4) are extracted from the local environment (the comparative figure for the Bashkurit Valley is 3.2 tonnes/capita), although only 12 of the 63 inhabitants are wholly or partly engaged in agriculture. The main harvest from agricultural land and forest is apples and pears (44%), maize (23%), grapes (13%) and fire wood (13%). Barley, wheat, sunflower, apricot, hay and straw account for less than two percent of the harvest.

Imports account for approximately 3.4 tonnes per person: 1.4 tonnes of biomass (food for the inhabitants' own consumption, seeds and piglets), 1.4 tonnes per person of fossil fuels for the operation of the vehicles and to a lesser extent for heaters, and 600 kg per hectare for agricultural inputs such as fertilizers or agents for weed and pest control (in Bashkurit 35 kg per hectare). Due to accounting conventions, electricity is not considered in the material flows, but it constitutes a significant part of the energy balance (ten per cent of energy consumption).

Of the locally extracted twelve tonnes of material per person, more than eight tonnes are exported. The farms are not geared to self-sufficiency, but to the supply of the nearby urban centres. Many inns and large customers are supplied with pure high quality fruit juice from these farms. Corn is sold primarily through the close by cooperative warehouse, and the wine obtained from the grapes is marketed directly or to large customers via the farmers' own sales channels. The four tonnes per person difference between extraction and export are partly residues from pressing fruit (mash) and partly self-supply. However, self-supply does not come from agriculture but in the form of fruits and vegetables from house gardens or extensive orchards behind the farm buildings, and as firewood from own forests, which is partly fired in modern heating systems.

**Table 4.** Per capita material flows and consumption (import plus imports minus exports) in 2001 [28].

| Category (kg Per'Person and Year) | Extraction | Imports | Exports | Consumption |
|---|---|---|---|---|
| Biomass | 12,000 | 1400 | 8400 | 5000 |
| Mineral materials | 0 | 600 | 0 | 600 |
| Fossil energy carriers | 0 | 1400 | 0 | 1400 |
| Total | 12,000 | 3400 | 8400 | 7000 |

*5.3. Primary Energy, Societal Useful Energy and Human Labour*

A high amount of human labour is performed in Theyern, but with 2.5 h per person and day this is by far not as much as in the Bashkurit Valley. In terms of time use, it is striking that different households in Theyern are characterized by very different rhythms. For example, there is a peasant couple with an agriculture-dominated and tightly organized daily routine, a family of seven with various professions, or a household with a relatively young pensioner. In addition, the high mobility is noticeable. In Theyern, around 9000 km per year are covered per person. This is roughly 10 times that of the kilometres per capita travelled in the Bashkurit valley. In view of the different travel speeds of walking (five kilometres per hour) and driving (50 km per hour), the average time required per person is about the same in Theyern and the Bashkurit valley (30 min per day and person).

In Theyern, an average of almost 2.5 h per day is spent on wage labour (in the Bashkurit Valley 5). This low number of labour hours per capita is above all since young and old people do work very little. The interviews in Theyern show that young people provide a very small labour time input to farming or other economic activities. The same applies to older persons who play workwise a minor role in agriculture but are important in the household chores. This can also be explained by the fact that in the years before the survey, many of the farms were converted from so-called mixed farms (coined by cattle, fields and extensive orchards) to specialized farms (fruit orchards) and therefore the experience of older farmers was not appropriate any longer. For example, an older farmer expressed in an interview the common norm in previous times that the good apples were selected for eating, the bad ones were sorted out to make juice. Today, however, the good apples are selected for juice making. He concluded that he does not want to join in in such practices which run counter to his inner understanding how things should be done well. Considering this withdrawal of older farmers from agricultural activities, the average 2.5 h per day can be regarded as a realistic estimate.

7 to 8% of the total primary energy is provided in the form of food (Table 5). In relation to the primary energy spent, the useful energy of human labour is negligible (due to the low numbers of hours worked and the assumption that the physical work delivered per hour is in accordance to [20] due to the used machinery much lower in industrial than in agrarian societies). The following energy flow calculations are done per inhabitant since (a) the purpose of a social system's functioning ultimately is to reproduce a human population within a territory [26], and (b) it is necessary for a meaningful comparison of different social systems with different population sizes.

Thus, per inhabitant, the few farm animals consume a slightly larger portion of biomass than humans. Farm animals do almost no work but are mostly used to produce food, which is only to a small extent consumed locally. Most the animal products are marketed and bring income. The work of the animals is only due to the horses, which a family keeps for riding in their free time. The useful energy is here the drive force provided during the rides. Primary energy consumption for the sum of privately and agriculturally used fleet of vehicles is the largest amount of Theyern's total primary energy consumption. Due to the low conversion efficiency, however, this contributes only about 12% to the total useful energy consumed in Theyern. The room heat is next largest area of primary energy consumption. This is a result of the considerable amount of building stock which the residents in Theyern inhabit. Compared to Bashkurit, of course, the different climatic conditions are decisive (annual average temperature of Theyern about ten degrees Celsius, Bashkurit valley about 20 degrees Celsius), but also the different comfort requirements play a role as well as the limited capacity of households to acquire more fuels (local fuel is scarce and there are no heaters at all for burning fossil fuels).

In total, about 40% of the primary energy used is from renewable energy sources (biomass and hydroelectric power) and 60% from fossil energy. While in the Bashkurit Valley, 30% of the total useful energy was still provided by humans, the share in Theyern is diminishingly small (about 0.1%). The consumption of useful energy in agriculture is much higher compared to the Bashkurit valley. Farmers can dispose of approximately 5300 Megajoule of work per inhabitant of Theyern (including persons not working in agriculture). If we divide this useful energy only amongst persons involved in agriculture, this is 20,000 Megajoule per person. In the Bashkurit Valley, this is 420 Megajoule per person including all and 530 Megajoule per person engaged in agriculture.

Conversion efficiencies are very low for human labour and even lower for animal labour. This is not surprising since humans are well nourished but in this energy regime there is little need for their physical work capacity. The same applies for farm animals, which are kept for milk, eggs and meat and therefore do not provide useful energy. Technology using biomass, hydropower and fossil fuels has far higher conversion efficiencies. Amongst them drive force has the lowest due to low efficiencies of the combustion engines and losses during the supply chain. In the case of space heating, due to reasons based in the second law of thermodynamics, all forms of energy like mechanical, electrical or chemical can be completely converted into heat. This contrasts with other energy conversions and enables the highest efficiency in practice.

The conversion efficiency for lighting benefits from the fact that it is partly based on hydropower, which has lower losses during conversion and thermal power plants which partly make use of the heat for district heating. Altogether, the total conversion efficiency in Theyern is by a factor 10 higher than in the Bashkurit valley, due to both the higher share of fossil fuels in the energy mix and modern conversion technologies.

**Table 5.** Per capita primary energy (including imports and exclusive exports) and useful energy for Theyern in MJ per person and year in 2001 (Source: [28,30], own assumptions).

| Primary Energy (MJ Per Person and Year) | | Useful Energy (MJ Per Person and Year) | | Conversion Efficiency |
|---|---|---|---|---|
| Food | 15,000 | Human labour | 50 | 0.0030 |
| Fodder | 20,000 | Animal labour | 5 | 0.0003 |
| Fuel for tractors and trucks | 55,000 | Driving power agriculture | 5300 | 0.0960 |
| Fuel for cars and motor bikes | 30,000 | Driving power private use | 2800 | 0.0930 |
| Energy for cooking | 8000 | Heat for cooking | 2400 | 0.3000 |
| Oil (30%) + Biomass (70%) for space heating | 64,000 | Space heating | 37,000 | 0.5780 |
| Hydro power & Fossil energy carriers (Austrian mix) | 20,000 | Lighting, appliances, electronic devices | 6000 | 0.3000 |
| Total | 212,000 | | 53,555 | 0.2520 |

## 6. Changes in the Societal Metabolism-Labour Nexus across Energy Regimes

Based on the two case studies, we have portrayed the societal metabolism-labour nexus for an agrarian and an industrial rural community. Despite the cultural differences, both cases can be regarded as illustrative for the respective energy regime.

Figure 5 compares the two cases with the bandwidth of indicators for agrarian and industrial societies as presented in Haberl et al. [5]. The Bashkurit Valley is in terms of material consumption at the lower end of agrarian societies. Theyern is outside the bandwidth. For this there is a simple explanation. Material flows at the country level include flows for building and maintaining large infrastructures like dams, highways and power plants, which can't be found in local rural cases like Theyern. So, while the domestic material consumption of mineral materials (metals and non-metallic minerals) in Theyern was 0.6 tonnes per capita and year (see Figure 6), the material consumption of mineral materials was about 15 tonnes per capita and year in Austria in 2001. About 80% of these materials is used for construction activities [52]. This difference between local rural cases and the country level is in agrarian societies less pronounced than in industrial societies due to the lower level of urbanisation [21]. In terms of primary energy both cases are within the bandwidth, albeit in the lower range of it out of the same reasons as for material consumption. Moreover, in spite of great cultural differences, the biophysical basic characteristics of the Ethiopian case also correspond to those of a historical reconstruction of the material and energy flows of the village of Theyern in Austria around 1830, i.e., at a time when Theyern can still be described as an agrarian society [28]. Thus, basic indicators like material consumption and primary energy display significant differences between the two cases in line with the differences at country level. Altogether, these comparisons show that each case can be regarded as illustrative for the energy regime it belongs to. This allows to further interpret differences regarding more detailed material and energy indicators as well as labour time as differences of the energy regime.

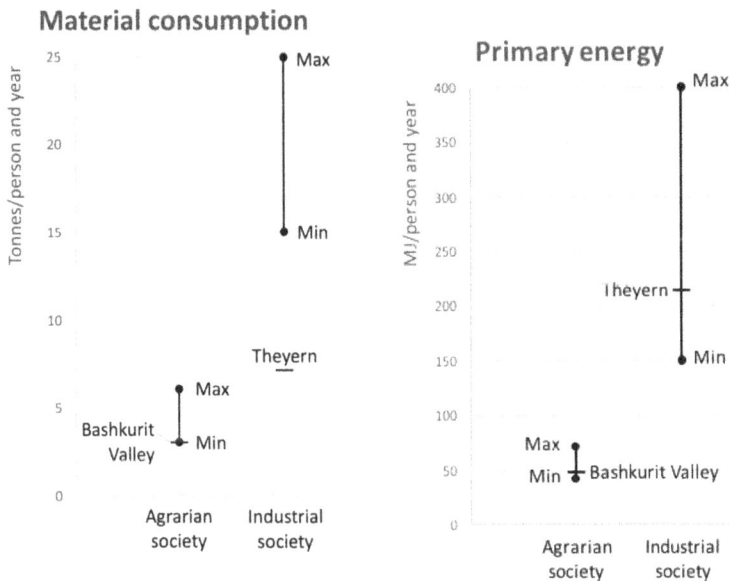

**Figure 5.** Bandwidth (Min-Max) of domestic material consumption and primary energy per capita and year for agrarian and industrial societies according to literature [5] in comparison to the local rural cases Bashkurit Valley, Ethiopia and Theyern, Austria.

Material flows in both cases are dominated by biomass (see Figure 6). Domestic material consumption of biomass differs almost by a factor two, which is because people in Theyern are better nourished (three times higher calories per capita) and dispose partly over fuel-wood based space heating, which is missing in the Bashkurit Valley. However, other material flow indicators display even bigger differences. Biomass extraction is in Theyern four times higher than in the Bashkurit Valley. While in Bashkurit trade flows are very moderate, with exports of 0.3 tonnes of biomass per capita and year, Theyern can export 8.4 tonnes of biomass, which is the same level as the overall material consumption in Theyern. This low surplus in Bashkurit Valley shows the limited capacity of rural villages in agrarian societies to support an urban population, while Theyern's high surplus as net exporter of biomass can contribute to sustain high population numbers not involved in agriculture (this is 94% in Austria) [48]. This is made possible by the availability of fossil fuel carriers. In tonnes, they are not so significant (1.4 tonnes per capita and year), albeit together with technologies they entail far reaching consequences for labour. This becomes visible in the energy flows of Figure 6. The primary energy is in Theyern five times higher than in the Bashkurit Valley. While the primary energy for humans and animals together is the same amount in both cases, primary energy for technology is in Theyern almost 20 times higher than in the Ethiopian case. The estimate for labour in terms of societal useful energy reveals the drastic differences of the two local systems, with Theyern disposing over 54 times more useful energy than the Bashkurit Valley. The increase in disparities from primary to useful energy between the two cases is due to the higher conversion efficiencies of fossil fuel-based technologies, which are dominant in Theyern but at low levels in Bashkurit Valley (only cooking stoves and petroleum lamps). Technologies provide in Theyern 92 times more useful energy than in the Bashkurit Valley, while useful energy provided by human labour is in Theyern a fifth (0.2) and labour hours are a half of the values in the Bashkurit Valley (see Figure 7). Thus, physical work of humans is both supported and substituted by using fossil energy powered machines in Theyern. Furthermore, since useful energy is not scarce anymore in Theyern, they could build up stocks like buildings, machinery and household artefacts 92 times heavier than in the Bashkurit Valley.

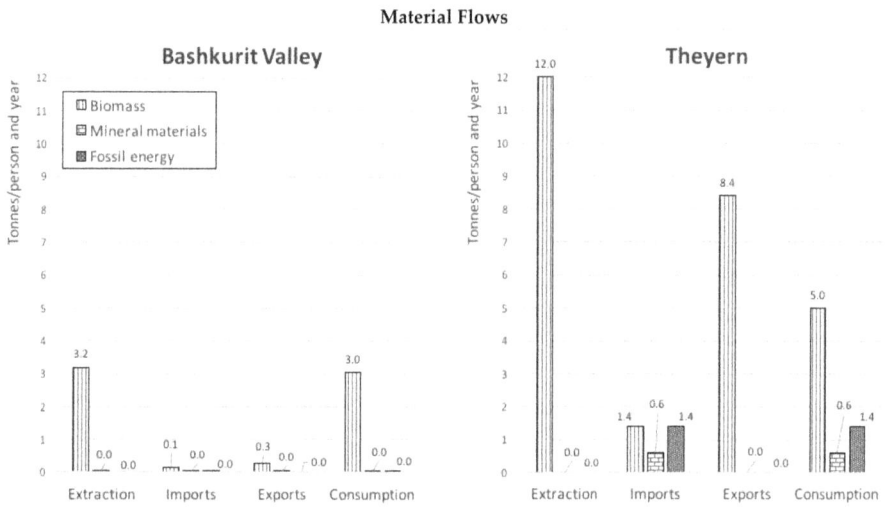

**Figure 6.** *Cont.*

**Energy flows**

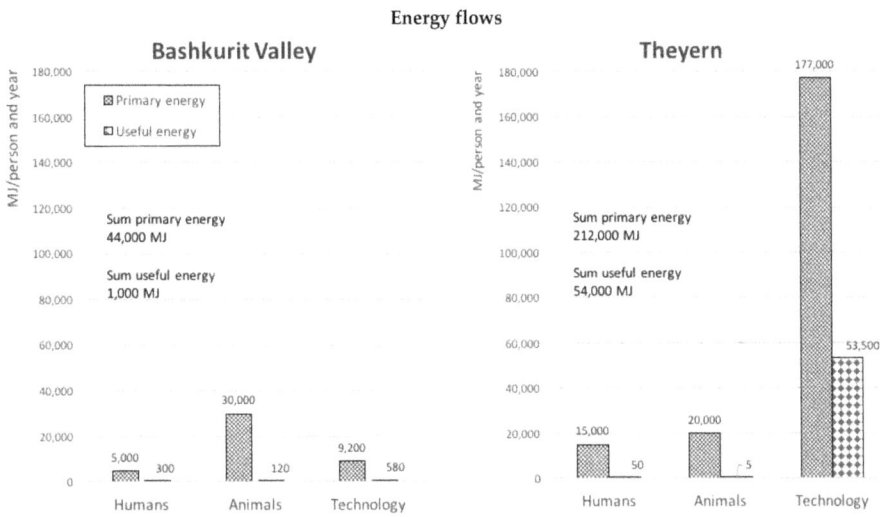

**Figure 6.** Material and energy flows displayed for both cases. On top the material flows extraction, imports, exports and consumption is shown for biomass, mineral materials and fossil energy in tonnes per capita and year for the Bashkurit Valley (**left**) and for Theyern (**right**). Mineral materials include metals here as well. The two lower graphs show primary and useful energy as delivered by humans, animals and technology in MJ per capita and year for the Bashkurit Valley (**left**) and for Theyern (**right**).

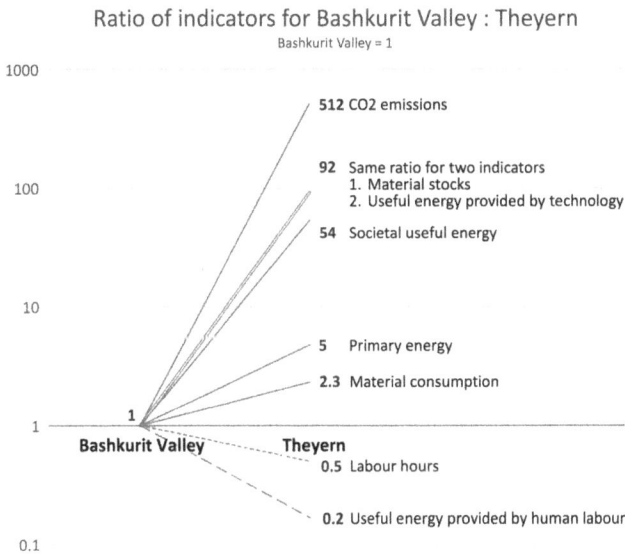

**Figure 7.** Ratio of indicators: Bashkurit Valley is indexed with 1 and Theyern is by a factor of x bigger or smaller. Thus, CO2 emissions are 512 times higher in Theyern than in the Bashkurit Valley.

After discussing the sustainability problem of declining food returns for a growing population in the Bashkurit Valley, we want to highlight one of the critical sustainability problems related to Theyern. As part of an industrial society Theyern has increased its land use intensity by use of fertilizers and machinery powered by fossil fuels. Further, fossil fuels are used in the private consumption of households. This adds up to the consumption of 1400 kg of fossil energy carriers per person and year plus a consumption of 190,000 kWh of electricity partly generated by burning fossil fuels in caloric power plants. In the Bashkurit Valley 3 kg of petroleum was used and no electricity at all. In terms of $CO_2$ this means 11 kg in Bashkurit Valley and 5600 kg per person and year in Theyern, which means about 500 times higher $CO_2$ emissions in Theyern than in the Bashkurit Valley.

Summing up on the societal metabolism-labour nexus across energy regimes, in the industrial case people work less in terms of both useful energy they provide and labour hours they spend, are better nourished, dispose over far more useful energy mainly provided by fossil fuel-powered technology and can export far higher food supplies than it is possible in the agrarian case. Thus, industrial societies have a high share of the labour force freed from the need of food provisioning allowing for all different forms of labour as known today [6]. This can be assessed as massive change in the societal metabolism-labour nexus based on the abundant availability of energy as provided by fossil fuels. However, high carbon emissions of industrial societies as the ones in Theyern call for low carbon development strategies. The necessary abandoning of fossil energy carriers means withdrawing from the very same energy base that enabled todays multiplicity how human labour appears today. While the new energy transition will not mean a return to agrarian societies, it is very likely to entail drastic consequences for the societal metabolism-labour nexus and consequently for human labour. Changes promise to be as far-reaching for human labour as in past transitions from one energy regime to another. However, the topic how human labour might change during the next transition to a post fossil society is still an under-researched area with high relevance for all.

**Acknowledgments:** The case study on Theyern was part of the research project GEWIN that was funded within the Austrian landscape research program of the Austrian federal ministry of science and education. The case study of the Bashkurit valley was made possible by a doctoral scholarship for Hailemariam Birke Andarge (Appear grant of the Austrian Development Cooperation). The socioecological perspective on labour was developed by Marina Fischer-Kowalski together with Willi Haas during the FP7 EU project NEUJOBS. The authors thank an anonymous reviewer for valuable comments. No funds for covering the costs to publish in open access were received.

**Author Contributions:** Willi Haas has provided the overall framework of the article. The case study of Theyern was conducted by a research team led by him. Hailemariam Birke Andarge provided the Ethiopian case.

**Conflicts of Interest:** The authors declare no conflict of interest.

## References

1. Rockström, J.; Steffen, W.; Noone, K.; Persson, Å.; Chapin, F.S.; Lambin, E.F.; Lenton, T.M.; Scheffer, M.; Folke, C.; Schellnhuber, H.J.; et al. A safe operating space for humanity. *Nature* **2009**, *461*, 472–475. [CrossRef] [PubMed]
2. Crutzen, P.J.; Stoermer, E.F. The "Anthropocene". *IGBP Newsl.* **2000**, *41*, 17–18.
3. Fischer-Kowalski, M.; Krausmann, F.; Pallua, I. A sociometabolic reading of the Anthropocene: Modes of subsistence, population size and human impact on Earth. *Anthr. Rev.* **2014**, *1*, 8–33. [CrossRef]
4. Sieferle, R.P. *The Subterranean Forest: Energy Systems and The Industrial Revolution*; White Horse Press: Cambridge, UK, 2001.
5. Haberl, H.; Fischer-Kowalski, M.; Krausmann, F.; Martinez-Alier, J.; Winiwarter, V. A socio-metabolic transition towards sustainability? Challenges for another Great Transformation. *Sustain. Dev.* **2011**, *19*, 1–14. [CrossRef]
6. Fischer-Kowalski, M.; Haas, W. Exploring the transformation of human labour in relation to socio-ecological transitions. In *Let's Get to Work! The Future of Labour in Europe*; Beblavy, M., Maselli, I., Veselková, M., Eds.; Centre for European Policy Studies (CEPS): Brussels, Belgium, 2014; Volume 1, p. 250.
7. Fischer-Kowalski, M.; Weisz, H. The Archipelago of Social Ecology and the Island of the Vienna School. In *Social Ecology: Society-Nature Relations across Time and Space*; Springer: Cham, Switzerland, 2016; pp. 3–28.
8. Fischer-Kowalski, M.; Weisz, H. Society as hybrid between material and symbolic realms. *Adv. Hum. Ecol.* **1999**, *8*, 215–251.
9. De Molina Navarro, M.G.; Toledo, V.M. *The Social Metabolism: A Socio-Ecological Theory of Historical Change*; Springer: Cham, Switzerland, 2014.

10. Fischer-Kowalski, M.; Haberl, H. Tons, joules, and money: Modes of production and their sustainability problems. *Soc. Nat. Resour.* **1997**, *10*, 61–85. [CrossRef]
11. Matthews, E.; Amann, C.; Bringezu, S.; Hüttler, W.; Ottke, C.; Rodenburg, E.; Rogich, D.; Schandl, H.; Van, E.; Voet, D.; et al. *The Weight of Nations-Material Outflows from Industrial Economies*; World Resources Institute: Washington, DC, USA, 2000.
12. Marx, K. Die Produktion des absoluten Mehrwerts. 5. Kap. Arbeitsprozess und Verwertungsprozess. *K Marx F Engels Werke* **1968**, *23*, 192–213.
13. Haberl, H.; Zangerl-Weisz, H. Kolonisierende Eingriffe: Systematik und Wirkweise. In *Gesellschaftlicher Stoffwechsel und Kolonisierung von Natur: Ein Versuch in Sozialer Ökologie*; Fischer-Kowalski, M., Fischer Kowalski, M., Eds.; Overseas Publ. Association (OPA): Amsterdam, The Netherlands, 1997.
14. Pimentel, D.; Acquay, H.; Biltonen, M.; Rice, P.; Silva, M.; Nelson, J.; Lipner, V.; Giordano, S.; Horowitz, A.; D'Amore, M. Environmental and Economic Costs of Pesticide Use. *BioScience* **1992**, *42*, 750–760. [CrossRef]
15. Tilman, D.; Cassman, K.G.; Matson, P.A.; Naylor, R.; Polasky, S. Agricultural sustainability and intensive production practices. *Nature* **2002**, *418*, 671–677. [CrossRef] [PubMed]
16. Markussen, M.; Østergård, H. Energy Analysis of the Danish Food Production System: Food-EROI and Fossil Fuel Dependency. *Energies* **2013**, *6*, 4170–4186. [CrossRef]
17. Pimentel, D.; Hurd, L.E.; Bellotti, A.C.; Forster, M.J.; Oka, I.N.; Sholes, O.D.; Whitman, R.J. Food Production and the Energy Crisis. *Science* **1973**, *182*, 443–449. [CrossRef] [PubMed]
18. Sieferle, R.P. *Rückblick auf Die Natur: Eine Geschichte des Menschen und Seiner Umwelt*; Luchterhand: München, Germany, 1997.
19. Fischer-Kowalski, M.; Haberl, H. *Socioecological Transitions and Global Change: Trajectories of Social Metabolism and Land Use*; Edward Elgar Publishing: Cheltenham, UK, 2007.
20. Haberl, H. The Energetic Metabolism of Societies: Part II: Empirical Examples. *J. Ind. Ecol.* **2001**, *5*, 71–88. [CrossRef]
21. Fields, G. Urbanization and the transition from agrarian to industrial society. *Berkeley Plan. J.* **1999**, *13*, 102–128.
22. Fischer-Kowalski, M.; Haas, W. Toward a Socioecological Concept of Human Labor. In *Social Ecology: Society-Nature Relations across Time and Space*; Haberl, H., Fischer-Kowalski, M., Krausmann, F., Winiwarter, V., Eds.; Springer: Cham, Switzerland, 2016; pp. 259–276.
23. Eurostat. *Economy-Wide Material Flow Accounts (EW-MFA)—Compilation Guide 2013*; Eurostat: Luxembourg, 2013; p. 87.
24. Krausmann, F.; Weisz, H.; Eisenmenger, N.; Schütz, H.; Haas, W.; Schaffartzik, A. *Economy-Wide Material Flow Accounting Introduction and Guide*; Institute of Social Ecology: Vienna, Austria, 2015.
25. Fischer-Kowalski, M.; Krausmann, F.; Giljum, S.; Lutter, S.; Mayer, A.; Bringezu, S.; Moriguchi, Y.; Schütz, H.; Schandl, H.; Weisz, H. Methodology and indicators of economy-wide material flow accounting. *J. Ind. Ecol.* **2011**, *15*, 855–876. [CrossRef]
26. Singh, S.J.; Ringhofer, L.; Haas, W.; Krausmann, F.; Lauk, C.; Fischer-Kowalski, M. *Local Studies Manual: A Researcher's Guide for Investigating the Social Metabolism of Rural Systems*; Social Ecology Working Paper; Institute of Social Ecology: Vienna, Austria, 2010.
27. Andarge, H.B. People, Energy, and Environment—A Socio-Ecological Analysis in Bashkurit Watershed, Ethiopia. Ph.D. Dissertation, Alpen Adria Universitaet, Klagenfurt, Austria, 2014.
28. Haas, W.; Krausmann, F. Transition-related changes in the metabolic profile of an Austrian rural village. *Work. Soc. Ecol.* **2015**, *153*, 42.
29. Zabalza Bribián, I.; Valero Capilla, A.; Aranda Usón, A. Life cycle assessment of building materials: Comparative analysis of energy and environmental impacts and evaluation of the eco-efficiency improvement potential. *Build. Environ.* **2011**, *46*, 1133–1140. [CrossRef]
30. Haberl, H. The energetic metabolism of societies: Part I: Accounting concepts. *J. Ind. Ecol.* **2001**, *5*, 11–33. [CrossRef]
31. Brockway, P.E.; Barrett, J.R.; Foxon, T.J.; Steinberger, J.K. Divergence of Trends in US and UK Aggregate Exergy Efficiencies 1960–2010. *Environ. Sci. Technol.* **2014**, *48*, 9874–9881. [CrossRef] [PubMed]
32. Rant, Z. Exergie, ein neues Wort fur "Technische Arbeitsfaehigkeit"(Exergy, a new word for technical availability). *Forsch. Auf Dem Geb. Ingenieurwesens A* **1956**, *22*, 36–37.

33. Smil, V. Agricultural energy costs: National analyses. In *Energy Farm Production*; Fluck, R.C., Ed.; Elsevier: Amsterdam, The Netherlands, 1992; pp. 85–100.

34. Smil, V. *Energy Transitions: History, Requirements, Prospects*; ABC-CLIO: Santa Barbara, CA, USA, 2010.

35. Hammond, G.; Stapleton, A. Exergy analysis of the United Kingdom energy system. *Proc. Inst. Mech. Eng. Part J. Power Energy* **2001**, *215*, 141–162. [CrossRef]

36. Geller, H.S. Cooking in the ungra area: Fuel efficiency, energy losses, and opportunities for reducing firewood consumption. *Biomass* **1982**, *2*, 83–101. [CrossRef]

37. $CO_2$-Calculator. Available online: http://www.energyglobe.com/de_at/haus-wohnung/energiesparen-im-haushalt/online-checks-zum-thema/co2-rechner/ (accessed on 1 May 2017).

38. FAOSTAT Macro Indicators. Available online: http://www.fao.org/faostat/en/#data/MK (accessed on 21 April 2017).

39. Asrat, F. *Ethiopia Country Report for the 2014 Ministerial Conference on Youth Employment*; Federal TVET Agency, Ministry of Education: Addis Ababa, Ethiopia, 2014.

40. Mayer, A.; Schaffartzik, A.; Haas, W.; Rojas-Sepulveda, A.; World Rainforest Movement; GRAIN; Brad, A.; Pichler, M.; Plank, C.; Andarge, H.B. *Patterns of Global Biomass Trade—Implications for Food Sovereignty and Socio-Environmental Conflicts*; EJOLT Report No. 20; ICTA, Universitat Autonoma de Barcelona: Barcelona, Spain, 2014.

41. Salomon, J.A.; Wang, H.; Freeman, M.K.; Vos, T.; Flaxman, A.D.; Lopez, A.D.; Murray, C.J. Healthy life expectancy for 187 countries, 1990–2010: A systematic analysis for the Global Burden Disease Study 2010. *Lancet* **2013**, *380*, 2144–2162. [CrossRef]

42. Wiedenhofer, D.; Steinberger, J.K.; Eisenmenger, N.; Haas, W. Maintenance and Expansion: Modeling Material Stocks and Flows for Residential Buildings and Transportation Networks in the EU25: Stocks and Flows in the EU25. *J. Ind. Ecol.* **2015**, *19*, 538–551. [CrossRef] [PubMed]

43. FAO/WHO/UNU. *Expert Consultation Human Energy Requirements 2001*; Food and Agriculture Organization of the United Nations: Rome, Italy, 2011.

44. Ringhofer, L.; Fischer-Kowalski, M. Method Box: Functional Time Use Analysis. In *Social Ecology: Society-Nature Relations across Time and Space*; Haberl, H., Fischer-Kowalski, M., Krausmann, F., Winiwarter, V., Eds.; Springer: Cham, Switzerland, 2016; pp. 519–522.

45. Fischer-Kowalski, M.; Singh, S.J.; Lauk, C.; Remesch, A.; Ringhofer, L.; Grünbühel, C.M. Sociometabolic transitions in subsistence communities: Boserup revisited in four comparative case studies. *Hum. Ecol. Rev.* **2011**, *18*, 147–158.

46. Utlu, Z.; Hepbasli, A. Estimating the energy and exergy utilization efficiencies for the residential–commercial sector: An application. *Energy Policy* **2006**, *34*, 1097–1105. [CrossRef]

47. Boserup, E. *The Conditions of Agricultural Growth. The Economics of Agrarian Change under Population Pressure*; Aldine/Earthscan: Chicago, IL, USA, 1965.

48. Statistik Austria Erwerbsstatistik Nach Wirtschaftssektoren. Available online: http://www.statistik.at/web_de/services/statcube/index.html (accessed on 21 April 2017).

49. Statistik Austria Bevölkerung nach Lebensunterhalt und Geschlecht seit 1994. Available online: http://www.statistik.at/web_de/statistiken/menschen_und_gesellschaft/arbeitsmarkt/erwerbsstatus/index.html (accessed on 9 January 2017).

50. Statistik Austria Bevölkerung am 1.1.2016 nach Ortschaften. Available online: http://www.statistik.at/wcm/idc/idcplg?IdcService=GET_PDF_FILE&RevisionSelectionMethod=LatestReleased&dDocName=103419 (accessed on 13 January 2017).

51. Statistik Austria Ortsverzeichnis Niederösterreich. 2005. Available online: https://www.statistik.at/wcm/idc/idcplg?IdcService=GET_PDF_FILE&RevisionSelectionMethod=LatestReleased&dDocName=007110 (accessed on 13 January 2017).

52. Eisenmenger, N.; Fridolin, K.; Milota, E.; Schaffartzik, A. *Ressourcennutzung in Österreich–Bericht 2011*; Bundesminist. Für Land- Forstwirtsch. Umw. Wasserwirtsch. Bundesminist. Für Wirtsch. Fam. Jugend Wien: Wien, Austria, 2011.

*sustainability*

MDPI

*Article*

# More Than a Potential Hazard—Approaching Risks from a Social-Ecological Perspective

**Carolin Völker** [1,2,*], **Johanna Kramm** [1,*], **Heide Kerber** [1], **Engelbert Schramm** [1,2], **Martina Winker** [1] and **Martin Zimmermann** [1]

[1]   ISOE—Institute for Social-Ecological Research, 60486 Frankfurt/Main, Germany; kerber@isoe.de (H.K.); schramm@isoe.de (E.S.); winker@isoe.de (M.W.); zimmermann@isoe.de (M.Z.)

[2]   Senckenberg Biodiversity and Climate Research Centre (BiK-F), 60325 Frankfurt/Main, Germany

[*]   Correspondence: voelker@isoe.de (C.V.); kramm@isoe.de (J.K.)

Received: 30 January 2017; Accepted: 27 May 2017; Published: 26 June 2017

**Abstract:** Risks have been classically understood as a probability of damage or a potential hazard resulting in appropriate management strategies. However, research on environmental issues such as pollutants in the aquatic environment or the impacts of climate change have shown that classical management approaches do not sufficiently cover these interactions between society and nature. There have been several attempts to develop interdisciplinary approaches to risk that include natural as well as social science contributions. In this paper, the authors aim at developing a social-ecological perspective on risk by drawing on the concept of societal relations to nature and the model of provisioning systems. This perspective is used to analyze four cases, pharmaceuticals, microplastics, semicentralized water infrastructures and forest management, with regard to risk identification, assessment and management. Finally, the paper aims at developing a perspective on risks which takes into account non-intended side-effects, system interdependencies and uncertainty.

**Keywords:** pharmaceuticals; microplastics; semicentralized water infrastructures; forest management; provisioning system; normal operation

---

## 1. Introduction

The detection of anthropogenic micropollutants in the water cycle or the environmental changes caused by climate change are examples of current risks, which point to the relation between society and nature, as well as to the relational character of risk; put in a simple way, this is a relation of cause and (often undesired or adverse) effect. The term "risk" enjoys great popularity, though at the same time it comes with a variety of definitions [1]. A definition often referred to says: the "term 'risk' denotes the likelihood that an undesirable state of reality (adverse effects) may occur as a result of natural events or human activities" [2] (p. 50). A distinction is made between risk and hazard. While hazards are defined as a potential source of harm (e.g., toxicity of a chemical substance), a risk emerges when there is a likelihood that the hazard will produce harm [2]. The technocratic understanding is an equation involving the possibility of an adverse effect and the potential damage, while emphasis in the social sciences, for instance is laid on perception and decision-making [3]. Different concepts and theories of risk have been developed. They vary from those found in cultural studies and the social sciences, which embrace a constructivist approach to nature, and the technical and natural sciences which adopt an objectivist approach (for an overview see [4,5]). Social science theories comprise system theory following Luhmann, which stresses ways of risk communication [6], cultural theories working on cultural assumptions on risk and risk perception [7], and psychological theory centering around subjective judgment of the extent and character of risks [8]. In the natural sciences, for instance ecotoxicology, environmental risk assessment considers a cause deriving from a toxic substance (e.g., chemical, which has effects on an organism). An "environmental risk" is then a product of exposition, toxicity and sensitivity. Classical risk analysis calculates the possibility of an adverse event and the potential damage, for instance an assessment of the ecotoxicity of hazardous substances based on dose-response relationships. Similar accounts for engineering sciences. Here, risks are calculated for socio-technical systems such as energy or water infrastructures which are susceptible to hazards.

There has been a debate in risk research on how to consider society-nature interactions in risk analysis [1,9]. This paper takes up this debate and suggests a social-ecological perspective on risks which focuses on the interactions between society and nature. The concept of societal relations to nature [10] and the model of social-ecological systems (SES) as provisioning systems [11,12] (Hummel et al. 2017 in this special issue) are used to characterize society-nature interactions. The aim of the paper is to elaborate a social-ecological perspective on risks by drawing on four case studies analyzed with the concepts of societal relations to nature and provisioning systems.

The next section gives an overview of the current risk research. The subsequent section introduces the four case studies and in a second step discusses them, deploying the analytical framework of provisioning systems. In a third section, aspects of the studies are discussed to develop conceptual notes on a social-ecological risk perspective. It is further discussed how this perspective helps to identify risk management strategies and how to govern risks.

## 2. State of the Art: Risk and the Social-Ecological Perspective

As outlined in the introduction, the conceptualization and the understanding of "risk" vary highly. A common and quite popular way of classifying different risk concepts has been a disciplinary ordering. Several reviews [2,13,14], notably work by Ortwin Renn, have presented the concept as used in natural and technical sciences, economics, psychology and social and cultural studies.

Examples for technical and natural science concepts of risks comprise actuarial analysis of the insurance business, in which the expected value—an equation of the probability of an event multiplied by the insurance sum of the event—is the reference point; toxicological and epistemological risk assessments using causal models of risks to identify substances, like carbon monoxide or benzene, that may cause harm to humans or the environment [15] and probabilistic risk assessments which attempt to predict the probability of undesirable events within complex technical systems [16]. In sum, they "anticipate potential physical harm to human beings, cultural artefacts, or ecosystems, average these events over time and space, and use relative frequencies or estimated probabilities (observed or modelled) as a mean to specify likelihood" [2] (p. 52). These kinds of risk analyses are useful for identifying causes and predicting events with undesirable effects and therefore help decision-makers in accident management and emergency planning. They are important to improve safety of technical systems [2]. The technical and quantitative risk analyses have been subject to criticism from social sciences [14,17]. Among other critique, the objectivity of the identified risks is questioned and it is emphasized that risk analysis cannot be considered as "value-free scientific activity" [2] (p. 52). Furthermore, critics point to the subjective perceptions of people with under- or overestimating the probability of certain adverse events and the role of the media [18,19].

Renn clustered risk approaches in social sciences along two axes from individualistic to structuralist approaches and from constructivist to realistic ones [2]. In this taxonomy, he classifies postmodern theory [20], cultural theory [21] and Luhmann´s system theory [22] (constructivist and structuralist), critical theory [23] (realistic and structuralist), rational choice theories [24] (realistic and individualistic) and theories of reflexive modernization [25] (individualistic and constructivist). Overall, the approaches have in common that risks are related to decision-making under uncertainty [26].

Aven [27] (p. 34) suggests to overcome a classification of risk which is based on "objective reality" vs. "societal phenomenon". He argues for a new way of classifying risks and develops six risk perspectives based on six development paths. By this, he develops perspectives which are crossing scientific disciplines, but he remarks that the "risk perspective chosen strongly influences the way risk is analyzed and hence it may have serious implications for the risk management and decision making" [27] (p. 42). There have been further attempts to develop an integrative approach to risk [28]. The discussed approaches comprise the social amplification of risk concept [18,29], the risk types according to WBGU [30] and finally the IRGC´s risk governance framework [31]. The model of the International Risk Governance Council (IRGC), which in contrast to the Red Book model to the co-evolutionary approach [32] integrates the assessment sphere, the management sphere and the communication sphere and by this accommodates the technical/natural science and the social dimensions of risk.

This paper elaborates a further perspective, which builds on the integrative aspects of risk research and at the same time takes into account the society-nature interactions. For developing a social-ecological risk perspective, we draw therefore on the concept of societal relations to nature [10] and the model of social-ecological systems (SES) as provisioning systems [11,12].

The concept of "societal relations to nature" frames society-nature interactions in a non-dualistic way as patterns of regulation aiming at the reproduction of societies. Particular sets of characteristic societal relations to nature can be represented in a provisioning system [12] (Hummel et al. 2017 in this issue). Provisioning systems are "developed [and regulated] by societies [to] provide goods and services such as food, water, or energy [and] based on ecosystems and their geophysical environments" [11] (p. 11). Resource usage in a provisioning system is regulated by practices, knowledge, technologies and institutions. This regulation of provisioning systems can cause intended and non-intended effects on ecological and biophysical processes and structures (ecosystems or human health), as well as on social processes and structures (economies, etc.) within a system and on other provisioning systems [11]. These effects can cause feedback processes that might call for new decisions, or have effects on cultural and social processes. Such feedback processes can be immediate or buffered; they can be anticipated with a proactive reaction, or non-anticipated with a reactive response, and they can take place within a system as well as affecting other systems. Thus, these systems are linked. Taking feedback processes and unintended effects into account, it is important to be aware that every action associated with risk management and problem solving can cause non-intended effects and risks; this is called self-referentiality [33].

Who or what is "at risk"? From a social-ecological perspective, societies as well as ecosystems can be at risk, ranging from "material things", such as infrastructure, buildings, etc., to humans and non-humans, like organisms and whole ecosystems, to social-ecological systems, like the operation of a provisioning system. The causes of risks lie mostly in human activities, since most bio-physical processes and natural resources are regulated by societies [11].

## 3. Social-Ecological Perspective in Risk Research

The four case studies were taken from projects conducted at the ISOE—Institute for Social-Ecological Research and were selected because they all address interactions between society and nature and represent a broad spectrum of risk themes. The case studies are on (a) pharmaceutical residues in the aquatic environment, (b) microplastics in the aquatic environment, (c) semicentralized water infrastructures, and (d) forest management. The case of pharmaceuticals was derived from studies conducted within several projects, including start (2005–2008) and SAUBER+ (2011–2015), the microplastics example is covered in the project PlastX (2016–2021), semicentralized water infrastructures are dealt with in the project Semizentral (2013–2017), and forest management within the BiK-F project (2008–2021).

The cases were analyzed by identifying (1) possible negative effects, (2) the affected (provisioning) systems, and (3) the causes of the potential effects. The following questions were used as guide for discussing the results:

What is the scope of the negative effects and how might they affect other linked systems?

How do these systems interplay?

Which risk management strategies exist and do they produce competing interests?

*3.1. Pharmaceutical Residues in the Aquatic Environment*

3.1.1. Case Outline

First detections of pharmaceutical residues in the aquatic environment date back to the 1970s [34–37]. Nowadays, pharmaceutical residues can be detected in nearly all compartments of the water cycle. Their existence has been demonstrated in surface waters, seawaters, groundwater and even at the nano-level in drinking water in a few cases [38,39].

The original cause of pharmaceutical residues in the water cycle lies in human medical care: Some pharmaceutical substances used for human therapy are only slightly transformed by the human body, excreted, and afterwards not completely eliminated in wastewater treatment plants (WWTPs). Finally they are discharged into the aquatic environment. Besides urban WWTPs, discharges from manufacturing, animal husbandry, and aquacultures are further emission pathways [40].

The introduction of some of these pharmaceuticals into the environment poses a risk to aquatic wildlife, since negative effects have been demonstrated at environmentally relevant concentrations, for instance in the case of endocrine disruptors such as steroid hormones [39]. Uncertainty exists regarding the risks for whole ecosystems due to knowledge gaps in respect of the mixture toxicity of pharmaceuticals, as well as chronic and subtle effects that might have consequences at the population level [39]. Acute effects for human beings due to the occurrence of pharmaceutical residues in drinking water do not exist; the daily intake via drinking water is far from reaching therapeutic doses [41]. However, since pharmaceuticals are continuously released into the environment ("pseudopersistent"), serving to sustain long-term exposure, precautionary measures are discussed to reduce risks in advance. In most cases, these measures comprise end-of-pipe solutions, e.g., modifications of WWTPs in order to improve removal efficiencies.

3.1.2. Risks as Side-Effects of Normal Operation

The social-ecological perspective on the issue of pharmaceutical residues in the aquatic environment clearly points to a risk involving the "water services" provisioning system as pharmaceutical residues are detectable in aquatic ecosystems as well as in drinking water. The "water services" provisioning system comprises the disposal of wastewater, the connected aquatic ecosystems, and the water supply, including drinking water. As wastewater constitutes the risk-producing part of the provisioning system, measures to prevent the emission of pharmaceuticals predominantly focus on WWTPs and advanced treatment technologies.

However, by including interconnected provisioning sytems as well, the mode of risk production comes into focus: in the case of the "water services" provisioning system, the risk for the aquatic ecosystem and consequently for human health is produced by wastewater emission but it originates from the use of pharmaceuticals to treat and prevent diseases, and hence from the provisioning system "human health care" (Figure 1). What is striking is the fact that the use, excretion and disposal of pharmaceuticals is organized in a way that these agents are intentionally released into the water cycle. Consequently, the health care system in its "normal mode of operation" produces a risk for the "water services" provisioning system [42].

It becomes clear that measures within this provisioning system, such as an additional treatment step in wastewater treatment plants, are not enough. Such measure must be complemented by innovations in the health care system to reduce the load of pharmaceuticals in water bodies. This means that as a part of precautionary measures, the "human health care" provisioning system needs to be readjusted. If possible, the consumption of pharmaceuticals should be reduced, and supported by measures such as different prescribing practices and disease prevention, proper handling of pharmaceuticals, and the development of environmentally friendly pharmaceuticals [43].

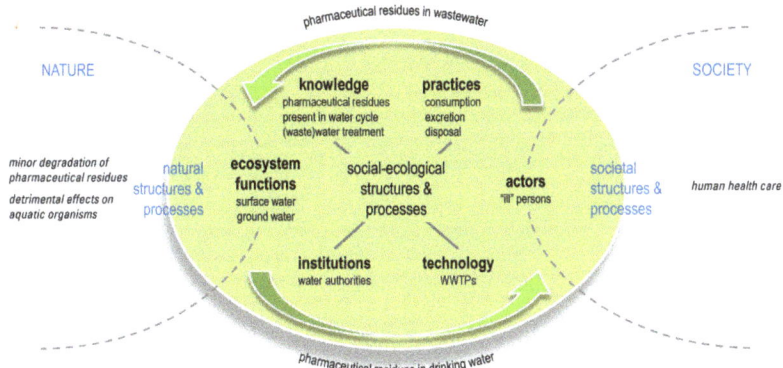

**Figure 1.** Impact of human health care on the "water services" provisioning system (own draft based on [11]).

In conclusion, the social-ecological perspective does not consider extreme events or disasters in order to develop risk management strategies. By drawing on the concept of provisioning systems, risks are discovered which arise through "processes of self-endangerment brought about by modern, highly interconnected societies" [42] (p. 357). Therefore, this analytical perspective leads to risk management strategies which target the reorganization of the involved provisioning systems [42].

### 3.2. Microplastics in the Aquatic Environment

#### 3.2.1. Case Outline

The term "microplastics" emerged in the 2000s to describe small plastic fragments detected in seawaters since the 1970s [44]. Meanwhile, a rising number of studies has discovered the vast extent of microplastic pollution reaching from deep-sea sediments to freshwater environments [45,46].

The causes of microplastic pollution are as variable as the material types introduced into the environment. So-called "primary" microplastics are manufactured for their application in specific products, such as cosmetics or air-blasting technology, but they also involve virgin plastic production pellets [47]. These primary microplastics mainly enter the aquatic environment via domestic and industrial WWTPs [48]. "Secondary" microplastics derive from the breakdown of larger plastic items, such as plastic waste introduced into the environment, but also chemical fibers washed from fabrics, or tire dust from road traffic discharged by WWTPs [47]. Secondary microplastics are considered to be the main source of microplastic pollution [49].

Uncertainties exist regarding the risks these microplastic particles pose to the ecosystems where they persist [50]. Studies have shown that microplastics are ingested by various aquatic organisms including organisms that play a role in human consumption (such as shellfish and fish) [51,52]. So far, biological effects have only been detected in the laboratory at particle concentrations that have limited environmental relevance [53]. Furthermore, risk assessment is complicated, as microplastics might also transfer plastic-associated chemicals (such as additives or adsorbed pollutants) to organisms, with negative impacts on their health. Currently, researchers argue that the existing knowledge gaps are too large to properly assess the risks of microplastics for ecosystems [54]. The same holds true in respect of the risks for human health of microplastics-contaminated food [55]. Nevertheless, precautionary measures are discussed which mostly focus on primary microplastics. These measures comprise advanced wastewater treatment and regulatory approaches, such as a ban (in the United States) on microplastics in specific cosmetic products [56]. Furthermore, voluntary commitments have been made by plastic processing industries addressing zero pellet loss or phase-out of microplastics in cosmetics [56].

3.2.2. Societal Significance of Risks

The microplastics example shows some parallels to pharmaceutical residues in the water cycle. A chemical agent is introduced into aquatic ecosystems via WWTPs, involving the "water services" provisioning system. Microplastics from cosmetic products or fibers washed from fabrics also enter the environment as side effects of the normal wastewater processing strategy.

From a social-ecological perspective, the case of microplastics cannot be considered in isolation from plastic waste in a larger context, as the pathway via WWTPs accounts for only a relatively small amount of the pollution. As mentioned above, most plastic particles in the environment stem from the degradation of larger plastic waste. The origin of this plastic waste cannot be attributed to one specific provisioning system that produces the risk and can be readjusted. Since plastic products, for instance plastic packaging, are an integral part of modern societies, a variety of provisioning systems come into focus, ranging from food supply to human health care. These systems meet when it comes to the disposal of plastics and proper waste management. In this case, it can be argued that if the systems set up for correct disposal of plastics operated properly, plastics would not reach the environment in relevant amounts. However, the huge amount of plastics leaking into the environment worldwide throws doubts on the existence of "normal" operating waste management systems. It seems that the current mode, with its leakages, is the "normal" mode. Nevertheless, risk management strategies should include proper waste management, but also—and especially in countries which lack waste disposal systems—target the reorganization of the respective systems regarding sustainable production and consumption patterns to reduce the production of plastic waste.

However, the microplastics example shows further aspects which come into focus by adopting a social-ecological perspective. In this case, societal perception and the mobilization potential of the society are important drivers for risk management. So far, ecotoxicological studies have not been able to conclusively characterize the risk posed by microplastics to aquatic ecosystems due to the above-mentioned complex material characteristics. Despite the absence of clear scientific evidence, it is common sense in most societies that plastics should not end up in the environment. In many cases, there are aesthetic reasons for this, which have political, social, and economic consequences, for example when there is a risk of losing tourism revenues due to beaches littered with plastic debris [57–59]. Furthermore, the public debate, triggered by the media [60,61], centers around the potential abundance of microplastics in human food suggesting human health risks. These reports are only based on single observations, and as yet there is no evidence of negative effects on human health. Nevertheless, microplastics are perceived as a threat to the "food" provisioning system and thus to human health [62]. The social-ecological approach to the issue of microplastics in the aquatic environment considers both technical risk calculations, here the ecotoxicological approach, and the societal significance of the risk. This makes it possible to address the ecological and the social consequences. Moreover, communication processes involving public interest groups help to identify resilient strategies for coping with uncertainties and thus managing the risk [63].

*3.3. Semicentralized Water Infrastructures*

3.3.1. Case Outline

Technical innovations such as treatment and reuse of partial wastewater streams at their point of origin in combination with new management approaches result in radical innovations of the water infrastructure [64–67]. These concepts lead to a different design: decentralized or semicentralized components cause a higher flexibility due to a modular architecture and enable more options for water and energy reuse. Wastewater streams of different quality (e.g., blackwater from toilets and greywater from showers, sinks, kitchens, washing machines) can be treated separately by using dual sewers. In addition, greywater can be recycled to produce service water for toilet flushing or the irrigation of green space. Through this modular architecture, urban water services can be adjusted flexibly to rapidly growing (or shrinking) population numbers. Furthermore, the reuse of water (and heat or other resources) contributes to an efficient management of natural resources.

Water supply and disposal systems are categorized as critical infrastructures, i.e., they are essential for the functioning of a society and its economy. The vulnerability of novel infrastructures must be analyzed from this perspective, where vulnerability is understood as the susceptibility and resilience of the system's components to certain hazards [68]. The potential causes of a failure of the water infrastructure, or of parts of it, can be internal or external. They include natural hazards (heavy rainfall, floods, droughts, earthquakes) and human or technical failure (pipe bursts, sabotage). The failure of the water infrastructure, in turn, might endanger the users' health or surrounding ecosystems, for instance if the user has contact with contaminated service water, or if untreated wastewater is discharged into the environment.

Risk analyses for wastewater utilities are usually based on technical safety management. The probability of occurrence of potential hazards (threats) to the infrastructure is quantified based on statistical data or estimations. In addition, the damage to subjects of protection (e.g., users, technical infrastructure, ecosystems) is quantified in monetary terms. High probabilities of occurrence and high possible damage indicate a high level of risk and necessitate precautionary measures to prevent the predicted damage.

### 3.3.2. Integrative Risk Assessment

The case of semicentralized water infrastructures differs from the previous examples of pharmaceuticals and microplastics as it has a strong spatial dimension by focusing on a specific technological system. Furthermore, this system is already able to combat existing problems within the "water services" provisioning system, coping for instance with water scarcity by implementing improved water efficiency. The focus in this case is on the hypothetical risks posed by hazards, or a failure of interconnected provisioning systems, such as (drinking) water supply, electricity, or the disposal of food waste. The aim is to minimize risks which do not occur in the context of normal operation, but which are due to negative events which may have a natural origin (e.g., extreme weather, landslides) or a socio-technical one (e.g., vandalism, faulty connections). The hazards can be multiple and lie within the system itself (human or technical failure) or be exogenous (weather). At risk is the safe operation of the water infrastructure guaranteeing an adequate service water supply, as well as wastewater management, including safe discharge of the treated wastewater into the environment. In a worst case scenario, the effects may lead to a breakdown of the infrastructure, resulting in health hazards for the users due to an inadequate service water supply, and/or environmental pollution due to the disposal of wastewater which is untreated or inadequately treated.

To avoid these consequences, risk analyses serve to assess the vulnerability of the system to these hazards, and to take adequate measures for risk management. In addition to the conventionally applied risk assessment approach, the social-ecological approach integrates the interdependencies between the infrastructure and connected provisioning systems and ecosystems. By adopting this systemic perspective, differences in the appropriation of natural resources become apparent, since the semicentralized system is more resource efficient than conventional infrastructures, for example in terms of groundwater resources for (drinking) water supply purposes. Furthermore, the users become a significant part of the analysis. On the one hand, users might be influenced by malfunctioning of provisioning systems that are dependent on the functioning of the water infrastructure. On the other hand, the kind of usage (routines, usage patterns, misuse) influences the functioning of the infrastructure, which is why the users' practices and their handling of the water infrastructure are integrated into the analysis. The described integrative take on technological risk or vulnerability analysis hence contributes to a broader understanding of social-ecological interdependencies, and offers more promising options for action in respect of risk management. This approach also considers the diversity and complexity of the actors (e.g., users, operators, craftsmen), as well as their practices.

Consequently, the actors do not have to adapt their actions to the infrastructure's requirements and its assigned management regime, but rather the water system is adjusted to the actors' needs and behavior.

*3.4. Sustainable Forest Management*

3.4.1. Case Outline

Central European forests play an important role in timber production, but also serve as a place for outdoor recreation, for regeneration of water, or as a protection against mudflows and rockfalls (in Alpine regions). Global climate change directly and indirectly affects the growth and productivity of Central European forests. Besides forest biome shifts, an immediate risk is the loss of forests at specific locations. If forest ecosystems are degraded, or even destroyed, their recultivation is challenging. Effects that point to this risk can already be observed in form of the destabilization of existing forests, for example in specific regions in Germany (Hessian Ried): oak forests have been severely depleted due to climate change, a lowering of the groundwater level, and cockchafer epidemics [69]. However, no reliable statements can be made on which of the observed effects were caused by climate change and which occurred due to silvicultural mistakes, excessive numbers of game animals, or other causes [70].

To date, Central European forests have mostly been cultivated at a profit. But in future, substantial declines in earnings caused by climate change are expected by forest economists [71]. Experts predict a decline in timber yields [71], as well as a decline in forest functions [72]. Because of the loss of retention functions offered by alluvial forests, residential areas near rivers are increasingly threatened by floods. And the regeneration of ground water is endangered when the soil loses its ability to retain water or to decompose contaminants, due to a loss of forest vegetation [73].

So far, risk management has focused on the reduction of operational risks of the forest enterprises [74]. Most foresters have decided on a "no-change" strategy because of their doubts whether active management is possible. Smaller numbers tend to follow a "trial and error" strategy, or to plant more climate resistant tree species [75]. Therefore, the challenge for forest policy and science is to identify the tree species with a predicted optimum yield in future [76].

3.4.2. Integrative Management Approach

The example of Central European forests also has a strong spatial dimension. In this case, forestry systems are at risk due to climate change, and risk management concentrates on their future cultivation. Drawing on the concept of provisioning systems, the described forestry systems represent a characteristic case of societal relations to nature. These forests have been cultivated for specific purposes for a long time, and provide specific ecosystem services, such as timber, provision of clean water and air, water retention and flood protection, but also cultural services, like space for recreation. If the functioning of these forestry systems fails, the connected provisioning systems which benefit from the ecosystem services provided will bear the consequences.

At the moment, the regional characteristics of climate change and subsequent effects on forests are difficult to predict. Nevertheless, climatic extremes (storms, droughts, torrential rains, frost events) are likely to occur more often in the future, and have been addressed in forestry practice, at least in Germany. In this regard, an adaptive management approach is pursued: forests are restructured by planting tree species which are less susceptible to environmental changes, including climate resistant species from North America such as the Douglas fir. However, this approach solely addresses economic losses for forest enterprises, as forest management concentrates on profit-yielding species and the ecosystem service "timber production" [74]. In contrast to forest practitioners, conservationists and recreational visitors demand sites with a portfolio of different trees. In this regard, an adaptive management strategy should consider how to maintain the diverse ecosystem services of forestry systems. The social-ecological approach embraces interconnected provisioning systems and, therefore, the "multifunctionality" of forests. This approach thus addresses the needs and demands of linked provisioning systems and seeks to negotiate between the different stakeholders involved. Moreover, a tree portfolio maintaining the multifunctionality of forests for provisioning systems simultaneously addresses the functioning of the ecosystem. The social-ecological perspective, therefore, helps to develop management practices that are sustainable and maintain the characteristic societal relations to these ecosystems.

## 4. Discussion

In accordance with newer approaches to risk governance [18,29–31], the social-ecological perspective aims at integrating technical and social contributions to risk assessment, evaluation, and management in order to comprehensively address complex environmental issues. However, by drawing on the concepts of societal relations to nature and provisioning systems, the social-ecological perspective focuses on specific aspects which are outlined in the following:

- The social-ecological perspective does not consider natural hazards such as earthquakes or volcanic events that might have catastrophic effects and require crisis management. The same holds true for risks that can be understood as "normal accidents" (cf. [77]) but have solely social implications, for instance working conditions or health risks such as smoking. In social-ecological research, the focus is set on risks that arise through specific society-nature interactions which are regulated in a non-sustainable way and affect ecological as well as social processes and structures. By framing risks as a product of the "normal" operation of a provisioning system, the analytical perspective focuses on the mode of risk production and, therefore, leads to the conception that the involved provisioning systems need to be reorganized, as shown by the case of pharmaceutical residues in the water cycle. The social-ecological analysis includes the fact that risks can be transboundary, traveling across the border of risk producing systems into other linked (provisioning) systems [78]. Thus, these risks can be characterized by a specific vibrancy which affects other linked entities or systems, as demonstrated in the case of pharmaceuticals and microplastics. On the one hand, these linkages result from a physical connection of the involved provisioning systems, like the connection of the health care system and the "water services" system through wastewater flows. On the other hand, these linkages emerge through social processes (like communication, practices) affecting different social, political and economic spheres, for instance societal aversion to beached plastic debris inducing income losses in the tourism sector. Thus, the social-ecological approach includes both so-called objective approaches, trying to assess risks with probability assessments, and constructivist approaches that consider how the risk is socially constructed and perceived.
- The social-ecological perspective takes into account that risk management approaches can always have (non-intended) side effects and affect other linked entities or systems, as shown in the cases of novel water infrastructures and forest management. In this regard, a social-ecological approach integrates interdependencies of involved systems with the respective stakeholders from the beginning. Thus, knowledge from the technical and natural sciences is combined with knowledge from the social sciences on everyday practices, for instance the behavior of users in the case of novel water infrastructures.

From a social-ecological perspective, the starting point of risk analysis depends on the guiding question (this question might come up within a scientific community or from discussions in society); the question can focus on a single agent and its potential impacts on a system (Figure 2a) or on a system (with its subsystems) that is at risk due to several stressors (Figure 2b). In the case of pharmaceuticals and microplastic particles, the focus is on specific chemical substances introduced into the aquatic environment resulting in certain risks for the ecosystems involved. By contrast, the cases of water infrastructures and forests focus on specific systems that are at risk due to natural and anthropogenic causes. In the first case (Figure 2a), the analysis aims at making a hazard assessment of the substance and its possible reduction and removal, while in the second case (Figure 2b) it is centered on adaptation of the system's functions to cope in an enhanced way with its stressors. Further perspectives may evolve by considering other cases.

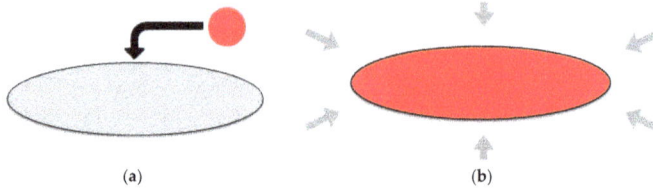

(a)                                                    (b)

**Figure 2.** Different foci for risk analysis. (**a**) Focus on a single agent and its potential impact on a system; (**b**) focus on a system that is at risk due to several stressors.

As argued above, it is important to consider linkages and interdependencies in the social-ecological system. In the first case, it is important to analyze the system from which the agent derives, i.e., the causing or risk producing system. Furthermore, effects of the agent on other systems are crucial for understanding the cause-effect relationships in depth. In the second case, understanding the interdependencies of related systems is imperative for integrated management, as shown by the interplay of groundwater resources, timber production, space for human recreation, etc., in the case of forest management.

In accordance with other integrative risk governance approaches [28,79], the widening of the analytical focus subsequently has implications for management of the identified risks. These integrative management approaches include the technical as well as the social dimension of risks [79]. Table 1 gives an overview of risk management strategies frequently applied to the cases outlined in this article, and their extension through an integrated, social-ecological perspective.

**Table 1.** Approaches to risk management.

|  | Frequently Applied Management Approaches | Management Approaches Derived from a Social-Ecological Analytical Framework |
| --- | --- | --- |
| Pharmaceutical residues | Advanced wastewater treatment (end-of-pipe solution) | Integrative sustainability policies [43] including<br>• a transformation of the health care system (improved doctor–patient communication, preventive measures to reduce the prevalence of risk factors related to certain diseases, etc.)<br>• encouragement for the introduction of environmentally friendly pharmaceuticals<br>• measures in the area of environmental engineering |
| Microplastics | Regulations and measures focusing on primary microplastics | Integrative strategies for sustainable production, usage and disposal of plastics including<br>• transformations of practices involving short-lived plastic products in various provisioning systems<br>• the implementation of measures towards a circular economy<br>• encouragement for the introduction of environmentally friendly polymers |
| Semicentralized water infrastructures | Safe operation of the system by technical safety management | Widening the scope to the social and ecological environment of the technical system by taking into accountthe<br>• actors of the system and their behavior<br>• interdependencies with connected provisioning systems<br>• impacts on the appropriation of natural resources |
| Forest management | Adaptive management focusing on timber production | Adaptive management considering the "multifunctionality" of forests by<br>• identifying a diverse tree portfolio able to maintain multiple ecosystem functions<br>• integrating the needs and demands of different stakeholders and required ecosystem services |

As in the case of pharmaceuticals, but also in the other cases, the social-ecological approach to these issues aims at transforming the "normal" operation mode of provisioning systems. This is certainly a challenging endeavor, because it is not possible to apply this type of management within a single provisioning system. A comprehensive management approach is necessary which takes all linked systems into account, especially if the system´s transformation is dependent on simultaneous measures in other systems. As shown in Table 1 for the case of pharmaceuticals, a transformation of the health care system is desirable, allowing for synergies like a reduction of pharmaceuticals in water bodies and improvements in the field of disease prevention. If the dynamics in coupled systems are not sufficiently addressed, risk management approaches will remain one-dimensional and will only deal with the symptoms (cf. [43,73]). But a "multifunctional" perspective, as described in the case of forest management, can make it possible to identify risks for other, linked provisioning systems [80].

A key challenge of managing risks is dealing with uncertainty [81]. Uncertainty arises because, as shown in the case studies, the risks are often complex and characterized by a lack of knowledge on cause-effect linkages regarding time, scope and spatiality. A high degree of uncertainty, in turn, impedes the development of effective risk management strategies. Therefore, risk assessment aims to resolve these uncertainties by closing the existing knowledge gaps. In most cases, however, a knowledge base that eliminates any uncertainty can never be achieved. For instance, in the cases of pharmaceuticals and microplastics, it is practically impossible to gain scientific evidence showing the effects on different organisms of every possible combination of substances, dosages, exposure duration, etc. [50]. More scientific evidence broadens the knowledge base, but some uncertainty will always remain (see Table 2). Furthermore, more scientific research may reveal new questions and non-knowledge, leading to an increased complexity and even more uncertainty. Therefore, the management of risks is always accompanied by uncertainty, and it is not necessary to understand all possible cause-effect linkages in order to develop solution strategies, but rather the question is how much scientific knowledge is necessary for legitimizing informed action.

**Table 2.** Different types of uncertainty encountered within the case studies.

|  | Reason for Uncertainty | Degree of Uncertainty |
|---|---|---|
| Pharmaceutical residues | Lack of knowledge on subtle, chronic, and mixture effects | Knowledge base concerning specific adverse effects can be broadened, but never entirely be established. Some uncertainty will remain. |
| Microplastics | Lack of knowledge on biological effects | Knowledge base concerning specific adverse effects can be broadened, but never entirely be established. Some uncertainty will remain. |
| Semicentralized water Infrastructures | Lack of experience of the systems' processes and operations | Knowledge base and experience of the systems' processes and operations will develop. Uncertainty will be low within the system boundaries. |
| Forest management | Lack of predictability of climate change effects | Cause and effect are indeterminable and only visible in retrospect, uncertainty remains high. |

Furthermore, the lack of evidence on clear cause-effect linkages can cause different ways of seeing and perceiving the risks [82]. In the case of forest management, the lack of secured knowledge enhances competing views on management strategies and leads to conflicts of different interest groups. This ambiguity may arise in connection with risk management strategies, but also in connection with how scientific evidence is interpreted [50,82].

To properly address uncertainty and ambiguity in risk management, a debate on the tolerability of risks involving a wide range of experts is a prerequisite. For determining tolerability, not only scientific evidence is needed, but societal judgment and shared social values also play a crucial role [81]. Transdisciplinary social-ecological approaches are suitable for triggering debates on tolerability, as well as on the precautionary principle and other possible measures aiming at risk distribution. Finally, a critical assessment of the relevance of risks is required which centers not only on scientific evidence, but also on ethical considerations, addressing the normative dimensions of societal relations to nature and their sustainable transformation.

## 5. Conclusions

In this paper, a social-ecological perspective on risks has been elaborated by drawing on the concepts of societal relations to nature and provisioning systems. Based on four case studies it has been shown that the social-ecological approach enables a holistic view on risks by (a) considering linkages and interdependencies between the involved provisioning systems, (b) unveiling the mode of risk production and (c) combining knowledge from natural and social sciences. As a consequence, the social-ecological perspective broadens the scope for risk management strategies and suggests a reorganization of the involved provisioning systems. By pursuing a transdisciplinary approach involving a wide range of experts and the public, the social-ecological perspective allows for the development of informed risk management strategies.

**Acknowledgments:** The authors wish to thank Thomas Kluge for his helpful hints and comments. The case studies on pharmaceuticals, microplastics, and semicentralized water infrastructures are based on results accomplished within projects funded by the Federal Ministry of Education and Research (BMBF): "start: Management strategies for pharmaceutical residues in drinking water" (2005–2008; grant number: 07VPS16); "SAUBER+: Innovative concepts for wastewater from public health sector facilities" (2011–2015; grant number 02WRS1280B); "PlastX—Plastics as a systemic risk for social-ecological supply systems" (2015–2021; grant number: 01UU1603A); "Semizentral: Resource-efficient and flexible supply and waste disposal infrastructure systems for rapidly growing cities of the future—Phase 2" (2013–2017; grant number: 02WCL1266G). The case study on forest management was derived from the Senckenberg Biodiversity and Climate Research Centre (BiK-F) which was funded by the Hessian initiative for scientific and economic excellence (LOEWE) from 2008 to 2014.

**Author Contributions:** Carolin Völker and Johanna Kramm wrote the manuscript, contributed the microplastics example (Section 3.2), the literature review and the conceptual framing. Heide Kerber contributed to the pharmaceuticals example (Section 3.1). Engelbert Schramm contributed to the example of forest management (Section 3.4). Martina Winker contributed to the examples of pharmaceuticals (Section 3.1) and semicentralized water infrastructures (Section 3.3). Martin Zimmermann contributed to the example of semicentralized water infrastructures (Section 3.3). All authors compiled the figures and tables and contributed to the discussion.

**Conflicts of Interest:** The authors declare no conflict of interest.

## References

1.  Neisser, F. 'Riskscapes' and risk management: Review and synthesis of an actor-network theory approach. *Risk Manag.* **2014**, *16*, 88–120. [CrossRef]
2.  Renn, O. Concepts of risk. An interdisciplinary review. Part 1: Disciplinary risk concepts. *GAIA Ecol. Perspect. Sci. Soc.* **2008**, *17*, 50–66. [CrossRef]
3.  Renn, O.; Dreyer, M.; Klinke, A.; Schweizer, P.-J. Systemische Risiken: Charakterisierung, Management und Intergration in eine aktive Nachhaltigkeitspolitik. In *Jahrbuch Ökologische Ökonomik 5 (Soziale Nachhaltigkeit)*; Beckenbach, F., Hampicke, U., Leipert, C., Meran, G., Minsch, J., Nutzinger, H.G., Pfriem, R., Weimann, J., Wirl, F., Witt, U., Eds.; Metropolis: Marburg, Germany, 2007; pp. 157–158.
4.  Schramm, E.; Lux, A. *Klimabedingte Biodiversitätsrisiken. Ein neues Forschungsgebiet für BiK-F*; Knowledge Flow Paper, Biodiversität und Klima Forschungszentrum No. 16; LOEWE Biodiversität und Klima Forschungszentrum (BiK-F): Frankfurt am Main, Germany, 2014.
5.  Birkmann, J. Regulation and coupling of society and nature in the context of natural hazards. In *Coping with Global Environmental Change, Disasters and Security: Threats, Challenges, Vulnerabilities and Risks*; Brauch, H.G., Oswald Spring, Ú., Mesjasz, C., Grin, J., Kameri-Mbote, P., Chourou, B., Dunay, P., Birkmann, J., Eds.; Springer: Berlin/Heidelberg, Germany, 2011; pp. 1103–1127.
6.  Luhmann, N. *Risk. A Sociological Theory*; Aldine de Gruyter: New York, NY, USA, 1995.
7.  Douglas, M. *Risk Acceptability According to the Social Sciences*; Sage: New York, NY, USA, 1985.
8.  Eiser, R.; Miles, S.; Frewer, L. Trust, perceived risk and attitudes towards food technologies. *J. Appl. Soc. Psychol.* **2002**, *32*, 2423–2434. [CrossRef]
9.  Renn, O.; Schweizer, P.-J.; Dreyer, M.; Klinke, A. *Risiko. Über den Gesellschaftlichen Umgang mit Unsicherheit*; Oekom: München, Germany, 2007.

10. Becker, E.; Jahn, T. (Eds.) *Soziale Ökologie. Grundzüge einer Wissenschaft von den Gesellschaftlichen Naturverhältnissen*; Campus Verlag: Frankfurt am Main, Germany, 2006.

11. Hummel, D.; Jahn, T.; Schramm, E. *Social-Ecological Analysis of Climate Induced Changes in Biodiversity. Outline of a Research Concept*; Knowledge Flow Paper, Biodiversität und Klima Forschungszentrum No. 11; LOEWE Biodiversität und Klima Forschungszentrum (BiK-F): Frankfurt am Main, Germany, 2011.

12. Hummel, D.; Jahn, T.; Keil, F.; Liehr, S.; Stieß, I. Social Ecology als Critical, Transdisciplinary Science— Conceptualizing, Analyzing, and Shaping Societal Relations to Nature. *Sustainability* **2017**. submitted.

13. Hampel, J. Different concepts of risk: A challenge for risk communication. *Int. J. Med. Microbiol.* **2006**, *296*, 5–10. [CrossRef] [PubMed]

14. Taylor-Gooby, P.; Zinn, J.O. Current directions in risk research: New developments in Psychology and Sociology. *Risk Anal.* **2006**, *26*, 397–411. [CrossRef] [PubMed]

15. Haddad, S.; Béliveau, M.; Tardif, R.; Krishnan, K. A PBPK Modeling-Based Approach to Account for Interactions in the Health Risk Assessment of Chemical Mixtures. *Toxicol. Sci.* **2001**, *63*, 125–131. [CrossRef] [PubMed]

16. Bedford, T.; Cooke, R. *Probabilistic Risk Analysis. Foundations and Methods*; Cambridge University Press: Cambridge, UK, 2001.

17. Egner, H.; Pott, A. (Eds.) *Geographische Risikoforschung. Zur Konstruktion Verräumlichter Risiken und Sicherheiten*; Franz Steiner Verlag: Stuttgart, Germany, 2010.

18. Kasperson, R.E.; Renn, O.; Slovic, P.; Brown, H.S.; Emel, J.; Goble, R.; Kasperson, J.X.; Ratick, S. The Social Amplification of Risk A Conceptual Framework. *Risk Anal.* **1988**, *8*, 177–187. [CrossRef]

19. Rohrmann, B.; Renn, O. Risk Perception Research. In *Cross-Cultural Risk Perception: A Survey of Empirical Studies*; Renn, O., Rohrmann, B., Eds.; Springer: Boston, MA, USA, 2000; pp. 11–53.

20. Tagg, J. *Grounds of Dispute. Art History, Cultural Politics and the Discursive Field*; University of Minnesota Press: Minneapolis, MN, USA, 1992.

21. Douglas, M.; Wildavsky, A. *Risk and Culture. An Essay on the Selection of Technological and Environmental Dangers*; University of California Press: Berkeley, CA, USA, 1983.

22. Luhmann, N. *Communication and Social Order. Risk: A Sociological Theory*; Fourth Printing; Transaction Publishers: New Brunswick, NJ, USA, 2008.

23. Habermas, J. *Moral Consciousness and Communicative Action*; MIT Press: Cambridge, MA, USA, 1990.

24. Jaeger, C.C.; Webler, T.; Rosa, E.A.; Renn, O. *Risk, Uncertainty and Rational Action*; Routledge: Oxon, NY, USA, 2001.

25. Beck, U. *Risk Society. Towards a New Modernity*; Sage Publication: Los Angeles, CA, USA, 2010.

26. Egner, H.; Pott, A. Risiko und Raum: Das Angebot der Beobachtungstheorie. In *Geographische Risikoforschung: Zur Konstruktion Verräumlichter Risiken und Sicherheiten*; Egner, H., Pott, A., Eds.; Franz Steiner Verlag: Stuttgart, Germany, 2010; pp. 9–31.

27. Aven, T. The risk concept–historical and recent development trends. *Reliab. Eng. Syst. Saf.* **2012**, *99*, 33–44. [CrossRef]

28. Renn, O. Concepts of risk: An interdisciplinary review. Part 2: Integrative approaches. *GAIA Ecol. Perspect. Sci. Soc.* **2008**, *17*, 196–204. [CrossRef]

29. Kasperson, R.E.; Renn, O.; Slovic, P.; Kasperson, J.X.; Emani, S. Social amplification of risk: The media and public response. In *Waste Processing, Transportation, Storage and Disposal, Technical Programs and Public Education. Volume 1: High-Level Waste and General Interest*; Post, R.G., Ed.; Arizona Board of Regents: Tucson, AZ, USA, 1989; pp. 131–135.

30. WBGU—German Advisory Council on Global Change. *World in Transition. Strategies for Managing Global Environmental Risks: Annual Report 1998*; German Advisory Council on Global Change: Heidelberg, Germnay, 2000.

31. International Risk Governance Council. *An Introduction to the IRGC Risk Governance Framework*; IRGC: Geneva, Switzerland, 2007.

32. Millstone, E. Science, risk and governance: Radical rhetorics and the realities of reform in food safety governance. *Res. Policy* **2009**, *38*, 624–636. [CrossRef]

33. Hummel, D.; Kluge, T. Regulationen. In *Soziale Ökologie: Grundzüge einer Wissenschaft von den Gesellschaftlichen Naturverhältnissen*; Becker, E., Jahn, T., Eds.; Campus Verlag: Frankfurt am Main, Germany, 2006; pp. 248–258.

34. Tabak, H.H.; Bunch, R.L. Steroid hormones as water pollutants. I. Metabolism of natural and synthetic ovulation-inhibiting hormones by microorganisms of activated sludge and primary settled sewage. *Dev. Ind. Microbiol.* **1970**, *11*, 367–376.

35. Norpoth, K.; Nehrkorn, A.; Kirchner, M.; Holsen, H.; Teipel, H. Untersuchungen zur Frage der Löslichkeit und Stabilität ovulationshemmender Steroide in Wasser, Abwässern und belebtschlamm. *Z. Bakteriol. Mikrobiol. Hyg.* **1973**, *156*, 500–511.

36. Hignite, C.; Azarnoff, D.L. Drugs and drug metabolites as environmental contaminants: Chlorophenoxyisobutyrate and salicylic acid in sewage water effluent. *Life Sci.* **1977**, *20*, 337–341. [CrossRef]

37. Garrison, A.W.; Pope, J.D.; Allen, F.R. GC/MS analysis of organic compounds in domestic wastewaters. In *Identification and Analysis of Organic Pollutants in Water*; Ann Arbor Sicence Publisher Sicence Publisher, Inc.: Ann Arbor, MI, USA, 1976; pp. 517–556.

38. Heberer, T. Occurrence, fate, and removal of pharmaceutical residues in the aquatic environment: A review of recent research data. *Toxicol. Lett.* **2002**, *131*, 5–17. [CrossRef]

39. Fent, K.; Weston, A.A.; Caminada, D. Ecotoxicology of human pharmaceuticals. *Aquat. Toxicol.* **2006**, *76*, 122–159. [CrossRef] [PubMed]

40. Weber, F.; Bergmann, A.; Hickmann, S.; Ebert, I.; Hein, A.; Küster, A. Pharmaceuticals in the environment— Global occurrences and perspectives. *Environ. Toxicol. Chem.* **2015**, *34*, 823–835.

41. Webb, S.; Ternes, T.; Gibert, M.; Olejniczak, K. Indirect human exposure to pharmaceuticals via drinking water. *Toxicol. Lett.* **2003**, *142*, 157–167. [CrossRef]

42. Keil, F.; Bechmann, G.; Kümmerer, K.; Schramm, E. Systemic risk governance for pharmaceutical residues in drinking water. *GAIA Ecol. Perspect. Sci. Soc.* **2008**, *17*, 355–361. [CrossRef]

43. Brandmayr, C.; Kerber, H.; Winker, M.; Schramm, E. Impact assessment of emission management strategies of the pharmaceuticals Metformin and Metoprolol to the aquatic environment using Bayesian networks. *Sci. Total Environ.* **2015**, *532*, 605–616. [CrossRef] [PubMed]

44. Thompson, R.C.; Olsen, Y.; Mitchell, R.P.; Davis, A.; Rowland, S.J.; John, A.W.G.; McGonigle, D.; Russell, A.E. Lost at sea: Where is all the plastic? *Science* **2004**, *304*, 838. [CrossRef] [PubMed]

45. Van Cauwenberghe, L.; Vanreusel, A.; Mees, J.; Janssen, C.R. Microplastic pollution in deep-sea sediments. *Environ. Pollut.* **2013**, *182*, 495–499. [CrossRef] [PubMed]

46. Mani, T.; Hauk, A.; Walter, U.; Burkhardt-Holm, P. Microplastics profile along the Rhine river. *Sci. Rep.* **2015**, *5*, 17988. [CrossRef] [PubMed]

47. Cole, M.; Lindeque, P.; Halsband, C.; Galloway, T.S. Microplastics as contaminants in the marine environment: A review. *Mar. Pollut. Bull.* **2011**, *62*, 2588–2597. [CrossRef] [PubMed]

48. Derraik, J.G.B. The pollution of the marine environment by plastic debris: A review. *Mar. Pollut. Bull.* **2002**, *44*, 842–852. [CrossRef]

49. Jambeck, J.R.; Geyer, R.; Wilcox, C.; Siegler, T.R.; Perryman, M.; Andrady, A.; Narayan, R.; Law, K.L. Plastic waste inputs from land into the ocean. *Science* **2015**, *347*, 768–771. [CrossRef] [PubMed]

50. Kramm, J.; Völker, C. Understanding risks of microplastics: A social-ecological risk perspective. In *Freshwater Microplastics: Emerging Contaminants?* Wagner, M., Lambert, S., Eds.; Springer: New York, NY, USA, 2017; in press.

51. Von Moos, N.; Burkhardt-Holm, P.; Köhler, A. Uptake and effects of microplastics on cells and tissue of the blue mussel *Mytilus edulis* L. after an experimental exposure. *Environ. Sci. Technol.* **2012**, *46*, 11327–11335. [CrossRef] [PubMed]

52. Van Cauwenberghe, L.; Janssen, C.R. Microplastics in bivalves cultured for human consumption. *Environ. Pollut.* **2014**, *193*, 65–70. [CrossRef] [PubMed]

53. Duis, K.; Coors, A. Microplastics in the aquatic and terrestrial environment: Sources (with a specific focus on personal care products), fate and effects. *Environ. Sci. Eur.* **2016**, *28*, 1. [CrossRef] [PubMed]

54. Dris, R.; Imhof, H.; Sanchez, W.; Gasperi, J.; Galgani, F.; Tassin, B.; Laforsch, C. Beyond the ocean: Contamination of freshwater ecosystems with (micro-) plastic particles. *Environ. Chem.* **2015**, *12*, 539–550. [CrossRef]

55. Federal Institute for Risk Assessment. Mikroplastikpartikel in Lebensmitteln. Stellungnahme Nr. 013/2015. 2015. Available online: http://www.bfr.bund.de/cm/343/mikroplastikpartikel-in-lebensmitteln.pdf (accessed on 30 May 2017).

56. Girard, N.; Lester, S.; Paton-Young, A.; Saner, M. *Microbeads: "Tip of the Toxic Plastic-berg"? Regulation, Alternatives, and Future Implications*; Institute for Science, Society and Policy: Ottawa, ON, Canada, 2016.

57. Ballance, A.; Ryan, P.G.; Turpie, J.K. How much is a clean beach worth? The impact of litter on beach users in the Cape Peninsula, South Africa. *S. Afr. J. Sci.* **2000**, *96*, 210–230.

58. Free, C.M.; Jensen, O.P.; Mason, S.A.; Eriksen, M.; Williamson, N.J.; Boldgiv, B. High-levels of microplastic pollution in a large, remote, mountain lake. *Mar. Pollut. Bull.* **2014**, *85*, 156–163. [CrossRef] [PubMed]

59. Jang, Y.C.; Hong, S.; Lee, J.; Lee, M.J.; Shim, W.J. Estimation of lost tourism revenue in Geoje Island from the 2011 marine debris pollution event in South Korea. *Mar. Pollut. Bull.* **2014**, *81*, 49–54. [CrossRef] [PubMed]

60. Der Spiegel. Plastikteilchen Verunreinigen Lebensmittel. Unterschätzte Gefahr, 2013. 17 November 2013. Available online: http://www.spiegel.de/wissenschaft/technik/winzige-plastikteile-verunreinigen-lebensmittel-a-934057.html (accessed on 30 September 2016).

61. NDR. Mikroplastik in Mineralwasser und Bier. 2014. Available online: http://www.ndr.de/ratgeber/verbraucher/Mikroplastik-in-Mineralwasser-und-Bier,mikroplastik134.html (accessed on 30 September 2016).

62. Federal Institute for Risk Assessment. *BfR Consumer Monitor 02 | 2016*; Federal Institute for Risk Assessment: Berlin, Germany, 2016.

63. Renn, O. The Role of Risk Perception for Risk Management. *Reliab. Eng. Syst. Saf.* **1998**, *59*, 49–62. [CrossRef]

64. Winker, M.; Schramm, E.; Schulz, O.; Zimmermann, M.; Liehr, S. Integrated water research and how it can help address the challenges faced by Germany's water sector. *Environ. Earth Sci.* **2016**, *75*, 1226. [CrossRef]

65. Otterpohl, R.; Albold, A.; Oldenburg, M. Source control in urban sanitation and waste management: Ten systems with reuse of resources. *Water Sci. Technol.* **1999**, *39*, 153–160. [CrossRef]

66. Larsen, T.A.; Alder, A.C.; Eggen, R.I.L.; Maurer, M.; Lienert, J. Source separation: Will we see a paradigm shift in wastewater handling? *Environ. Sci. Technol.* **2009**, *43*, 6121–6125. [CrossRef] [PubMed]

67. Schramm, E.; Kerber, H.; Trapp, J.H.; Zimmermann, M.; Winker, M. Novel urban water systems in Germany: Governance structures to encourage transformation. *Urban Water J.* **2017**. accepted for publication.

68. Birkmann, J. Measuring vulnerability to promote disaster-resilient societies: Conceptual frameworks and definitions. In *Measuring Vulnerability to Natural Hazards: Towards Disaster Resilient Societies*; Birkmann, J., Ed.; United Nations University Press: New Delhi, India, 2006; pp. 7–54.

69. Sutmöller, J.; Spellmann, H.; Fiebiger, C.; Albert, M. Der Klimawandel und seine Auswirkungen auf die Buchenwälder in Deutschland The effects of climate change on beech forests in Germany. *Ergeb. Angew. Forsch. Buche* **2008**, *3*, 135.

70. Yousefpour, R.; Jacobsen, J.B.; Thorsen, B.J.; Meilby, H.; Hanewinkel, M.; Oehler, K. A review of decision-making approaches to handle uncertainty and risk in adaptive forest management under climate change. *Ann. For. Sci.* **2012**, *69*, 1–15. [CrossRef]

71. Hanewinkel, M.; Cullmann, D.A.; Schelhaas, M.-J.; Nabuurs, G.-J.; Zimmermann, N.E. Climate change may cause severe loss in the economic value of European forest land. *Nat. Clim. Chang.* **2013**, *3*, 203–207. [CrossRef]

72. Kromp-Kolb, H.; Lindenthal, T.; Bohunovsky, L. Österreichischer Sachstandsbericht Klimawandel 2014. *GAIA Ecol. Perspect. Sci. Soc.* **2014**, *23*, 363–365.

73. Schramm, E.; Holland, V. Climate change management in Central European forestry and the perspective of risk sharing. *For. Policy* **2017**. submitted.

74. Schramm, E. Gesellschaftliche Wahrnehmung klimabedingter Biodiversitätsveränderungen in der Forstwirtschaft. In *Klimawandel und Biodiversität: Folgen für Deutschland*; Mosbrugger, V., Brasseur, G.P., Schaller, M., Eds.; WBG Wissenschaftliche Buchgesellschaft: Darmstadt, Germany, 2012; pp. 374–376.

75. Yousefpour, R.; Hanewinkel, M. Forestry professionals' perceptions of climate change, impacts and adaptation strategies for forests in south-west Germany. *Clim. Chang.* **2015**, *130*, 273–286. [CrossRef]

76. Schramm, E.; Litschel, J. Stakeholder-Dialoge—Ein Instrument zur Bearbeitung von Konflikten um Biodiversität in mitteleuropäischen Wäldern. *Nat. Landsch.* **2014**, 478–482. [CrossRef]

77. Perrow, C. *Normal Accidents. Living with High Risk Technologies*; Princeton University Press: Princeton, NJ, USA, 1999.

78. Klinke, A.; Renn, O. Systemic risks as challenge for policy making in risk governance. *Forum Qual. Soc. Res.* **2006**, *7*, Art.33. Available online: http://www.qualitative-research.net/index.php/fqs/article/viewArticle/64/131 (accessed on 30 May 2017).

79. Beisheim, M.; Rudloff, B.; Ulmer, K. *Risiko-Governance. Umgang Mit Globalen und Vernetzten Risiken*; SWP Berlin: Berlin, Germany, 2012; Volume 8.

80. Suda, M.; Pukall, K. Multifunktionale Forstwirtschaft zwischen Inklusion und Extinktion (Essay). *Schweiz. Z. Forstwes.* **2014**, *165*, 333–338. [CrossRef]

81. Klinke, A.; Renn, O. A new approach to risk evaluation and management: Risk-based, precaution-based, and discourse-based strategies. *Risk Anal.* **2002**, *22*, 1071–1094. [CrossRef] [PubMed]

82. Shaxson, L. Structuring policy problems for plastics, the environment and human health: Reflections from the UK. *Philos. Trans. R. Soc. B Biol. Sci.* **2009**, *364*, 2141–2151. [CrossRef] [PubMed]

![sustainability logo] *sustainability*

MDPI

*Article*

# Extractive Economies in Material and Political Terms: Broadening the Analytical Scope

Anke Schaffartzik * and Melanie Pichler

Institute of Social Ecology (SEC), Alpen-Adria University, A-1070 Vienna, Austria; melanie.pichler@aau.at
* Correspondence: anke.schaffartzik@aau.at; Tel.: +43-1-5224000-409

Received: 3 February 2017; Accepted: 19 May 2017; Published: 26 June 2017

**Abstract:** In order to curb environmental impact, absolute resource use reductions are urgently needed. To reach this goal, multi-scalar synergies and trade-offs in global resource use must be effectively addressed. We propose that better understanding the role of extractive economies—economies that extract raw material for export—in global resource use patterns is a prerequisite to identifying such synergies and trade-offs. By combining a system-wide environmental accounting perspective with insights from political ecology and political economy research, we demonstrate that (1) the extractivist expansion may be the corollary of reduced immediate environmental impact in the industrialized countries; and (2) the material flow patterns on which this result is based do not suffice to identify the mechanisms underlying extractivist development and its role in global resource use. Our work on extractive economies illustrates that, in order to supply transformative knowledge for sustainability transformation, biophysical and socio-political conceptualizations of society-nature relations must be more strongly integrated within the interdisciplinary sustainability sciences in general and social ecology in particular.

**Keywords:** extractive economies; international trade; material flow accounting; political ecology; social ecology

---

## 1. Introduction

If we are to have any chance at sustainability transformations, we must reduce our global resource use in absolute terms [1]. Some of the richest countries in the world seem to have made progress towards such reductions: Since the 1970s, countries in Europe, North America, and Asia that industrialized early have experienced stagnating or even declining material resource use [2,3] as reflected in their domestic material consumption (DMC). Under the economy-wide material flow accounting framework, DMC corresponds to material extraction (from agriculture, forestry, and mining) plus material imports minus material exports and covers all the material which an economy integrates into stocks (buildings, infrastructure, and durable goods) and/or either immediately or eventually turns into waste or emissions [4]. When trade plays a dominant role in aggregate material use, the DMC indicator may not offer sufficient information on underlying material flow patterns. What is not directly captured in aggregate material use trajectories is the importance of other countries as sources of raw materials for the environmental progress of the industrialized countries. We postulate that extractive economies which extract raw material for export (as primary goods) [5,6] play a pivotal role in both the stagnation of material resource use in the industrialized countries and in the growth of material use at the global level. By supplying energy- and material-intensive primary resources to the global market, extractive economies enable other countries to concentrate on the addition of value in the secondary or even the tertiary sectors. These sectors tend to be, in direct terms, less material- and energy-intensive and feature high-value output. Imports, rather than domestic extraction, are an increasingly important source of resources for industrialized economies. Extractive economies, in contrast, may often be dependent on imports of secondary commodities in meeting their final demand [7,8]. The win-win situation in the industrialized countries (experiencing increasing affluence and decreasing resource use) may actually be a lose-lose situation globally (unsustainable income and increasing resource use). As a source of income, the extraction of raw materials, even of 'renewable' biomass, does not offer unlimited growth opportunities: mineral resources are extracted at rates much higher than the rates at which they are naturally replenished while biomass harvest is ultimately limited by available land area (and subject to more confining boundaries when nutrient depletion, water cycles, and environmental impact are taken into account). Yet, the expansion of extractive activities, especially in low- and medium-income countries, is currently being encouraged through a series of political instruments. The European Raw Materials Initiative [9], for example, is aimed at ensuring global supply of raw materials to the European Union while simultaneously encouraging the extraction of raw materials within its borders to improve security of supply [10], not only for minerals but also for biotic resources extracted through agriculture and forestry as expounded in the bioeconomy strategy [11]. The UN holds that extractive industries can ultimately contribute to achieving the Sustainable Development Goals [12,13].

In promoting the extractive activity on which high-consuming and growing economies so heavily depend, Norway, Canada, and Australia are commonly cited as best-practice examples in which significant gains in affluence have been made through expanding extractive activity [13]. All three are both important global suppliers of resources and affluent, economically thriving countries. This general observation as well as our empirical work, with which we also seek to identify to what extent the grounds for this success of a small number of countries may be replicated by other economies, refers to the aggregate national level. Underlying the development at this level of scale are subnational disparities that form part of the staple theory of growth [14], developed in the early 20th century for Canada and considering the role of what we might refer to as subnational core and peripheral regions in raw materials-based growth. While high-income countries in which extraction figures prominently may appear to form a contrast to those countries that exchange low-price primary commodity exports for high-price secondary commodity imports (e.g., Chile, Colombia, Ecuador, Mexico, and Peru [15–18]), it is possible that they share communalities at the subnational level.

The extractivist expansion is potentially a critical barrier to sustainability transitions. In order to come to a better understanding of extractive economies, we propose that it is necessary to (1) identify resource use trajectories of these economies and their role in global resource use patterns and (2) integrate insights from such empirical work within a conceptual framework allowing for the identification of functional mechanisms in this particular form of societal organization of resource use. This is, of course, a tall order and not one we will fulfill with this article. Rather than a full-fledged socio-ecological analysis of extractive economies, we offer an exploration of quantitative material flow data that can be used to characterize extractive economies and a discussion of the 'bridges' to political ecology and political economy research that identify qualitative and space-sensitive understandings of the factors and mechanisms shaping resource use in extractive economies and, by extension, globally. The investigation of resource use trajectories is based on a system-wide perspective and internationally harmonized and comparable indicators [4] which allows us to take synergies and trade-offs across levels of scale into account. The mechanisms underlying these resource use patterns cannot be identified based on quantitative information alone, but require the conceptual, functional understanding that can be gained from solid theoretical underpinning.

We develop a framework of socio-metabolic and socio-economic indicators (biophysical resource extraction, trade, monetary income) and apply it to a global sample of 142 countries in the year 2010. We enrich the emerging biophysical characteristics with those that political ecology and political economy research have identified as potentially important for extractive economies. We propose that social ecology's conceptualization of society as simultaneously socio-cultural and biophysical [19] allows this research field to serve as an umbrella under which the socio-metabolic and political-economic analyses of extractive economies can enter into productive dialogue. This conceptualization of society stems from social ecology as developed at the Institute of Social Ecology in Vienna where both authors of this article are based. The editorial for this special issue of Sustainability provides a differentiated discussion of this and other approaches to social ecology.

We discuss major research strands on extractive economies in Section 2 and present a framework of socio-metabolic and socio-economic indicators to quantitatively distinguish different types of extractive economies in Section 3. In Section 4, we present our results—an overview of extractive economies according to material flows and types and national income. In Section 5, we discuss insights from political ecology and political economy research that may enrich the data-driven and quantitative analysis of extractive economies. We propose that by crossing these bridges, we may arrive at a better understanding of the role of extractive economies in the global economy. Such an understanding is essential not only for assessments of this particular development trajectory but is also a prerequisite to evaluating the sustainability of the world's wealthiest economies.

## 2. Extractive Economies in the Global Economy

Our research on extractive economies builds on important previous work, especially from development studies, (ecological) economics, and political ecology, aimed at providing a general narrative on the role of resource provision and consumption in the global economy. After the Second World War, economists noted that, in contrast to countries specialized in manufacturing, primary commodity exporters had experienced deteriorating terms of trade [20,21]. The observation of prices for primary commodities declining compared to the prices of secondary commodities in the long-run came to be known as the Prebisch-Singer thesis [22] and influenced the rise of dependency theory [23]. World-systems theory has observed similarly skewed relations between core and periphery countries [6,24]. Based on these conceptualizations, the theory of ecologically unequal exchange holds that countries which incur high environmental impacts in the extraction and preliminary processing of raw materials for comparatively low-price exports do not benefit accordingly in financial or material terms [25,26]. In development economics, the theory of the "resource curse" has been developed which holds that countries with an abundance of natural resources 'underperform' economically compared to those countries with a lower resource endowment [27–29]. Resource-rich countries may even be more strongly prone to suffer from armed conflict [30]. For our research, the resource curse hypothesis introduces two important points of departure: (1) The entailed critique of assumed socio-economic benefits of extractivist expansion, a pillar of past and current development policy, to which we add the resource use perspective; and (2) The question whether the observed negative impacts of abundant resource endowment are a function of that endowment or perhaps more accurately of mechanisms of societal organization governing the use of resources. For our research, it is additionally noteworthy that the development of the resource curse hypothesis—and even its critique [31,32]!—relied on monetary measurement of natural resources, either as natural resource abundance or dependence (monetary value of primary exports divided by national income) or as natural wealth [33]. More recent research on these interrelated topics (core-periphery relations, ecologically unequal exchange, and the resource curse) has commonly sought to correct this monetary bias. Especially in ecological economics and political ecology, the relationship between *physical* and *monetary* flows has been studied in order to determine whether countries benefit from trade or not. Studies empirically testing for patterns of ecologically unequal exchange have not reached agreement in their conclusions [34–39], also because an acceptable monetary amount of revenue for the social and environmental impacts of resource extraction for export cannot be defined [40]. Recently, the research agenda has been developed to more strongly focus on biophysical flows in identifying ecologically unequal exchange and (potential for) environmental conflict (e.g., [40–42]).

Data on biophysical trade flows alone do not sufficiently characterize an extractive economy as socio-economically very different countries such as Australia, Azerbaijan, Bolivia, and Canada would then be grouped together [17]. These countries are net-exporters, and yet, a qualitative distinction between their extractive activities must be made. Within a high-income economy, for example, extractive sectors are not necessarily characterized by low capital inputs or lack of skilled labor [43]. Recent contributions from political ecology and political economy have provided differentiated analyses of resource-based development with special emphasis on the access to and control over resources, (transnational) production and consumption networks, and inequality and justice with regard to burdens and benefits of extraction. In Section 5, we further elaborate on these contributions in light of the results of our quantitative analysis of extractive economies.

### 3. Defining Extractive Economies in Quantitative Terms

Our empirical work focuses on insights into extractive economies that can be gained from system-wide material flow data. While this information is pivotal to our understanding of the role of extractive economies in the global economy and global environmental change, we do not expect it to allow for a full-fledged understanding of the observed resource use patterns. On the basis of political ecology and political economy research, we further deliberate these limitations in Section 5.

In material terms, we understand extractive economies to be characterized by the domestic extraction of materials for export and propose two data-driven types of operationalization of this understanding. The first allows us to identify economies in which exports biophysically dominate material use by accounting for more than 50% of all material input (from domestic extraction and imports). The second isolates countries that are net-exporters, i.e., they export more than they import. The decision-making flow chart in Figure 1 illustrates how we implemented this understanding with the help of standard indicators of material flow accounting: export per direct material input (DMI = domestic extraction plus imports) or physical trade balance (PTB = imports minus exports, i.e., net imports) and domestic extraction (DE) per DMI. Of course, the two categories potentially overlap: Countries can be export-dominated and net-exporters.

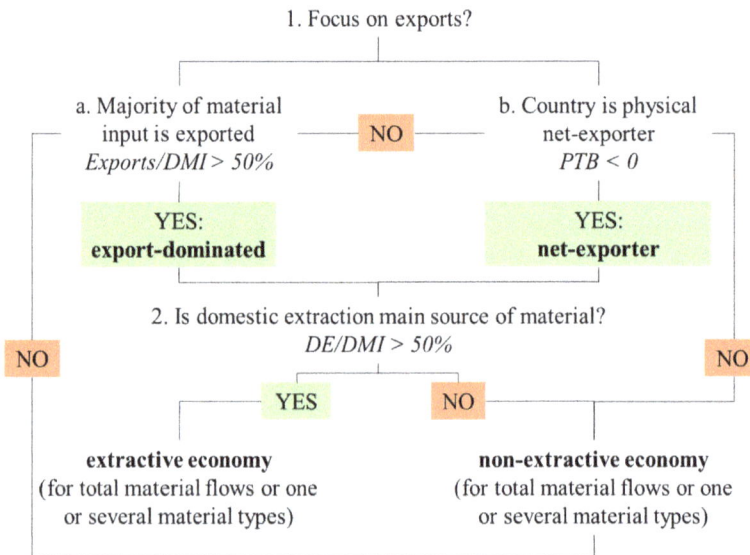

**Figure 1.** Decision-making flow chart for identification of extractive economy (with regard to total material flows or flows of one or several material types) based on the material flow indicators exports, direct material input (DMI), physical trade balance (PTB), and domestic extraction (DE).

The distinction between export-dominated and net-exporting economies is a priori driven by discernible patterns in the material flow data. Based on potential drivers of such distinct material use patterns, we also expect the two types of extractive economies to be politically and economically distinct. The fact that export-dominated economies provide the major share of the material available to them to the global economy may constitute, if underlying production structures are considered, an advantage rather than a loss. *Export-dominated economies* may be highly efficient in transforming raw materials and primary products into commodities for export, they may strike a beneficial balance in the international division of labor concerning their material inputs, and they may be, in the short- to medium-term, partially protected from international raw material price volatility. For many raw material types, the processing stages that precede export involve so much waste generation that the share of exports in DMI does not rise above 50%. The specific combination of material types, material sources, and technology in export-dominated economies must allow them to process material input into export with comparatively low material losses. For example, processing of crude oil for export involves significantly lower material losses than is the case for copper ore. Resource-rich countries with small populations (e.g., countries with high per capita agricultural land availability) require relatively less of their resources for the consumption of their population (e.g., agricultural biomass) than do countries that are more populous. Export-dominated economies may additionally be extractive economies of the second degree in that they not only process domestically extracted raw materials but also import primary products for export. Especially the high-income extractive economies tend to be importers of primary products for further processing. By re-exporting materials that have already undergone some degree of processing, economies require relatively less material and energy input in the production process than economies that rely exclusively on domestically extracted raw material. In the international division of labor, they act not only as resource suppliers but also rely on extractive sectors in other countries. This mixed sourcing of raw materials and primary products, however, may allow them to benefit from international price volatility for primary commodities and offer some degree of protection from the so-called "specialization trap" [44,45].

While we expect very few economies to fulfill the export-dominated criteria, many more countries and even entire country groupings and regions are known to be biophysical net-exporters [3]. *Net-exporters* are global suppliers of material and provide more to the international markets than they require from them. From the national perspective, these net-exports often constitute an important source of income. In the international context, however, net-exporters may be relatively small contributors to global supply, offering them little leverage and making them economically and politically vulnerable as a result of their own specialization in extraction and production for export.

In order to be considered as extractive economies, both the export-dominated countries and the net-exporters had to domestically extract the majority of the total material available to them (DE/DMI >50%). While, as previously stated, we were interested in considering as extractive those economies that complement resource extraction with the import of primary commodities for further processing, we did not want to include countries relying chiefly on the re-export of imported materials as extractive. Countries may, for example, specialize in the production of high-grade steel for export but not engage in extraction of iron ore. Instead, they import iron concentrate for further processing. While this constitutes an interesting form of international exchange, it does not, in our understanding, characterize an extractive economy.

We analyzed exports, net imports, DE, and DMI and the relationships among them as totals as well as for material types to allow for economies behaving in an extractive manner with regard to one or several types of materials but not with regard to others. The indicators were processed in physical units (tonnes) for a global sample of 142 countries in the year 2010. The underlying material flow data were extracted from the database maintained at the Institute of Social Ecology, Vienna [3]. This view of the global economy in 2010 affords the advantage of data coverage for a large number of countries. By choosing a static perspective, we are able to gain insight into resource patterns at one point in time which can be grasped through the conceptualization we propose. While this provides us with the opportunity to test and discuss the potential and limits of our definition of extractive economies, it does not take development pathways into account. They are not addressed within the scope of this article but we do consider such pathways (e.g., of decreasing or increasing material extraction, exports, income) to be very important in advancing research on extractive economies.

We based our primary definition of extractive economies on purely biophysical indicators in order to be able to identify those economies that dominantly exported domestically sourced resources. Of course, there is a 'hidden' socio-economic component to these indicators in the sense that they do not reflect purely geomorphological and/or climatic endowment with resources but also the socio-economic feasibility of their extraction and trade. Additionally, past resource depletion determines whether countries in 2010 meet the criteria we have defined.

We wanted to determine whether the resource use patterns of extractive economies, defined according to biophysical criteria, coincided in any way with economic performance. In a first approach, we used gross national income (GNI) expressed in 2010 US dollars according to the Atlas method [46]. Based on the per capita income level, we further distinguished countries and country groupings by four income categories: high income (above 12,275 US $/cap), upper middle income (above 3976 US $), lower middle income (above 1005 US $/cap), and low income [47].

Previous socio-ecological studies on the role of trade for resource use patterns have commonly relied on the physical trade balance (PTB = imports − exports), an indicator derived from material flow accounting [4]. Countries, regions or country groupings have been distinguished according to whether they are net-suppliers (i.e., net-exporters) or net-consumers (i.e., net-importers) of materials [34,48]. Extractive economies were generally characterized by a negative PTB, i.e., as net-exporters. However, even countries that are not physical net exporters may be extracting/producing for foreign demand. It has been shown for the case of Chile, for example, that a significant share of the country's very high level of material consumption consists of waste rock extracted during mining of metals for export in a concentrated or pure form [49,50]. This waste rock itself is not exported and yet must be extracted so that copper may be exported. In order to generate a more detailed picture of international trade patterns, the material flow accounting (MFA) community is currently developing methods which allow for the quantification of the amount of material used for the production of exports (and hence not used to satisfy domestic final demand). These *upstream material requirements* include the mass of the trade flows upon crossing the border as it is traditionally covered in MFA (see [51]) and additionally allocate the material used in the production of exports but not incorporated in the flow itself. Accounting for upstream material requirements currently relies on approximations via monetary input-output data which reflect the structure of production and consumption [52] because data in physical units denoting the material use for production of exports are not available [53]. In the study of extractive economies, however, information on material and economic trade patterns is needed independently so that interrelations between both types of development may be considered. This criterion is not met by upstream material accounts based on monetary input-output tables [53], for a discussion of the implications, also see [54]. We have therefore used material flow data from biophysical accounting for our analysis.

## 4. Comparing Extractive Economies

In the following, we apply our two-part, data-driven definition of extractive economies (Figure 1) to our data sample to provide a quantitative distinction among extractive economies, which can be productively discussed in light of political ecology and political economy advances.

### 4.1. Extractive Economies in Global Trade by Income Groups

Material flow data aggregated by national income shows that some of the world's richest countries rely on the rest of the world (home to the majority of the global population) to supply them with natural resources. Since the Second World War, early industrializers in Europe, North America, and Asia have been, as a country grouping, net-importers of material. Since the 1970s, fossil fuels (mainly coal, petroleum, and natural gas) have accounted for more than half of the imports of the early industrializers [3]. With the exception of Asia in the most recent decades, all other world regions

(Sub-Saharan Africa, Middle East and North Africa, Latin America and the Caribbean, and the countries which emerged from the dissolution of the Soviet Union) at all other points in time between 1950 and 2010 were net-exporters of material.

In approaching the economic performance of extractive economies, we grouped countries in 2010 according to their income level and calculated the PTB by material categories for each country grouping. The early industrializers largely coincide with the high-income countries that were the grouping with the highest net imports in 2010: Their imports exceeded their exports by 500 million tons (Figure 2). In 2010, the high-income countries (dominated, for most metals, by Australia and Canada) accounted for 25% or more of the global extraction of iron ore, bauxite, zinc, lead, gold, ilmenite, and rutile [3]. While the high-income countries were net-exporters of metals and biomass (from agriculture and forestry), they relied very strongly on imports of fossil fuels (mainly coal, oil, and natural gas) and, in much lower amounts, of construction minerals (such as sand and gravel), industrial minerals (e.g., fertilizers), and complex multi-material products that consist of several different types of material. By displaying trade flows by income groupings, we can determine that these high net imports stem from the middle-income countries. Fossil fuels flow into the high-income countries from, for example, Nigeria, Yemen, and Indonesia. Net-exports of metals and biomass flow from high-income (e.g., Australia for metals, many European countries for biomass) and lower-middle-income countries (especially in Sub-Saharan Africa and Southeast Asia as well as India, for example) to the upper-middle-income countries (among which China has become an important importer). Overall, the middle-income countries supplied approximately 900 million tons of material to the global economy in 2010, with the larger share stemming from the upper-middle-income countries. The low-income countries are characterized by low integration into the global economy and their net imports are approximately one tenth of the net imports of the high-income countries.

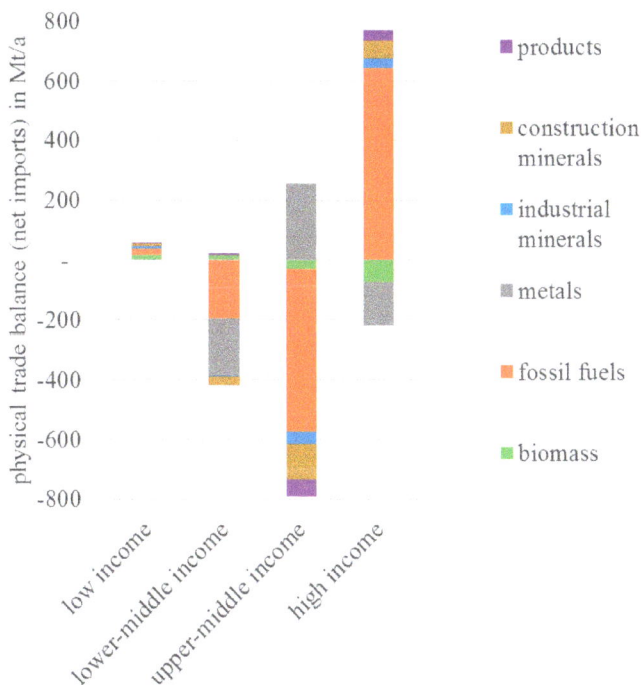

**Figure 2.** Physical trade balances (imports minus exports) in Megatons per year (Mt/a) by material categories for income-based country groupings [47]. Source of material flow data: [3].

*4.2. Extractive Economies' Material Flows and Income*

We found a small and inhomogeneous group of seven countries to meet the criteria for *export-dominated extractive economies* (Table 1). The seven countries were Norway, the Republic of the Congo, Qatar, Trinidad and Tobago, Oman, Kuwait, and Australia. Given the size of this group and the exogenously defined cut-off threshold (50%) in our extractive economies definition (Figure 1), we do not propose to conduct a systematic analysis of this result. What does stand out, however, is the dominant role of domestic sources in fossil energy supply for all of these countries (and for those countries approaching but not meeting the 50% exports/DMI criterion). In these countries, more than 80% of fossil fuel DMI stems from domestic extraction; in some cases, this share approaches 100%. All export-dominated extractive economies were net-exporters of fossil fuels and fossil fuels were the most important export flow in monetary terms. With the exception of Australia, 50% or more of monetary exports of these countries consisted of crude and refined petroleum and petroleum gas [55]. For Australia, coal contributed significantly to monetary fossil energy carrier exports and this material category was closely followed by metals (especially iron ore but also copper) in monetary terms. This material evidence supports the assumption we initially made with regard to the efficiency with which export-dominated economies must transform extracted raw materials into exports. This efficiency is, in general, higher for fossil fuels than, for example, for metals. In addition, the availability of (comparatively cheap and stably priced) energy may be a decisive factor in biophysically sustaining export-dominated economies. Together, these seven economies accounted for almost 14% of all physical exports and for less than 1% of the world population. They generated almost 3% of the global income, and all but one (the Republic of the Congo) were high-income countries; their average income in 2010 amounted to over 43,000 US \$/cap (with 12,275 US \$/cap being the threshold for high income that year). A small number of countries appear to be thriving economically based on the extraction of high amounts of resources for export coupled with domestic fossil energy availability. This extractivism-income link, however, neither applies to the majority of countries engaged in extractive activity nor does it represent a trajectory that can be easily replicated as it appears to rely, at least in part, on the biophysical availability of fossil fuels and the economic feasibility of their extraction.

**Table 1.** Overview of extractive economies in 2010. Share in global exports indicates percentage of global biophysical exports in each material category originating from the group of countries classified as extractive with regard to that material. Income level distinguishes high (H), upper middle (UM), and lower middle (LM) income. None of the groups of economies defined as extractive experienced average low income as per the World Bank [47] definition. Sources of data: [3,46].

| | Export-Dominated | | | | Net-Exporters | | | |
|---|---|---|---|---|---|---|---|---|
| | n(1) | Share in Global Exports | Average Income (GNI/cap) | Average Income Level | n(2) | Share in Global Exports | Average Income (GNI/cap) | Average Income Level |
| Total | 7 | 13.5% | 43,228 | H | 49 | 59.0% | 6813 | UM |
| Biomass | 1 | 0.1% | 36,670 | H | 57 | 68.3% | 10,136 | UM |
| Fossil Fuels | 25 | 52.0% | 6787 | UM | 38 | 76.0% | 6110 | UM |
| Metals | 11 | 47.1% | 3359 | LM | 28 | 62.6% | 4176 | UM |
| Industrial Minerals | 14 | 32.3% | 10,673 | UM | 26 | 61.0% | 7484 | UM |
| Construction Minerals | 3 | 3.8% | 8301 | UM | 45 | 78.3% | 7150 | UM |

Many more countries (49) were *net-exporters* of total materials. This group of extractive economies accounted for almost 60% of total global physical exports in 2010, approximately 24% of the population, and 19% of global income. With an average income of 6813 US \$/cap (less than 16% of the average income of export-dominated extractive economies), this country grouping could be classified as upper middle income. Within this group of global suppliers of material resources, those countries with higher national income tend to have higher net exports per capita (Figure 3). Countries that structurally dedicate the resources available to them to export appear to profit economically if they are able to export relatively more of those resources. Accordingly, those extractive economies with the highest per capita net exports are also those with the highest average income. Given this relationship, the low level of integration of low-income economies into global material trade (Figure 3), may play a functional role in their extractivism-income link. The expansion of primary exports, however, cannot be considered as the key to sustainable economic growth: the possibility for expansion is limited based on resource endowment (location and size of the country) and extraction and processing capabilities. While the latter can be improved to some extent through investment, the former is subject to physical limits and to trade-offs with other forms of land use. At the same time, high average income may be associated with high domestic inequality in income (see Section 5.3). Therefore, the progression of countries from low income and low exports to high income and high exports in Figure 3 cannot be understood as a potential development trajectory for an individual country.

The relationship between exports and GDP is a long-standing topic in development economics, with the identified positive correlation (of the monetary indicators) constituting one of the pillars of export-led development policies [56,57]. As opposed to the analysis based purely on monetary indicators, our results allow us to make two important distinctions: (1) the positive correlation between income and physical exports holds true only for net-exporting countries, i.e., cannot be verified for a general relationship between exports and income; and (2) among the net-exporters, any possible contribution of exports to income also comes at an environmental cost (more material must be exported).

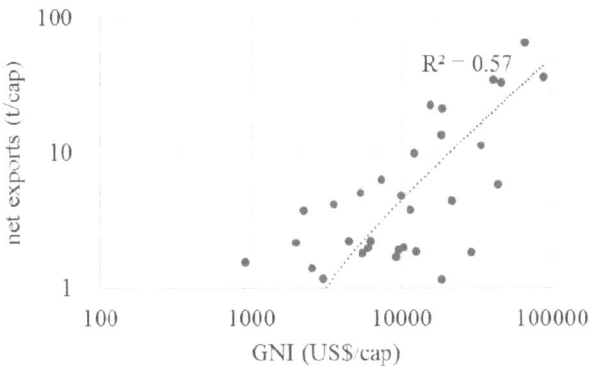

**Figure 3.** Gross national income (GNI) per capita and net exports in tons per capita (t/cap) for 49 extractive economies (export-dominated and net-exporters) in 2010. The linear relationship is highly significant ($p < 0.001$). Please note that both axes are logarithmically scaled. Sources of data: [3,43].

*4.3. Extractive Economies by Material Types*

The number of export-dominated countries is very small for biomass (Iceland) and construction minerals (Norway, Bhutan, and the Bahamas) and stands in strong contrast to the number of net-exporters for these same materials: 57 countries for biomass, corresponding to over 68% of global biomass exports, and 45 countries for construction minerals (78% of global exports). Most countries domestically consume the bulk of biomass and construction minerals extracted. This does not necessarily mean that the use of these materials is directly driven by domestic final demand: For example, construction minerals may be used in factories for the production of exports and grazed biomass and fodder crops may be used to feed livestock and produce animal products for export. For fossil fuels, 25 countries were export-dominated and 38 were net-exporters. Fossil fuel extractive economies accounted for more than $\frac{3}{4}$ of all fossil fuel exports in 2010. Countries of all income categories were represented in this group with an average income between 6000 and 7000 US $/cap corresponding to upper middle income. For metals, 11 economies were export-dominated and 28 were net-exporters. Eleven countries accounted for almost 50% of global metal exports, the 28 net-exporters accounted for approximately 15% more. In spite of their significant contribution to global supply, the export-dominated economies for metals feature the lowest average income (below 3500 US \$/cap, lower middle income) out of all the groupings presented in Table 1. In contrast to the extractive economies for all other material types, the average income is lower here for the export-dominated economies than for the net-exporters. For metals, in particular, the amount of material loss (waste rock) between extraction and export can be very large; the metal content of extracted ore may be below 1%. For industrial minerals, the contribution to global exports is more evenly divided between the extractive economies with the 14 export-dominated countries contributing 32% and the 26 net-exporters contributing slightly less than twice that share (61%).

The positive relationship between income and net exports for total material flows (Figure 2) does not hold true for all of the material categories. Even at the very high level of aggregation by five material categories, which resources a country extracts for export makes a difference in that country's economic performance. A positive correlation between income and per capita net exports was only found for fossil fuels ($R^2 = 0.47$, $p < 0.001$). For all other materials, the analysis did indicate a positive relation ($R > 0$) but with a low goodness-of-fit and/or low significance of the determined relationship.

## 5. Insights from Political Ecology and Political Economy

The quantitative economic and material flow data provides important insights on extractive economies in aggregate terms of inhomogeneous material categories and at the national level. To better understand the potential prospects and risks of extractive economies, we bring these results into dialogue with insights from political ecology and political economy research. These case studies have closely examined extractive activity for specific resources and countries as well as at different scales to gain insight into the role of political and economic mechanisms that shape extractive economies. From the wealth of theoretical underpinning and empirical evidence applicable to our research on extractive economies, we have selected two fundamental mechanisms by which resource extraction and use is societally organized: (1) the access to land as a precondition to any type of extractivist expansion; and (2) the integration of the economy in question in transnational networks of production and consumption as decisive for the role which export-led extraction may play. Both these mechanisms are related to inequality and justice with direct implications for the distribution of benefits and burdens associated with extractivist development.

### 5.1. Contested Access to and Control Over Land and Resources

Extractive activity requires access to and control over territories in order to extract one particular material or resource. Vast areas of land are typically appropriated; water and energy constitute other important inputs. Political ecology research has identified the pivotal role of control over land in extractive economies [58–61]. Extractive industries require exclusive control over land that often excludes existing users and uses from these lands. The state and power relations within (state) institutions play a crucial role in defining and formalizing exclusive property rights which precede the extraction of resources from specific territories (both for state and private companies) [62]. The resulting land-use competition [63] or competing claims to land [59] increasingly becomes manifest in conflicts over access to land, especially at the resource-use frontier where new claims are established [5,64,65]. Our study of material flows and economic indicators would lead us to expect *potential* for conflict wherever extraction expands [66], even in the high-income extractive economies. Comprehensive research into socio-ecological conflicts as represented by the *Environmental Justice Atlas* (compiled under the European FP7 project EJOLT (Environmental Justice Organizations, Liabilities and Trade): http://ejatlas.org/) empirically documents that the link is not as straight-forward as 'more extraction, more conflict'. Instead, *manifest* conflicts depend on specific power relations and institutional orders in the respective countries as well as on the specific integration of countries or local extraction sites in the world market [67,68]. Conflicts over water in the drought-plagued Murray-Darling basin in southeastern Australia, for example, have been carried out chiefly over economic issues, between farmers and miners and between governments at the sub-national level [69,70]. In many countries of the global South, in contrast, claims to subsistence and indigenous lands have pitted actors with much greater disparity in power against each other (also see Section 5.3). The recent opening of the economy for the exploitation of natural resources in Myanmar, for example, has given rise to resistance against land enclosure in the ethnically diverse uplands of the country [64]. Similarly, the rapid expansion of oil palm plantations in Indonesia has led to conflicts with indigenous peoples who challenge the territorialization and commodification of their lands [59]. What is observed as environmentally unequal exchange (Section 2) based on international inequalities between trade partners, may be directly linked to inequalities at the subnational level with regard to the access to resources and hence also the power to make decisions with regard to their use. In order to operationalize such inequality in biophysical resource use, aggregate national indicators do not suffice and current developments towards subnational or even spatially explicit data collection are imperative.

Through its role in securing access to and control over land, the state plays an important role in how and by whom rents from the extraction of natural resources are obtained. Governments at different levels may see their role in supporting extractive industries' access to land if rents, secured in essence by access to that land, constitute an important source of income for the national or regional economy [71,72]. This observation is potentially tied to one we have made based on material flow data: Countries that domestically extract and import raw and semi-processed materials for further processing before export are less dependent on rent as a source of income than those countries with no internal processing capacities. We suggested that, in general, the degree to which countries are able to build economic complexity surrounding their extractive industries may play an important role in national income. In light of the political ecology and economy research, we further propose that the share of rents in total national income would also play a role in how access to land is governed. The observed resource curse (Section 2) may depend less strongly on the endowment with natural resources (even if this were physically and not only monetarily captured) and more on how and to what end available resources are governed.

## 5.2. Transnational Production and Consumption Networks

In material terms, we were able to distinguish countries concentrating on the extraction of resources for export from those countries specializing in the extraction of individual material types (e.g., biomass, fossil fuels, metals). High-income extractive economies tend *not* to specialize in the extraction of only one type of material, although fossil fuels tend to be among the materials extracted (Section 4.2), and exhibit a high to intermediate economic complexity index [55]. Political economy research has linked the focus on the extraction of a small number of specific resources to the "specialization trap", in which countries are especially vulnerable to fluctuating world market commodity prices and ensuing boom-and-bust dynamics [44,45]. The reasons for such specialization may partially lie in resource endowment and can additionally be found in extractive industry's role in global value chains and world market integration [73,74]. Investment, governance, and ownership patterns in transnational production and consumption networks strongly influence the role of the extractive sector within the national economy. The national or regional advantage of the extractive sector may, therefore, not only depend on power relations between individual nation states but also on the relational ways in which companies, states, workers, and consumers are incorporated in production networks that transgress national boundaries [74]. Depending on these transnational relations, extractive sectors, especially in the Global South, tend to develop enclave economies. These enclaves are characterized by weak linkages between large-scale investments in extractive sectors and local companies or the national economy at large and are often led by international or foreign capital [74–76]. National governments and institutional structures may play a decisive role in creating linkage effects and contributing to the (economic) success (or lack thereof) of the extractive economy. An analysis of extractive economies has to consider the network of production and consumption to which an extractive industry contributes at the international (bilateral trade relations) and the transnational level (networks of corporations and businesses not apparent at the aggregate national scale), i.e., the conflicting interests and dependencies of companies, states, and other transnational players [74].

At the national level at which material flow data are currently available, we were able to identify the role of economies as global suppliers or consumers of resources. In addition, patterns in material extraction began to emerge, possibly providing some indication of biophysical manifestations of economic integration of the extractive activity. We found, for example, that extractive economies based on a mix of materials in their exports tended to feature higher income if they *also* extracted fossil fuels. The associated access to comparatively cheap energy for the extractive industries as well as for further processing may be decisive in the degree to which countries can economically benefit from their extractive activities. Yet, based on aggregate national material flow data, we cannot draw definite conclusions as to the integration of the extractive activity into the economy, which may be just as influential as the question of economic complexity.

*5.3. Inequality and Justice*

Linked to the access to and control over resources and the role of extractive sectors in (transnational) production and consumption networks are questions of inequality and justice [77]. This strand of research has centered on who stands to benefit economically and socially from resource extraction and who carries the socio-ecological costs, including, for example, pollution, loss of land, or health impacts [78].

Irrespective of average national income, extractive economies may face particular challenges with regard to intra-national inequality because increasing affluence at the national level (e.g., through the increase of rents from the export of primary commodities) is not necessarily linked to an equal share of this affluence by all social groups [71,76,79,80]. Even in countries that attempted to redirect increasing rents of natural resource exports towards marginalized groups in recent years (e.g., Ecuador, Bolivia, Argentina), the redistributional effects have remained marginal. In many cases, social policies were oriented towards conditional cash transfer programs to combat extreme poverty but have not taken up more structural efforts to reform the highly unequal tax system or land ownership structures that benefit the wealthy sectors of society [72].

Whereas political economy research has analyzed inequality and justice mainly from an economic and monetary perspective, political ecology has highlighted the political dimensions of justice [81,82]. In a multi-dimensional perspective, political representation and cultural recognition are incorporated in the analysis of socio-ecological justice with regard to resource extraction. Political representation asks *who* is included in decision-making processes and how rules and procedures for extractive economies are set. Cultural recognition demands a sensitivity towards alternative ways of living and forms of knowledge (e.g., subsistence farming, indigenous production systems) that challenge extractive sites and may be missed in economic analyses of inequality [66,81].

Systematically considering inequality along its environmental, social, and economic dimensions and in light of its sustainability implications and underlying drivers across levels of scale constitutes an important research frontier for social ecology. This frontier promises to be especially productive not only with regard to the specific role of inequality in extractive economies but also more generally in aiding social ecology to contribute to transformative knowledge for sustainability transformation.

## 6. Conclusions: Socio-Ecological Research on Extractive Economies

Starting from a data-driven socio-metabolic approach, this article has presented a comparison of extractive industries in quantitative terms. We found that the richest countries in the world have relied on the rest of the world to supply them with raw materials although they seem to have experienced declining material use in national extraction since the 1970s. Fossil fuels comprise the most important imports of primary commodities for these high-income countries. We distinguished between *export-dominated* countries that export more than 50% of their material input and *net-exporters* that provide the world market with more raw material than they import. Fossil fuels stand out, as all export-dominated countries are able to extract fossil fuels domestically. For most of them, fossil fuels are also the major export product. Whereas extractive economies are not necessarily linked to high or low levels of average national income, there are some interesting correlations. For countries that are net-exporters with regard to their total exports and imports, there is indeed a positive correlation with average national income, whereas metals export-dominated countries tend to exhibit comparatively low average income. For metals, in particular, the amount of material loss (waste rock) between extraction and export can be very large which might be an explanation for this pattern. By considering extractive economies in their aggregate, biophysical dimensions at the national level, we cannot conclude what causes the (economic) success or failure of such economies. We demonstrate that there are insights to be gained from this particular socio-ecological take on the extractive economy that provide potential for synergy with the political economy and political ecology research on associated socio-economic and political-institutional arrangements.

First, case studies have demonstrated how the question of access to and control over resources is decisive for biophysical availability of resources and emerging resource use patterns as well as for potential and manifest conflicts arising from extractive industries. Second, research on global value chains has highlighted the importance of transnational networks of production and consumption that transgress the national scale of analysis and have decisive consequences for the national and regional advantages (or lack thereof) of extraction sites. Overall, where extractivism is associated with higher average national income (as may be the case in the net-exporting countries), research on inequality and justice has shown that this rise in income is not necessarily linked to an equal share in this affluence by all social groups. Specific institutional arrangement and power relations matter, and monetary economic redistribution has to be accompanied by political and cultural dimensions of justice. Observed manifestations of the patterns in international resource trade such as the resource curse hypothesis and ecologically unequal exchange provide conceptual entry points for the type of integrated qualitative and quantitative research which we propose. At the same time, that research has the potential for further illuminating these concepts, especially by considering why they appear analytically valid yet not universally applicable. Here, the insights to be gained from material flow data on the composition of resource extraction and use constitute examples of how the mutual links between resource use patterns and their societal organization shape country-level development.

Any development trajectory based on the high and growing exploitation of natural resources will ultimately run into physical limits. These will initially be observed economically as rising costs of extraction but are ultimately biophysical in nature. In fact, some of the early industrializers have developed from extractive economies towards a stronger focus on the secondary and tertiary sectors. In the United Kingdom, for example, the domestic extraction of iron ore for the country's signature steel industry has long ceased to be economically viable. Instead, steel production has relied on imports of ores and concentrates. As, however, some of the world's most important extractive economies for iron began increasingly focusing on further processing of steel and on newly emerging markets, imports as a source of supply for the United Kingdom's industry dwindled. This 'other side of the coin' must be considered alongside the more common narrative of global glut and falling prices affecting the industrial motherland's ability to compete internationally in what was once one of its prime commodities [83,84]. Just as secondary processing relies on the extraction of primary material elsewhere, so does the tertiary sector rely to some extent on secondary (and by extension primary) economic activities. Continued growth, no matter which of the areas of the economy it relies on, is therefore ultimately unsustainable.

The development of an extractive economy occurs simultaneously in biophysical and in socio-political terms. As a result, research is required which allows us to "address social and natural structures and processes on an equal epistemological footing". Social ecology offers a conceptual framework of society-nature relations in which "human social and natural systems interact, coevolve over time and have substantial impacts upon one another" [19]. Within such a framework, the results on extractive economies' material flows and their socio-political arrangements must be further made mutually meaningful and applied in integrated socio-ecological research on extractive economies.

**Author Contributions:** A.S. analyzed the data, M.P. reviewed the literature, A.S. and M.P. wrote the paper.

**Conflicts of Interest:** The authors declare no conflict of interest.

## References

1.  Akenji, L.; Bengtsson, M.; Bleischwitz, R.; Tukker, A.; Schandl, H. Ossified materialism: Introduction to the special volume on absolute reductions in materials throughput and emissions. *J. Clean. Prod.* **2016**, *132*, 1–16. [CrossRef]
2.  Wiedenhofer, D.; Rovenskaya, E.; Haas, W.; Krausmann, F.; Pallua, I.; Fischer-Kowalski, M. Is there a 1970s Syndrome? Analyzing Structural Breaks in the Metabolism of Industrial Economies. *Energy Procedia* **2013**, *40*, 182–191.
3.  Schaffartzik, A.; Mayer, A.; Gingrich, S.; Eisenmenger, N.; Loy, C.; Krausmann, F. The global metabolic transition: Regional patterns and trends of global material flows, 1950–2010. *Glob. Environ. Chang.* **2014**, *26*, 87–97. [CrossRef] [PubMed]
4.  Fischer-Kowalski, M.; Krausmann, F.; Giljum, S.; Lutter, S.; Mayer, A.; Bringezu, S.; Moriguchi, Y.; Schütz, H.; Schandl, H.; Weisz, H. Methodology and Indicators of Economy-wide Material Flow Accounting. *J. Ind. Ecol.* **2011**, *15*, 855–876. [CrossRef]
5.  Bebbington, A. *Social Conflict, Economic Development and the Extractive Industry: Evidence from South America*; Routledge: London, UK, 2011.
6.  Bunker, S.G. *Underdeveloping the Amazon: Extraction, Unequal Exchange, and the Failure of the Modern State*; University of Chicago Press: Chicago, IL, USA, 1985.
7.  Eisenmenger, N.; Giljum, S. Evidence from societal metabolism studies for ecological unequal trade. In *The World System and the Earth System: Global Socioenvironmental Change and Sustainability since the Neolithic*; Hornborg, A., Crumley, C.L., Eds.; Left Coast Press: Walnut Creek, CA, USA, 2007; pp. 288–302.
8.  Schaffartzik, A.; Mayer, A.; Eisenmenger, N.; Krausmann, F. Global patterns of metal extractivism, 1950–2010: Providing the bones for the industrial society's skeleton. *Ecol. Econ.* **2016**, *122*, 101–110. [CrossRef]
9.  European Commission. *The Raw Materials Initiative. Meeting Our Critical Needs for Growth and Jobs in Europe*; European Commission: Brussels, Belgium, 2008.
10. European Commission. *Communication from the Commission to the European Parliament, the Council, the European Economic and Social Committee and the Committee of the Regions for a European Ind. Renaissance*; European Commission: Brussels, Belgium, 2014.
11. European Commission. *Innovating for Sustainable Growth. A Bioeconomy for Europe*; Publications Office of the European Union: Luxembourg, 2012.
12. United Nations. *Transforming Our World: The 2030 Agenda for Sustainable Development*; United Nations: New York, NY, USA, 2015; Volume A/RES/70/1.
13. UNDP Helen Clark: Speech at the Event "Extractive Industries and the Sustainable Development Goals—Enhancing Collaboration for Sustainability". Available online: http://www.undp.org/content/undp/en/home/presscenter/speeches/2015/09/27/helen-clark-speech-at-the-event-extractive-industries-and-the-sustainable-development-goals-enhancing-collaboration-for-sustainability-.html (accessed on 10 January 2017).
14. Watkins, M.H. A Staple Theory of Economic Growth. *Can. J. Econ. Political Sci.* **1963**, *29*, 141–158. [CrossRef]
15. Russi, D.; Gonzalez-Martinez, A.C.; Silva-Macher, J.C.; Giljum, S.; Martinez-Alier, J.; Vallejo, M.C. Material Flows in Latin America. *J. Ind. Ecol.* **2008**, *12*, 704–720. [CrossRef]
16. Gonzalez-Martinez, A.C.; Schandl, H. The biophysical perspective of a middle income economy: Material flows in Mexico. *Ecol. Econ.* **2008**, *68*, 317–327. [CrossRef]
17. Vallejo, M.C. Biophysical structure of the Ecuadorian economy, foreign trade, and policy implications. *Ecol. Econ.* **2010**, *70*, 159–169. [CrossRef]
18. Vallejo, M.C.; Pérez Rincón, M.A.; Martinez-Alier, J. Metabolic Profile of the Colombian Economy from 1970 to 2007. *J. Ind. Ecol.* **2011**, *15*, 245–267. [CrossRef]

19. Fischer-Kowalski, M.; Weisz, H. The Archipelago of Social Ecology and the Island of the Vienna School. In *Social Ecology. Society-Nature Relations across Time and Space*; Haberl, H., Fischer-Kowalski, M., Krausmann, F., Winiwarter, V., Eds.; Springer International Publishing: Cham, Switzerland, 2016; Volume 5, pp. 3–28.

20. Prebisch, R. *The Economic Development of Latin America and Its Principal Problems*; United Nations Economic Commission for Latin America: New York, NY, USA, 1949.

21. Singer, H.W. The Distribution of Gains between Investing and Borrowing Countries. *Am. Econ. Rev.* **1950**, *40*, 473–485.

22. Harvey, D.I.; Kellard, N.M.; Madsen, J.B.; Wohar, M.E. The Prebisch-Singer Hypothesis: Four Centuries of Evidence. *Rev. Econ. Stat.* **2010**, *92*, 367–377. [CrossRef]

23. Frank, A.G. *Latin America: Underdevelopment or Revolution: Essays on the Development of Underdevelopment and the Immediate Enemy*; Monthly Review Press: New York, NY, USA, 1969.

24. Wallerstein, I.M. *World-Systems Analysis: An Introduction*; Duke University Press: Durham, NC, USA, 2004.

25. Hornborg, A. Towards an ecological theory of unequal exchange: Articulating world system theory and ecological economics. *Ecol. Econ.* **1998**, *25*, 127–136. [CrossRef]

26. Muradian, R.; O'Connor, M.; Martinez-Alier, J. Embodied pollution in trade: Estimating the "environmental load displacement" of industrialised countries. *Ecol. Econ.* **2002**, *41*, 51–67. [CrossRef]

27. Sachs, J.D.; Warner, A.M. *Natural Resource Abundance and Economic Growth*; National Bureau of Economic Research: Cambridge, MA, USA, 1995.

28. Ross, M.L. The Political Economy of the Resource Curse. *World Politics* **1999**, *51*, 297–322. [CrossRef]

29. Karl, T.L. *The Paradox of Plenty: Oil Booms and Petro-States*; University of California Press: Oakland, CA, USA, 1997.

30. Collier, P.; Hoeffler, A. Greed and grievance in civil war. *Oxf. Econ. Pap.* **2004**, *56*, 563–595. [CrossRef]

31. Brunnschweiler, C.N.; Bulte, E.H. Linking Natural Resources to Slow Growth and More Conflict. *Science* **2008**, *320*, 616–617. [CrossRef] [PubMed]

32. Brunnschweiler, C.N.; Bulte, E.H. The resource curse revisited and revised: A tale of paradoxes and red herrings. *J. Environ. Econ. Manag.* **2008**, *55*, 248–264. [CrossRef]

33. World Bank. *Expanding the Measure of Wealth: Indicators of Environmentally Sustainable Development*; The World Bank: Washington, DC, USA, 1997; p. 1.

34. Giljum, S.; Eisenmenger, N. North-South Trade and the Distribution of Environmental Goods and Burdens: A Biophysical Perspective. *J. Environ. Dev.* **2004**, *13*, 73–100. [CrossRef]

35. Pérez-Rincón, M.A. Colombian international trade from a physical perspective: Towards an ecological "Prebisch thesis". *Ecol. Econ.* **2006**, *59*, 519–529. [CrossRef]

36. Moran, D.D.; Lenzen, M.; Kanemoto, K.; Geschke, A. Does ecologically unequal exchange occur? *Ecol. Econ.* **2013**, *89*, 177–186. [CrossRef]

37. Prell, C.; Feng, K.; Sun, L.; Geores, M.; Hubacek, K. The Economic Gains and Environmental Losses of US Consumption: A World-Systems and Input-Output Approach. *Soc. Forces* **2014**, *93*, 405–428. [CrossRef]

38. Weisz, H. Combining Social Metabolism and Input-Output Analyses to Account for Ecologically Unequal Trade. In *Rethinking Environmental History: World-System History and Global Environmental Change*; Hornborg, A., McNeill, J.R., Martinez-Alier, J., Eds.; Rowman Altamira: Lanham, MD, USA, 2007; pp. 289–306.

39. Dorninger, C.; Hornborg, A. Can EEMRIO analyses establish the occurrence of ecologically unequal exchange? *Ecol. Econ.* **2015**, *119*, 414–418. [CrossRef]

40. Hornborg, A.; Martinez-Alier, J. Ecologically unequal exchange and ecological debt. *J. Political Ecol.* **2016**, *23*, 328–333.

41. Dorninger, C.; Eisenmenger, N. South America's biophysical involvement in international trade: The physical trade balances of Argentina, Bolivia, and Brazil in the light of ecologically unequal exchange. *J. Political Ecol.* **2016**, *23*, 394–409.

42. Martinez-Alier, J.; Demaria, F.; Temper, L.; Walter, M. Changing social metabolism and environmental conflicts in India and South America. *J. Political Ecol.* **2016**, *23*, 467–491.

43. Schandl, H.; West, J. Material Flows and Material Productivity in China, Australia, and Japan. *J. Ind. Ecol.* **2012**, *16*, 352–364. [CrossRef]

44. Ekins, P.; Folke, C.; Costanza, R. Trade, environment and development: The issues in perspective. *Ecol. Econ.* **1994**, *9*, 1–12. [CrossRef]

45. Muradian, R.; Martinez-Alier, J. Trade and the environment: From a "Southern"perspective. *Ecol. Econ.* **2001**, *36*, 281–297. [CrossRef]

46. World Bank. *World DataBank. World Development Indicators*; World Bank: Washington, DC, USA, 2016.

47. World Bank. *World Bank GNI per Capita Operational Guidelines & Analytical Classifications*; World Bank: Washington, DC, USA, 2016.

48. Dittrich, M.; Bringezu, S. The physical dimension of international trade: Part 1: Direct global flows between 1962 and 2005. *Ecol. Econ.* **2010**, *69*, 1838–1847. [CrossRef]

49. Giljum, S. Trade, Materials Flows, and Economic Development in the South: The Example of Chile. *J. Ind. Ecol.* **2004**, *8*, 241–261. [CrossRef]

50. Muñoz, P.; Giljum, S.; Roca, J. The raw material equivalents of international trade. *J. Ind. Ecol.* **2009**, *13*, 881–897. [CrossRef]

51. Eurostat. *Economy-Wide Material Flow Accounts (EW-MFA)*; Eurostat: Luxembourg, 2012.

52. Wiedmann, T.; Wilting, H.C.; Lenzen, M.; Lutter, S.; Palm, V. Quo Vadis MRIO? Methodological, data and institutional requirements for multi-region input–output analysis. *Ecol. Econ.* **2011**, *70*, 1937–1945. [CrossRef]

53. Weisz, H.; Duchin, F. Physical and monetary input–output analysis: What makes the difference? *Ecol. Econ.* **2006**, *57*, 534–541. [CrossRef]

54. Hubacek, K.; Giljum, S. Applying physical input–output analysis to estimate land appropriation (ecological footprints) of international trade activities. *Ecol. Econ.* **2003**, *44*, 137–151. [CrossRef]

55. Simoes, A.; Hidalgo, C. *The Economic Complexity Observatory: An Analytical Tool for Understanding the Dynamics of Economic Development*; Massachusetts Institute of Technology: Boston, MA, USA, 2016.

56. Kavoussi, R.M. Export expansion and economic growth. *J. Dev. Econ.* **1984**, *14*, 241–250. [CrossRef]

57. Sheehey, E.J. Exports and growth: A flawed framework. *J. Dev. Stud.* **1990**, *27*, 111–116. [CrossRef]

58. Hall, D. Land grabs, land control, and Southeast Asian crop booms. *J. Peasant Stud.* **2011**, *38*, 837–857. [CrossRef]

59. Brad, A.; Schaffartzik, A.; Pichler, M.; Plank, C. Contested territorialization and biophysical expansion of oil palm plantations in Indonesia. *Geoforum* **2015**, *64*, 100–111. [CrossRef]

60. Ribot, J.C.; Peluso, N.L. A theory of access. *Rural Sociol.* **2003**, *68*, 153–181. [CrossRef]

61. Backhouse, M. Green grabbing—The case of palm oil expansion in so-called degraded areas in the eastern Brazilian Amazon. In *Political Ecology Agrofuels*; Routledge: London, UK, 2014; pp. 167–184.

62. Pichler, M. Legal Dispossession: State Strategies and Selectivities in the Expansion of Indonesian Palm Oil and Agrofuel Production. *Dev. Chang.* **2015**, *64*, 508–533. [CrossRef]

63. Niewöhner, J.; Nielsen, J.Ø.; Gasparri, I.; Gou, Y.; Hauge, M.; Joshi, N.; Schaffartzik, A.; Sejersen, F.; Seto, K.C.; Shughrue, C. Conceptualizing Distal Drivers in Land Use Competition. In *Land Use Competition*; Niewöhner, J., Bruns, A., Hostert, P., Krueger, T., Nielsen, J.Ø., Haberl, H., Lauk, C., Lutz, J., Müller, D., Eds.; Springer International Publishing: Cham, Switzerland, 2016; Volume 6, pp. 21–40.

64. Einzenberger, R. Contested Frontiers: Indigenous Mobilization and Control over Land and Natural Resources in Myanmar's Upland Areas. *Austrian J. South-East Asian Stud.* **2016**, *9*, 163.

65. Kelly, A.B.; Peluso, N.L. Frontiers of Commodification: State Lands and Their Formalization. *Soc. Nat. Res.* **2015**, *28*, 473–495. [CrossRef]

66. Martinez-Alier, J. Social metabolism, ecological distribution conflicts, and languages of valuation. *Capital. Nat. Soc.* **2009**, *20*, 58–87. [CrossRef]

67. Brand, U.; Dietz, K. (Neo-)Extraktivismus als Entwicklungsoption? Zu den aktuellen Dynamiken und Widersprüchen rohstoffbasierter Entwicklung in Lateinamerika. *Polit. Vierteljahresschr.* **2014**, *48*, 133–170.

68. Gellert, P.K. Extractive Regimes: Toward a Better Understanding of Indonesian Development: Extractive Regimes. *Rural Sociol.* **2010**, *75*, 28–57. [CrossRef]

69. NFF to Hold Urgent Talks on Water Crisis. *Syd. Morning Her.* **2007**. Available online: http://www.smh.com.au/national/nff-to-hold-urgent-talks-on-water-crisis-20070419-8kj.html (accessed on 24 May 2017).

70. Gale, M.; Edwards, M.; Wilson, L.; Greig, A. The Boomerang Effect: A Case Study of the Murray-Darling Basin Plan. *Aust. J. Public Adm.* **2014**, *73*, 153–163. [CrossRef]

71. Auty, R.M. Natural resources, capital accumulation and the resource curse. *Ecol. Econ.* **2007**, *61*, 627–634. [CrossRef]

72. Burchardt, H.-J.; Dietz, K. (Neo-)extractivism—A new challenge for development theory from Latin America. *Third World Q.* **2014**, *35*, 468–486. [CrossRef]

73. Staritz, C.; Gereffi, G.; Cattaneo, O. Shifting end markets and upgrading prospects in global value chains. *Int. J. Technol. Learn. Innov. Dev.* **2011**, *44*, 2.
74. Bridge, G. Global production networks and the extractive sector: Governing resource-based development. *J. Econ. Geogr.* **2008**, *8*, 389–419. [CrossRef]
75. Veltmeyer, H. The political economy of natural resource extraction: A new model or extractive imperialism? *Can. J. Dev. Stud. Rev. Can. D'études Dév.* **2013**, *34*, 79–95. [CrossRef]
76. Buur, L.; Monjane, C.M. Elite capture and the development of natural resource linkages in Mozambique. In *Fairness and Justice in Natural Resource Politics*; Pichler, M., Staritz, C., Küblböck, K., Plank, C., Raza, W., Peyré, F.R., Eds.; Routledge: London, UK, 2017; pp. 200–217.
77. Pichler, M.; Staritz, C.; Küblböck, K.; Plank, C.; Raza, W.; Peyré, F.R. *Fairness and Justice in Natural Resource Politics*; Routledge/Taylor & Francis Group: London, UK, 2016.
78. Martinez-Alier, J. *The Environmentalism of the Poor: A Study of Ecological Conflicts and Valuation*; Edward Elgar Publishing: Cheltenham, UK, 2003.
79. Kohl, B.; Farthing, L. Material constraints to popular imaginaries: The extractive economy and resource nationalism in Bolivia. *Political Geogr.* **2012**, *31*, 225–235. [CrossRef]
80. Andreucci, D.; Radhuber, I.M. Limits to "counter-neoliberal" reform: Mining expansion and the marginalisation of post-extractivist forces in Evo Morales's Bolivia. *Geoforum* **2015**, in press. [CrossRef]
81. Pichler, M. What´s democracy got to do with it? A political ecology perspective on socio-ecological justice. In *Fairness and Justice in Natural Resource Politics*; Pichler, M., Staritz, C., Küblböck, K., Plank, C., Raza, W., Peyré, F.R., Eds.; Routledge/Taylor & Francis Group: London, UK, 2016; pp. 33–51.
82. Brand, U.; Dietz, K.; Lang, M. Neo-Extractivism in Latin America-one side of a new phase of global capitalist dynamics. *Cienc. Política* **2016**, *11*, 125–159. [CrossRef]
83. Economist Tata Steel: Cast-Iron Arguments. *Economist* **2016**. Available online: http://www.economist.com/news/leaders/21696527-how-should-governments-cope-global-glut-steel-britain-depressing-case (accessed on 24 May 2017).
84. Economist Global Steel: Through the Mill. *Economist* **2016**. Available online: http://www.economist.com/news/business/21696556-it-hard-see-future-many-worlds-high-cost-steel-producers-britains-are-no (accessed on 24 May 2017).

*sustainability*

MDPI

*Article*

# Surplus, Scarcity and Soil Fertility in Pre-Industrial Austrian Agriculture—The Sustainability Costs of Inequality

Michael Gizicki-Neundlinger * and Dino Güldner

Institute of Social Ecology, Alpen-Adria-University, 1070 Vienna, Austria; dinoleon.gueldner@aau.at
* Correspondence: michael.neundlinger@aau.at; Tel.: +43-1-5224000-401

Academic Editors: Johanna Kramm, Melanie Pichler, Anke Schaffartzik and Martin Zimmermann
Received: 15 November 2016; Accepted: 2 February 2017; Published: 13 February 2017

**Abstract:** This paper takes a Long-Term Socio-Ecological Research (LTSER) perspective to integrate important aspects of social inequality into Socio-Ecological Metabolism (SEM) research. SEM has dealt with biophysical features of pre-industrial agricultural systems from a largely apolitical perspective, neglecting social relations and conditions of peasant production and reproduction. One of the politically and economically most important manorial systems in Early Modern Austria (Grundherrschaft Grafenegg) serves as a case study to reconstruct the unequal distribution of central resources between ruling landlords and subjected peasants. We show that peasant land use systems generated small surpluses only, whereas landlords enjoyed significant economies of scale. Furthermore, we explore what these conditions of landlord surplus and peasant scarcity implied for their respective agro-ecological sustainability. Finally, we argue that within pre-industrial agrarian systems sustainability costs of inequality were severely limiting margins for agricultural intensification and growth of peasant economies.

**Keywords:** long-term socio-ecological research; social inequality; land costs of sustainability; material and nutrient flow accounting; pre-industrial agriculture

## 1. Introduction

In recent years, scholars from different disciplinary backgrounds have studied the sustainability of pre-industrial agricultural systems under the category of "socio-ecological metabolism" (SEM) [1–3]. SEM research starts from the premise that every socio-economic system biophysically reproduces itself via a continuous, socially organized exchange of resources with its natural environment. In today's developing countries—and also in Europe's pre-industrial era—agriculture may be considered one of the core socio-metabolic strategies. Peasants invest labour to make use of land (and other natural resources) and continuously intervene into ecosystem dynamics to warrant steady flows of biomass for societal purposes (i.e., to maintain a certain socio-ecological metabolism). To investigate these metabolic interactions and their changes over time, Long-Term Socio-Ecological Research (LTSER) integrates perspectives from the social and the natural sciences [4,5]. Economic historians, social anthropologists and system ecologists have used indicators and concepts to investigate long-term trends in agrarian resource use at the national [6], regional [7] and local level [8]. Tello et al. [9,10] and Gingrich et al. [11] have intensively debated transitions between different energy regimes (from pre-industrial to fully industrialized agriculture). Tello et al. [12] and Marull et al. [13,14] have used SEM approaches to comprehensively study the evolution of land use changes and effects on local biodiversity. Also, flows of nutrients have been traced through agro-ecosystems to better understand issues of soil fertility [15–18]. Still, a vitally important aspect of (pre-industrial) socio-ecological systems has been widely under-researched. Even though Gonzalez de Molina and Toledo [3] have pointed out that an unequal distribution of materials, energy and nutrients within a specific socio-economic system may cause sustainability problems and lead to overexploitation of the resource base, the socio-economic conditions of peasant production and reproduction have largely been neglected in SEM research. Thus far, only a few authors [19–25] have tried to explicitly address the unequal distribution of resources within pre-industrial agriculture to open their SEM research for political analysis.

Within critical agrarian studies, analysis of social relations and inequalities determining pre-industrial peasant economic activities have a long tradition [26–29]. Some authors even claim that unfavourable property relations on land and labour, the extraction of agrarian surplus by landlord or state authorities and other processes of peasant exploitation and dispossession may be considered the prime determinants of agrarian change and the historical transition from feudalism to capitalism [30–34]. Drawing on Karl Marx' seminal ideas on the dialectics between capitalist agriculture and soil fertility, John Bellamy Foster added important ecological dimensions to the discussion on agrarian change. According to Foster, Marx had already formulated basic premises of the "metabolic rift" between growing industrial centres and their rural hinterlands: the large-scale capitalist agriculture of the 19th century led to a "material estrangement of human beings in capitalist society from the natural conditions of their existence" [35] (p. 383). This delocalization of agriculture and the alienation of the consumers from agricultural production had direct consequences for the "old ecological relations of production. In particular, the nutrient cycling of the old agrarian systems was disrupted" [36] (p. 126). Recently, Wittman [37], and Schneider and McMichael [38] proposed reworking and repairing the historic, capitalist metabolic rift under the auspices of "food sovereignty". Similar ideas of co-benefits between peasant autonomy and agro-ecological sustainability were put forward by several other scholars [39–44]. In the wake of the global land-grabbing processes of past years, critical agrarian studies have studied the disproportionate accumulation of "empty" and "cheap" land in the hands of a few globalized, corporate agro-businesses or national governments at the expenses of peasant communities, mostly in rural hinterlands of the Global South [45–47]. These new dynamics of peasant dispossession and land enclosures not only threaten food security, property relations and income opportunities of indigenous communities, but they also heavily intervene into local land use patterns and agro-ecosystem functions [48]. In a similar vein, scholars have investigated the biophysical burdens of ecologically unequal exchange related to international trade flows in the dominant world system. This concept refers to the externalization of ecological costs of production from the Western core nations to the production systems in the Global South [49–52].

In this paper, we try to integrate questions from critical agrarian studies into SEM research. Following Bernstein [53], we open the socio-ecological reading of our Austrian case study to some important aspects of inequality, explicitly focusing on the unequal distribution of agrarian resources and the extraction of agrarian surplus. Consequently, we present a first approach to use our empirical data on material and nutrient flows to support debates on accumulation, agrarian change and inequality from a biophysical perspective. Of course, the unequal distribution of land, biomass and nutrient resources between different agrarian agents represents a single dimension of social inequality only. Further research would be required to better understand the pivotal role of institutional settings, technological constraints and property relations that helped to establish, shape and maintain social inequality within pre-industrial agriculture. Here, we thus investigate landlord−peasant dialectics in one of the most politically and economically important manorial systems in Early Modern Austria (Grundherrschaft Grafenegg). Empirically, we analyse socio-ecological metabolisms of lords and peasants—in terms of biomass as well as plant nutrient flows—to scrutinize the unequal distribution of land, food and other resources between them and to explore implications of landlord surplus and peasant scarcity for the respective capacities to manage fertility of their soils.

To this end, first, we explain two central conditions of pre-industrial agriculture: the land costs of sustainability and the manorial regime. Second, we briefly introduce our case study region and the historical sources. Third, we reconstruct the unequal distribution of resources between lords and peasants and how it affected the sustainability of the respective agro-ecosystems. Finally, we show

how the maldistribution of socio-ecological capital limited margins for agricultural intensification and growth of the peasant economies. This is what we call the "sustainability costs of inequality".

## 2. Conditions of Pre-Industrial Agriculture in Central Europe

Agrarian sustainability is fundamentally linked to the maintenance soil fertility in the long run [54]. Soil fertility is determined by numerous biological (e.g., diversity of micro-organisms) or physical parameters (e.g., temperature, water and clay content of the soil), but primarily by the availability of the important macro-nutrients nitrogen (N), phosphorous (P), and potassium (K) in the agriculturally used soils [55]. In what follows, we focus on the availability of N exclusively; more data would be needed to reconstruct P and K dynamics in our case study. In fertile soils, the amount of nutrients extracted (via biomass harvest or grazing) or lost (via leaching and erosion) does not exceed the amount of nutrients replenished in the course of an agricultural year or rotation cycle. This holds true not only for pre-industrial agro-ecosystems as in our case study, but also in contemporary organic or industrial agricultural systems. An equilibrium state of nutrients is an important prerequisite to maintain ecosystem productivity and—therefore—a significant determinant of agricultural yields. Consequently, peasants have to ensure that sufficient quantities of nutrients are recovered each year in order to sustain food provision and security [56]. Nutrient replenishment in the soil is determined by various natural conditions, as well as through cultivation and management measures [15]. N deposition via rainfall is an important natural input into soil ecosystems. In Austria, annual precipitation ranges from 550 $L/m^2$ (lowland) to 1500 $L/m^2$ (pre-Alpine), adding up to 40% of the total N replenishment [57]. In regions with lower rainfall, other natural N inputs are of greater quantitative importance. For example, in the province of Lower Austria, non-symbiotic fixation (i.e., fixation of atmospheric N via free-living micro-bacteria) accounted for up to 35% of the annual nutrient inputs (see below).

In addition to these natural nutrient dynamics, peasants actively manage some socio-economic nutrient flows. Under the conditions of pre-industrial agriculture, chemically synthesized fertilizers were not available to replenish nutrients extracted or lost. Also, transportation opportunities were severely restricted, rendering the transfer of nutrients from remote agro-ecosystems relatively costly and inefficient [1]. Thus, for pre-industrial Central European agricultural systems—as well as for most agricultural systems in the Global South today—cultivation of leguminous crops, complex multi-annual rotation systems, irrigation measures or the application of manure may be considered the most important, local nutrient management strategy [58]. In quantitative terms, though, animal manure made up for the most significant nutrient input to the soil systems. Depending on the livestock density and the efficiency of the manure management, 30%–90% of the total extracted nutrients could be replenished [17,59]. One way of efficiently managing nutrients is livestock keeping and feeding. In Catalonia, for instance, the on-field burying of biomass constitutes another major N backflow into the soils [60]; however, this practice was not common in Austria. Livestock is able to convert biomass that is not suitable for human consumption (grass vegetation, bushes, stubble fields) into food products (and draught power) and it concentrates nutrients in a plant-available form—manure, which can be collected and applied on fields. Therefore, livestock keeping may be used to mobilize nutrients throughout the whole agricultural landscape, integrating different types of land uses at the local level and providing for a very important socio-economic nutrient backflow. In Austria—but also in other Central European regions—livestock was used to transfer nutrients from grassland or forests to more intensively used land, such as arable land, kitchen gardens or vineyards.

The necessity to support favourable livestock densities, however, did not come without costs. Sufficient land area was needed to feed animals either directly (from grazing areas) or indirectly with forage crops from arable land. In addition, livestock was regularly fed on cereal crops and their residues (mostly straw or bran). These "land costs of sustainability" may be considered an important condition of pre-industrial agriculture in Central Europe [1] and also in the Mediterranean [61,62]. High land costs may even push agro-ecosystems towards ecological disequilibrium. Feedstuff grown on cropland may directly compete with crops produced for human consumption. Also, nutrient depletion of grassland and forest due to intensive grazing may significantly reduce productivity of these ecosystems. Here, nutrient replenishment is mostly limited by natural processes (e.g., rainfall, fixation, formation of new soil), as peasant cultivation measures would entail relatively high labor requirements. And again, active nutrient replenishment in grassland and forest ecosystems would limit the availability of nutrients that can be transferred to intensively used croplands.

In order to navigate between favourable food production and overexploitation of soil nutrient resources, land costs needed to be actively managed. Evidently, this land management was embedded in distinct socio-economic contexts, unequal power relations and specific technological constraints. In what is Austria today, the "manorial regime" was a prime determinant of peasant livelihoods and land use practices from the early Middle Ages until the Liberal Reforms of 1848. The stark contrast between ruling landlords and subjected peasants was one of the most important structural elements of this historical era—not only in Austria, but all over Europe [63]. Under the conditions of the manorial regime, landlords formally owned the land and leased small, fragmented plots to their subjects ("rustic" lands). Rigorous land tenure excluded most of the population from large parts of the farmland. It should be noted that, within pre-industrial agriculture, the manorial regime represents one of the most important property systems on land. However, past tenurial relations and property rights were situated in specific socio-economic and historically contingent contexts. As Congost [64] has pointed out, many of the manorial rights were far from perfect. Also, dynamic Early Modern land markets may have created opportunities for peasants to increase their farm sizes [65]. Finally, common land resources without any formal property rights were a constitutive—and conflictive—element all over pre-industrial Europe [66]. To illustrate with an example, in one of our sample villages (Kamp), peasants had access to only 60% of the agriculturally productive land, whereas landlords controlled the remaining 40%. On average, small peasant families cultivated farms no larger than one hectare of cropland or small kitchen gardens, lacking the resources to keep more than a few heads of livestock. Also, peasant fields were situated in a complex amalgam of dispersed land tenure relations. On the contrary, landlord fields were geographically integrated to perform uninterrupted and efficient agricultural rotations [23]. In return for the land tenure, peasants were obliged to perform certain manorial services. On the one hand, they had to deliver some fractions of their surplus harvest as tithes or other forms of taxes. On the other hand, they were bound to perform corvée labour within their lord's domestic ("demesne") economy—bringing in the lord's harvest, working in the manorial manufacture, threshing corn, etc. Alternatively, peasants could compensate for their manorial obligations in money.

## 3. Lord and Peasant Agriculture in Grundherrschaft Grafenegg

In the first half of the nineteenth century, Lower Austria—one of the core provinces of the Austro-Hungarian Empire—was characterized by relatively strong manorial regimes [67]. The manor of Grafenegg in the central part of the province serves as our case study. Under the reign of the Breuner family (1730–1848), it became one of the most economically [68] and politically [69] influential manorial systems in Early Modern Austria. The manor comprised two demesne estates—Gut Grafenegg and Gut Neuaigen—which were located within the geographical boundaries of different peasant villages. In the nineteenth century, peasants were regularly embedded in broader, more complex networks of manorial obligations (from other manorial or ecclesial systems or the emerging state authorities). These other socio-economic systems, however, are beyond the scope of our study. Here, we assume that all villages containing Breuner demesne lands formally belonged to the manor of Grafenegg. In the historical sources, those villages were grouped together into two administrative regions—Augegend (Floodplain District) and Waldgegend (Forest District). Augegend data contains information on 9 out of 13 villages; Waldgegend data refers to 17 out of 30 villages. Information on the other villages was not available in the archives. Seven sample villages contained in the two regions were also investigated to represent different socio-economic and agro-ecological conditions of peasant agriculture in the manor, e.g., in the village of Kamp (Ka) we found a strong polarization between many smallholder families—with ploughing fields no larger than one hectare—and major demesne arable lands. In contrast, Untersebarn (U) hosted relatively large peasant holdings, and in Haitzendorf (H), land access was distributed very equally between the different village residents. In Etsdorf (E), we found many economically important peasant vineyards, whereas in Straß (St), almost all of the demesne wine production took place. Table 1 shows the main agrarian features of the two demesnes, the two districts and the seven villages. We can see the pronounced differences between lord and peasant economies—in terms of land use, population structure and livestock composition.

**Table 1.** Agrarian structure of the two demesne estates Gut Grafenegg (GG) and Gut Neuaigen (GN), of the two administrative regions Augegend (AG), Waldgegend (WG) and the seven sample villages of Kamp (Ka), Untersebarn (U), Grafenwörth (G), Haizendorf (H), Etsdorf (E), Sittendorf (Si) and Straß (St) in the first half of the nineteenth century.

| | GG | GN | AG | WG | Ka | U | G | H | E | Si | St |
|---|---|---|---|---|---|---|---|---|---|---|---|
| Area (km²) | 37 | 10 | 44 | 63 | 2.4 | 7.6 | 11.6 | 2.8 | 5.4 | 3.2 | 11.5 |
| Cropland (%) | 15 | 11 | 52 | 46 | 79 | 72 | 43 | 45 | 72 | 72 | 28 |
| Grassland (%) | 1 | 9 | 19 | 2 | 5 | 4 | 24 | 26 | 0 | 5 | 0 |
| Gardens (%) | 0 | 0 | 5 | 6 | 3 | 7 | 4 | 3 | 3 | 3 | 3 |
| Vineyards (%) | - | 4 | 1 | 29 | 6 | 2 | 1 | 1 | 22 | 18 | 40 |
| Forest (%) | 83 | 77 | 23 | 17 | 6 | 15 | 29 | 24 | 3 | 2 | 29 |
| Cereal Yields (kg·DM/ha × year) | 1004 | 1315 | 815 | 960 | 792 | 766 | 719 | 798 | 621 | 827 | 975 |
| Population (cap) | 138 | n. d. | 3124 | 7750 | 342 | 368 | 782 | 291 | 586 | 393 | 1072 |
| Density (cap/km²) | 3 | n. d. | 71 | 123 | 147 | 48 | 68 | 103 | 108 | 122 | 93 |
| No. Farms | 2 | 2 | 467 | 1099 | 42 | 54 | 123 | 41 | 76 | 55 | 158 |
| Farm Size (ha cropland) | 278 | 55 | 3 | 3 | 2 | 7 | 3 | 2 | 3 | 4 | 1 |
| Livestock (LU500) | 39 | 13 | 212 | 350 | 14 | 20 | 52 | 16 | 29 | 19 | 45 |
| Horses (%) | 16 | 6 | 22 | 16 | 14 | 10 | 20 | 15 | 21 | 9 | 10 |
| Bull (%) | — | — | 1 | 1 | — | — | 1 | — | — | 1 | 1 |
| Oxen (%) | 3 | — | 5 | 9 | 18 | 10 | - | 21 | 3 | 4 | — |
| Cows (%) | 7 | — | 54 | 55 | 47 | 57 | 59 | 54 | 53 | 64 | 78 |
| Heifers (%) | — | — | 3 | 1 | — | 3 | 2 | — | — | 2 | — |
| Pigs (%) | 1 | — | 7 | 6 | 10 | 4 | 11 | — | 8 | — | — |
| Sheep (%) | 63 | 91 | 8 | 12 | 11 | 5 | 7 | 9 | 15 | 19 | 11 |
| Poultry (%) | 10 | 3 | 0 | 1 | 1 | 0 | 0 | 0 | 0 | 1 | 1 |
| Density (LU500/km²) | 6 | 15 | 5 | 6 | 6 | 3 | 5 | 6 | 5 | 6 | 4 |

DM = dry matter; LU500 = livestock unit of 500 kg live weight; livestock density refers to manured land; n. d. = no data available; GG data refers to 1836, GN data refers to 1837 and village data refers to 1829–1830.

In terms of land use, the Breuner demesnes occupied almost 50 km², which was larger than the total land area of Augegend and almost as large as Waldgegend. These vast land areas enabled Breuner landlords to enjoy significant economies of scale. They primarily exploited forest resources—not only for timber exports, but also to fuel their brick manufacture. Even though soil conditions were generally favourable for intensive agriculture (chernozem and other humid black soils prevailed in the entire region), the respective share of cropland of the total demesne land was significantly lower compared to village arable land. Strikingly, cereal yields in the two demesne estates were higher than on the peasant fields and above the Austrian average of approximately 920 kg·DM/ha × year [59]. In the entire region, grazing opportunities were severely limited, except for some wet meadows next to the Danube River (in the southern part of the manor), and for a large communal pasture located in the village of Grafenwörth (G). Additionally, some parts of the wooded land provided another opportunity for communal grazing. In the northern hilly zones with favourable podzol loam soils, orchards and—the economically important—vineyards added to the land use portfolio of both lords and peasants.

The large demesne lands only hosted a few permanent residents, who were normally not directly involved in the agricultural cultivation (e.g., the nobility, their stewards, brewers, millers). The main agricultural workforce was recruited from the population of the subject villages, either within corvée or wage-labour relations. As shown in [23], small rustic land plots required less labour input than the large landlord economies. In some of the villages, smallholders accessed so little land that they only needed around half of their available labour time to cultivate it. Most of the population thus worked in service of the Breuner economy, generating some additional family income. Competition on the local labour market was probably high, as labour force was available in abundance. Landlords benefitted from this labour surplus situation, keeping wages at relatively low levels. On both landlord and peasant land, livestock was relatively scarce. Except for the estate Gut Neuaigen, livestock density was far below the Austrian average of 17 LU500/km$^2$ and was also lower than in other cropland regions in the province of Lower Austria, such as in the village of Theyern (about 16 km south of Grundherrschaft Grafenegg) where livestock density reached 24 LU500/km$^2$ [8]. The demesne livestock economy was dominated by large sheep herds reared for wool production for textile manufacturing. Peasants integrated a few larger ruminants (mostly oxen and cows) into their agricultural production system. In contrast to mono-functional sheep, cattle were held to serve multiple purposes. They provided draught power to pull agricultural machinery, produced livestock products (mostly meat and milk) and supplied heavily needed manure in greater quantities than sheep.

## 4. Historical Sources and Their Socio-Metabolic Reading

To efficiently organize their huge farming enterprises, Breuner landlords kept a complex administrative body. During the agricultural year, manorial bookkeepers meticulously compiled quantitative data on earnings and expenses, not only in monetary but also in physical terms. These accounting books comprise all relevant inputs and outputs to and from the manorial farmsteads and cover the most important agricultural activities in the demesne economy: production of cereals and legumes, production of hay and wine, harvest of fodder crops, livestock products, seeds applied, commodities exported, the collection of tithes and other taxes, imported goods such as salt, tools and candles, etc. [23]. Today, we can use the rich information contained in the accounting documents to reconstruct a fine-grained, biophysical picture of agricultural life on the Breuner demesnes. It should be pointed out that data generation from historical sources is a very sensitive, context-driven process. Gizicki-Neundlinger et al. [23], Gingrich et al. [57], and Krausmann [59] provide details on how to process historical data for socio-ecological analyses of 19th-century rural communities in Austria. Also, it is important to keep in mind to be aware of possible data uncertainty and keep a critical view on the sources. Historical sources are potentially biased and selective against their specific socio-economic background: "In the historical archives, accountancy records listing all the inputs and outputs involved in the agrarian cycle are never found" [70] (p. 174). For example, manorial accounting books do not contain any information on land ownership and tenure, on the occupational structure of the economy or on population and household composition. To close data gaps, we used as much site-specific information as possible, such as documents listing every single land parcel under direct Breuner cultivation [71]. For some of the more specific questions on ecological aspects of the demesne land use system, we had to rely on modelling assumptions derived from agro-ecology and socio-metabolic modelling (Güldner et al. [17], Krausmann [59], and Guzmán Casado et al. [72] give detailed descriptions of relevant conversion factors and modelling assumptions). Livestock grazing, for instance, is not reported within the accounting books. We used information on feed requirements of 19th-century farm animals, some quantitative data on feedstuff and local or regional feeding practices to estimate the share of grazed biomass within the animal diet. In other instances, the accounting books report seeds of certain fodder crops that were sown, but do not list any harvest of these crops. Here we considered that the crops were directly consumed by livestock as green fodder (e.g., millet and buckwheat).

To reconstruct peasant economies in the seven villages, we used one of the most important sources for Austrian agrarian and environmental history—the Franciscan cadaster [1,73,74]. From 1817 to 1856, expert commissions roamed the Habsburg Monarchy to undertake a comprehensive land tax survey of the whole country (530,000 km$^2$ in total). Numerous topographic maps and detailed descriptions of the land system were created, providing important information on peasant agricultural life.

The documents contain data on agricultural yields, livestock numbers, the demographic and occupational structure of the village, peasant diets, feeding practices, land tenure and ownership, etc. Overall, data from these records is considered very accurate [75]. Yields assembled within the cadaster refer to long-term averages and are therefore not biased by single-year anomalies [59]. Here, too, we used agro-ecological and socio-metabolic models to reconstruct important missing biomass flows, e.g., grazing, manure availability, etc. To close data gaps and to minimize data uncertainty in our study, we integrated and cross-checked information from the accounting books and the Franciscan cadaster. We found, for example, very high values for hay production for the villages on the alluvial floodplains. In light of the low livestock densities and the figures on landlord hay production, we assumed that the cadastral commission copied data from another region or year, and have corrected figures accordingly (Sandgruber [73] found similar practices in the province of Upper Austria). To clearly delineate landlord and peasant land use systems and to avoid double counting of land, we subtracted demesne plots found in the historical data of Gutsverwaltung Grafenegg [71] from the total village area provided in the cadaster. Yields were corrected correspondingly.

Based on the database established on the grounds of these sources, we derived a set of socio-metabolic indicators to investigate (1) the unequal distribution of crucial agrarian resources within the Breuner manorial regime and (2) the consequences for the sustainability of the associated agro-ecosystems. To compare the production of the lord and peasant economies at the level of the demesne estates and the two different regions, we calculated the respective amount of total biomass extraction per agricultural year (in metric tonnes of dry matter). This indicator comprises all agricultural products related to the production of food (crops, fruits, vegetables and wine), feed (straw, hay and grazed biomass) and wood. To compare the productivities of the landlord and peasant systems we related total biomass extraction to units of land area. At the intermediate level of manorial distribution, we estimated total biomass transferred (1) between lords and peasant land use systems and (2) on to local or regional markets, again at the level of the two estates and regions. This indicator comprises all the surplus extracted via tithes and taxes, backflows from lords to peasants (as rent in kind for specific manorial services) and trade relations—again in metric tonnes of dry matter per year and land unit. To assess the final biomass availability in each of the two systems, we related the biomass transfers to total biomass extracted. For the demesne economies, we added tithes and taxes as well as imports to the extraction, and subtracted rents in kind, exports and seeds. Final biomass availability for the peasant regions was estimated by reversing the calculations.

To elaborate on the aspect of food provision for both lords and peasants, we estimated the supply of plant- and animal-based food in terms of nutritional energy. Knowing the amount of food available for consumption—i.e., after all biomass transfers (and seed output as well as processing losses) have been deducted, expressed in gigajoule nutritional value per capita and year—allows exploring aspects of food provision under the conditions of manorial regimes. Using an average value of 3.4 GJ per capita and year as the minimum amount of energy required to sustain an individual metabolism [76,77], we can estimate nutritional surpluses and deficits in our case study. This is an average figure and probably on the higher end of actual average intake. The minimum metabolic requirement may vary substantially according to age, weight and occupation. To add an important economic dimension, we estimated the monetary value of the potentially marketable food. We assessed the amount of Kronen (Austrian currency of the time) to be gained by selling all available surplus food at the local markets—average contemporary prices for main agricultural products were taken from [78]. We compare these peasant income opportunities with the monetary obligations issued by the Breuner family—i.e., compensations for tithes and labour services found in the historical data of Gutsverwaltung Grafenegg [79,80].

Finally, to assess the complex issues of agrarian sustainability and soil fertility of the local agro-ecosystems, we calculated nitrogen (N) budgets for the two demesne estates, the two peasant regions Augegend and Waldgegend and for the seven sample villages. We reconstructed nutrient balances at the soil-surface scale, i.e., we assessed the total amount of annual natural and socio-economic N inputs and outputs to and from agriculturally used soils [81]. Positive budgets indicate that N is accumulating in the soil, whereas negative nutrient budgets indicate soil N depletion, pushing the land use system towards agro-ecological disequilibrium. Natural inputs (e.g., N deposited via rainfall or non-symbiotic fixation) and outputs (e.g., N losses due to denitrification, leaching and erosion) were quantified following [15,17], albeit translated into site-specific information for the province of Lower Austria. Landlord and peasant management of N fluxes was estimated according to land use information available in historical sources. We had to rely on assumptions from agro-ecology and socio-metabolic models to estimate some of the information required to perform the N balances. For instance, we followed Güldner, Krausmann, and Winiwarter [17] and used information on local feeding practices, the average number of days that animals were kept in stables, species-specific physiology and demographic change of the respective livestock species to estimate the amount of N retained in the livestock. After calculating N retention (and other losses) we arrived at the net amount of N available in manure, which we allocated to different types of land (mostly to cropland). As livestock density was comparably low in our case study region, we assumed that also human excreta were used to fertilize fields, at least next to the peasant houses. Therefore, we included an estimation of N contained in human excreta into our budgets. We modelled N content of human faeces and urine according to [82,83] and cross-checked our findings with rough estimates of N contained in the specific food intake. Here, we assumed large losses due to storage and transportation according to [84], arriving at similar values of net N availability in human excreta as [85,86].

## 5. Lord and Peasant Resource Use and Implications for Sustainability

Table 2 shows total biomass extraction, biomass transferred and final availability of biomass for the two Breuner demesnes Gut Grafenegg and Gut Neuaigen and for the two administrative districts Augegend and Waldgegend within one agricultural year. Comparing the biomass extracted per unit area, we see that the peasants used their resources more intensively than the landlords. In Waldgegend, biomass extraction per unit of land was greater than the demesne extraction by a factor of two and in Augegend by a factor of three. Also in the sample villages, where biomass extraction ranged between 1224 kg DM/ha × year in Sittendorf and 2563 kg DM/ha × year in Grafenwörth, peasant productivity greatly exceeded that of the landlord values. The same holds true if we compare demesne extraction to other case studies in early 19th-century Austria [57,59]. Looking at the composition of biomass extraction, we find that crop (and straw) production was important in the entire region, except for Gut Neuaigen. Viticulture seems unimportant in terms of physical (dry matter) extraction, but in the peasant districts, it provided an important source of monetary income. Grazed biomass represented higher shares of total biomass extraction in Gut Neuaigen—where the large manorial sheep herd was held—and in Waldgegend. Harvesting of hay was negligible in all the systems. Forestry played a dominant role in the demesne economy and in Augegend, where large alluvial floodplains were exploited by the peasant residents. Interestingly, in Waldgegend—the "forest district"—wood extraction was of least importance.

Comparing the amount of biomass transferred per unit area, we find that a relatively higher share of biomass was transferred within the manorial system. Here, between 145 and 290 kg DM/ha × year were mobilized as either inputs (tithes, taxes and imports) to or outputs (rent in kind and exports) from the demesne economy, accounting for approximately 25% and 65% of the annual demesne biomass extraction. In both Gut Grafenegg and Gut Neuaigen, exports for the local or regional markets constituted the most important output. Gut Grafenegg's tithes and Gut Neuaigen's imports of marketable goods accounted for the greatest inputs. Interestingly, the biomass flow from the landlords to the peasants (rent in kind) drastically exceeded their tithing obligations in Gut Neuaigen. Here, many peasant families worked in service of the Breuner economy, receiving agricultural goods in return. On the regional level, tithe and tax obligations made up for only 5% to 6% of the total physical produce. On the village level, manorial surplus extraction reached higher levels, e.g., in Kamp, Sittendorf or Haitzendorf [23]. In all cases, the biomass composition of tithe and rent flows was relatively similar.

**Table 2.** Total biomass extraction, transfers and availability in Gut Grafenegg (GG), Gut Neuaigen (GN), Augegend (AG) and Waldgegend (WG).

|  | GG | GN | AG | WG |
|---|---|---|---|---|
| **Total Biomass Extraction (t·DM/year)** | 1631 | 514 | 8898 | 7513 |
| **Biomass Extraction per land unit kg DM/ha × year** | 440 | 521 | 1704 | 1059 |
| Crops (%) | 16 | 6 | 15 | 23 |
| Fruits & Vegetables (%) | 0.3 | 0.05 | 2 | 1 |
| Wine (%) | 0.1 | — | 0.03 | 2 |
| Straw (%) | 26 | 11 | 25 | 27 |
| Hay (%) | 5 | — | 3 | 1 |
| Grazed Biomass (%) | 22 | 34 | 15 | 31 |
| Wood (%) | 31 | 50 | 40 | 15 |
| **Total Biomass Transferred (t·DM/year)** | 1075 | 143 | 280 | 40 |
| **Biomass Transferred per land unit (kg DM/ha × year)** | 290 | 145 | 54 | 6 |
| Tithe and Taxes (%) | 18 | 1 | 72 | 4 |
| Rent in Kind (%) | 7 | 35 | 28 | 96 |
| Import (%) | 8 | 16 | n. d. | n. d. |
| Export (%) | 67 | 48 | n. d. | n. d. |
| **Total Biomass Available (t·DM/year)** | 1078 | 412 | 8533 | 7230 |
| **Biomass Available per land unit (kg DM/ha × year)** | 291 | 418 | 1634 | 1019 |
| Food Available (GJ·nv/cap × year) | 17 | n. d. | 5.7 | 3.6 |
| Crops (%) | 73 | n. d. | 56 | 47 |
| Fruits & Vegetables (%) | 3 | n. d. | 11 | 3 |
| Wine (%) | 0.3 | n. d. | 0.4 | 11 |
| Livestock Products (%) | 24 | n. d. | 33 | 38 |
| Food needed for local demand (%) | 20 | n. d. | 62 | 97 |
| Marketable Food Potential (Kronen/cap × year) | 244 | n. d. | 60.7 | 2.9 |
| Taxes levied from Breuner (Kronen/cap × year) | — | n. d. | 1.4 | 3.4 |
| Tithes (%) | — | n. d. | 75 | 91 |
| Corvée (%) | — | n. d. | 25 | 9 |
| Balance (Kronen/cap × year) | 244 | n. d. | 59.3 | −0.5 |

DM = dry matter; GJ nv = Gigajoule nutritional value; Kronen = Austrian historical currency; n. d. = no data available; Gut Grafenegg (GG; 1837), Gut Neuaigen (GN; 1838), Augegend (AG; 1830) and Waldgegend (WG; 1829).

Food output per person greatly exceeded local demand in the landlord economy (at least for Gut Grafenegg; unfortunately, no population data was available for Gut Neuaigen). Surplus food was transported downstream the adjacent Danube River towards the capital city of Vienna, where it was profitably sold to feed a growing urban population [87]. In contrast, we find an almost negligible food surplus of 0.2 GJ per capita and year in the densely populated Waldgegend—pushing the peasant population towards the edge of agricultural subsistence—and a higher food surplus of 2.3 GJ per capita and year in the more intensively ploughed Augegend. Augegend peasants may have used their surplus food to participate in the local or regional markets, generating some modest monetary income of ca. 60 Kronen per capita. Sandgruber [88] found similar values for average smallholder families in the first half of the nineteenth century. If we compare the monetary value of the potential food surplus to tithing obligations issued by the Breuner family, we see that tax obligations were comparably small in Augegend. In Waldgegend, however, monetary obligations further aggravated subsistence pressure. Also, we have to bear in mind that under the conditions of the manorial regime, peasants were regularly embedded in a broader, more complex network of tithes and taxes (from other manorial or ecclesial systems or the emerging state authorities). Total tax burdens may have been substantially larger on the individual households than our figures suggest [89].

Therefore, within our case study region, peasant food provision may be considered very precarious (Waldgegend) or modest (Augegend), whereas the demesne economy regularly produced large surpluses of food and other resources (wood, wool, etc.). Under these conditions of relative subsistence pressure and local market incentives, the peasants used their land more intensively than the landlords did—as two indicators on output-related land use intensity suggest: Table 2 shows the estimates of biomass extraction per land unit in mass (kg·DM/ha), Table 3 shows plant nutrient extraction per land unit (kg N/ha). However, the production of more resources on the same land area may have created severe consequences for the agro-ecosystems, as agricultural intensification often comes at the cost of a long-term decrease in soil fertility [3,90]. Table 3 shows the total N budgets for the two demesne estates and the two peasant regions to empirically assess feedbacks between demesne surplus produce, peasant resource scarcity and the respective soil fertility.

**Table 3.** Nitrogen (N) balances for Gut Grafenegg (GG; 1837), Gut Neuaigen (GN; 1838), Augegend (AG; 1830) and Waldgegend (WG; 1829).

|  | GG | GN | AG | WG |
|---|---|---|---|---|
| **N Inputs (kg DM/ha × year)** | 14.3 | 13.7 | 19.2 | 20.5 |
| Rainfall (%) | 27 | 28 | 20 | 19 |
| Non-symbiotic fixation (%) | 33 | 35 | 21 | 18 |
| Symbiotic fixation (%) | 12 | 0.3 | 2 | 0.5 |
| Animal manure (%) | 25 | 35 | 36 | 37 |
| Human excreta (%) | 0 | 0 | 12 | 19 |
| Harvest Residues (%) | 1 | 1 | 3 | 2 |
| Seeds (%) | 2 | 1 | 6 | 5 |
| **N Outputs (kg DM/ha × year)** | 9.1 | 10.3 | 20.9 | 17.6/19.8 * |
| Harvest and grazing (%) | 48 | 45 | 71 | 62 |
| Denitrification (%) | 26 | 27 | 15 | 21 |
| Ammonia volatilization (%) | 25 | 27 | 13 | 18 |
| **Balance (kg DM/ha × year)** | 5.2 | 3.4 | −1.7 | 2.9/0.7 * |

\* Without access to forest grazing.

N balances indicate differences between the fertility of the demesne and rustic land. The two demesne economies and Waldgegend show positive N balances. In Augegend, slightly more N was extracted and lost than replenished during the entire agricultural year. Natural N dynamics (deposition via rainfall, non-symbiotic fixation, denitrification and volatilization) have to be considered when addressing any societal intervention. If we look into the N management practices of lords and peasants, however, we may find reasons for the differences in soil fertility. In our case study, landlords applied some of the decisive agrarian innovations of the time. On Gut Grafenegg, for instance, the cultivation of leguminous crops (beans, lentils, clover) played an integral role in soil fertility. Through symbiosis with rhizobium micro-bacteria, legumes are able to fix atmospheric N in the agricultural soil. Also, landlords integrated new fodder crops into their annual rotations. Former uncultivated fallow fields were now planted with roots and tubers (mostly potatoes and turnips in our case study). As a consequence, fodder availability was significantly raised within the demesne agricultural system. Abundant animal feedstuff—in combination with access to scarce grazing areas—was used to support large sheep herds and some larger ruminants, which supplied meat, milk and wool and considerable quantities of manure at the same time—either as on-field droppings or as manure collected in the stables. With these innovations, the relative ecological sustainability of the demesne agro-ecosystems was ensured and possible margins for agricultural intensification were created.

In the peasant regions, agrarian innovations were adopted more slowly. For example, clover was still very uncommon in the Lower Austrian peasant economies of the 19th century [88]. The low livestock density—compared to the Austrian average of 17 LU500/km$^2$, or to Gut Neuaigen—led to a relative scarcity of animal manure. We can thus assume that human excreta were also applied on intensively used land (orchards, vineyards and to a lesser extent, cropland) to substitute for manure of animal origin. In the 19th century, efficient management and application of human excreta was heavily debated among agronomists of the time [84]. Our reconstructions suggest that the application of human excreta may have considerably contributed to soil fertility of rustic lands—even though we do not find any explicit confirmation of this practice in the historical sources. According to our calculations, human excreta may have accounted for up to 20% of the total annual N replenishment. Still, in Augegend, the negative N balance indicates that N stocks in agricultural soils were slowly diminished. Here, market integration seems to be higher than in the other peasant regions. Accordingly, Augegend provides a vivid example of how metabolic rift processes may lead to N deficits, threatening the agro-ecological stability and food security in peasant land use systems. In Waldgegend, cultivation of common pool resources may be considered an important strategy to alleviate N scarcity. Here, communal forest grazing was the prime determinant of manure availability. But as Breuner landlords exerted strong, exclusive manorial rights to woodland areas, access to this vitally important N source may be considered extremely fragile and open to constant dispute [89]. If we consider that no forest grazing took place in Waldgegend, the total N balance drops to 0.7 kg·N/ha × year.

If we analyse the seven sample villages according to different land use types, we find a very similar picture of declining or stagnant levels of soil fertility for most peasant agro-ecosystems (see Figure 1). We find possible soil mining in four of them, whereas three villages show N balances around zero. In most of the cases, all land use types open to animal grazing (meadows, pastures and forests) and/or fodder production (cropland) show negative N balances. In addition, we can see that common pool resources may only restore soil fertility up to a certain threshold, as the case of Grafenwörth shows. Of the total grassland areas of the region, 30% was located in this village, and it shows a strongly negative N balance. The peasant livestock of all the neighbouring villages were regularly led to graze in Grafenwörth, transferring considerable amounts of N to their small plots of cropland. This drew on the fertility of the grazing areas (−5 kg·N/ha × year on meadows, −0.8 kg·N/ha × year on pasture and −1.1 kg·N/ha × year in the forests). Also, the amount of N removed via fallow grazing contributed to a highly negative balance on the cropland (−5.1 kg·N/ha × year) and village level (−3.6 kg·N/ha × year). The scarcity of fertile agricultural soils drove the peasant population of Grafenwörth towards forestry or the secondary sector, mostly artisanry and proto-manufacture. In Sittendorf, high levels of manorial surplus extraction led to a significantly negative N balance. Only in Haitzendorf and Straß was N successfully transferred from extensively used land (meadows, pasture and forests) towards intensively used land (cropland, vineyards and orchards), allowing for stable N balances of around 0.1 to 0.3 kg·N/ha × year at the village level.

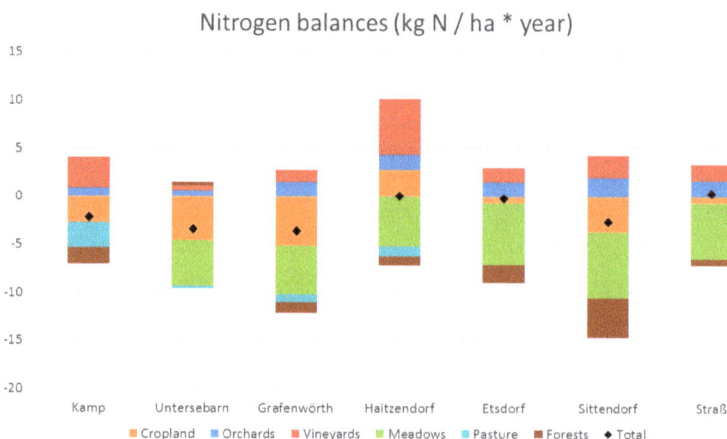

**Figure 1.** Nitrogen balances for the seven sample villages (1829–1830).

## 6. The Sustainability Costs of Inequality

Our empirical findings suggest that under the conditions of the manorial regime—characterized by surplus extraction via tithes and taxes, increased market integration and drastic scarcity of rustic land—a significant fraction of the peasant population was not able to cover the land costs of agrarian sustainability. Peasants could not maintain N transfers from extensively used grazing areas on to their small plots of cropland, as they were excluded from additional land and livestock resources. Rather, the accumulation of agrarian capital within the landlord economies—at the expenses of their subjects—led to a significant loss of sustainability of the peasant land use systems. In his seminal contribution from 1987, Marxist historian Robert Brenner already formulated the basic idea of what we call the sustainability costs of inequality.

> [B]ecause of lack of funds—due to landlords' extraction of rent and the extreme maldistribution of both land and capital, especially livestock—the peasantry was by and large unable to use the land they held in a free and rational manner. They could not, so to speak, put back what they took out of it. Thus the surplus-extraction relations of serfdom tended to lead to the exhaustion of peasant production per se; in particular, the inability to invest in animals for ploughing and as a source of manure led to deterioration of the soil, which in turn led to the extension of cultivation to land formerly reserved for the support of animals. This meant the cultivation of worse soils and at the same time fewer animals—and thus in the end of a vicious cycle of the destruction of the peasants' means of support [32] (p. 33).

This assumption is strongly corroborated by the results for our case study. The maldistribution of both land and capital in pre-industrial agriculture drove the peasant economies towards relatively unsustainable paths. Peasants were not only excluded from land and livestock, but also from vitally important nutrients, strongly limiting peasant capacities to cover the land costs of agricultural production. Small farm sizes—and the lack of grazing opportunities that came with it—led to comparably low agricultural yields within the peasant land use systems, which resulted in a precarious or relatively modest provision of food and also animal feedstuff. Again, the lack of feed resulted in low livestock densities and—therefore—manure scarcity, drastically limiting the amount of nutrients available to replenish annual N extraction and losses—even if we consider that human excrement was regularly applied on intensively used land. In turn, these nutrient deficits severely limited possibilities to intensify agricultural production. Figure 2 illustrates this "vicious cycle" of the sustainability costs of inequality.

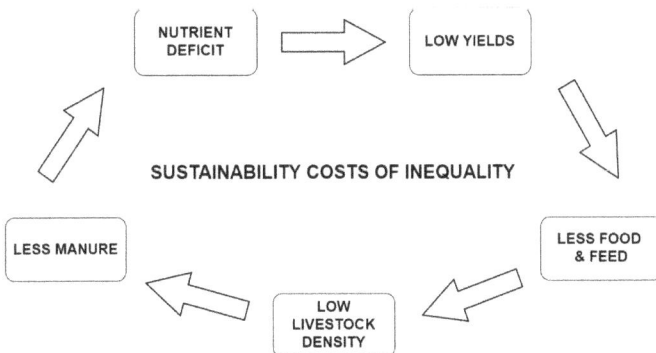

**Figure 2.** The vicious cycle limiting growth of peasant economies.

*Sustainability* **2017**, *9*, 265

Only beneficial N transfers from extensively used land to more intensively used land may have expanded the sustainability frontiers of peasant agriculture, i.e., providing food security while sustaining the agrarian resource base in the long run. In contrast, landlords were endowed with enough socio-ecological capital (land, food, livestock, revenues) and agrarian innovations (e.g., clover, potatoes) to compensate for the land costs of their demesne economies. In doing so, Breuner represents a typical "advanced organic economy" [91] of the 19th century, combining significant, market-oriented economies of scale with sound, sustainable resource management.

## 7. Conclusions

This paper shows how SEM research may benefit from critical questions on the socio-economic conditions of peasant production and reproduction and—alternatively—how debates on accumulation and inequality may benefit from biophysical approaches. In our case study on the Breuner demesne, we found that the huge sustainability costs of inequality drove peasants towards a struggle to meet their subsistence needs and to maintain fertility of their soils at the same time. Our reconstructions suggest that peasants used their resources more intensively (in terms of resource extraction per land unit) than their landlord counterparts. Ester Boserup described an increase in land use intensity as a common strategy to improve nutrient availability and agricultural yields under the conditions of population growth and land precariousness [92]. In the seven sample villages, sufficient labour time was available [23] to intensify land use on the small peasant plots and/or to include some land-saving strategies, e.g., cultivation of crops with higher nutritional values such as maize or potatoes [93]. Still, agricultural intensification seemed to be the exception rather than the norm, as the agro-ecological conditions of soil fertility did not allow for significant agrarian accumulation. Without structural solutions to cover the land costs—e.g., access to sufficient (common pool) grazing areas and the necessary livestock numbers, imports of feed and/or manure—peasants tended to slowly erode their soil resources. A second option to cope with the sustainability costs of inequality was to substitute subsistence crops for cash crops. Badia-Miro and Tello [21] and Parcerisas [22] have shown that, in Catalonia, vineyard specialization helped to alleviate subsistence pressure under pre-industrial conditions. In the seven sample villages, however, opportunities to shift to more intensive wine-growing seemed to be rather limited, as the historical sources indicate that significant amounts of manure were applied to vineyards already [94].

On the contrary, Breuner landlords created possible margins for agricultural intensification and economic growth while maintaining agro-ecological sustainability. In light of the relative fertility of their land, they could increase demesne production and further exacerbate accumulation of agrarian surplus. Therefore, the Breuner case shows how a systemic view at the intersection of SEM research and critical agrarian studies may help to better understand the maldistribution of agrarian capital and social inequalities at the onset of industrial transformation. Also, our LTSER case study shows how land-grabbing dynamics of today may intervene with fragile land use (and nutrient) equilibria of Global South production systems. New land enclosures and other processes of peasant dispossession may not only threaten property rights on land and food security of peasant communities, but also the agro-ecological sustainability of local agriculture.

**Acknowledgments:** We would like to thank Simone Gingrich and Fridolin Krausmann for comments on an earlier version of the manuscript, Enric Tello, Roc Padro and Sylvia Gierlinger for advice on the reconstructions of human excreta, the anonymous reviewers for helpful comments and Niederösterreichisches Landesarchiv and Gutsverwaltung Grafenegg for generously opening their (private) archives to us. This research was supported by the Canadian Social Sciences and Humanities Research Council, Partnership Grant 895-2011-1020.

**Author Contributions:** Michael Gizicki-Neundlinger conceived and designed the paper, performed data analysis and wrote the manuscript. Dino Güldner contributed analysis methods on the calculation of nutrient budgets. Both authors have read and approved the final manuscript.

**Conflicts of Interest:** The authors declare no conflict of interest.

## References

1. Sieferle, R.P. *Das Ende der Fläche: Zum Gesellschaftlichen Stoffwechsel der Industrialisierung*; Böhlau: Köln, Germany, 2006.
2. Arizpe, N.; Ramos-Martín, J.; Giampietro, M. An assessment of the metabolic profile implied by agricultural change in two rural communities in the North of Argentina. *Environ. Dev. Sustain.* **2014**, *16*, 903–924. [CrossRef]
3. De Molina Navarro, M.G.; Toledo, V. *The Social Metabolism: A Socio-Ecological Theory of Historical Change*; Springer: Cham, Switzerland, 2014.
4. Haberl, H.; Winiwarter, V.; Andersson, K.; Ayres, R.U.; Boone, C.; Castillo, A.; Cunfer, G.; Fischer-Kowalski, M.; Freudenburg, W.R.; Furman, E.; et al. From LTER to LTSER: Conceptualizing the Socioeconomic Dimension of Long-term Socioecological Research. *Ecol. Soc.* **2006**, *11*, 13. [CrossRef]
5. Singh, S.J.; Haberl, H.; Chertow, M.; Mirtl, M.; Schmid, M. (Eds.) *Long Term Socio-Ecological Research: Studies in Society: Nature Interactions across Spatial and Temporal Scales*; Springer: Dordrecht, The Netherlands, 2013.
6. Soto, D.; Infante-Amate, J.; Guzmán, G.I.; Cid, A.; Aguilera, E.; García, R.; de Molina, M.G. The social metabolism of biomass in Spain, 1900–2008: From food to feed-oriented changes in the agro-ecosystems. *Ecol. Econ.* **2016**, *128*, 130–138. [CrossRef]
7. Cunfer, G.; Krausmann, F. Adaptation on an Agricultural Frontier: Socio-Ecological Profiles of Great Plains Settlement, 1870–1940. *J. Interdiscip. Hist.* **2015**, *46*, 355–392. [CrossRef]
8. Krausmann, F. From energy source to sink: Transformations of Austrian agriculture. In *Social Ecology: Society-Nature Relations across Time and Space*; Haberl, H., Fischer-Kowalski, M., Krausmann, F., Eds.; Springer: Cham, Switzerland, 2016.
9. Tello, E.; Galán, E.; Cunfer, G.; Guzmán, G.I.; de Molina, M.G.; Krausmann, F.; Gingrich, S.; Sacristan, V.; Marco, I.; Padro, R.; et al. *A Proposal for a Workable Analysis of Energy Return on Investment (EROI) in Agroecosystems: Part I: Analytical Approach*; Social Ecology Working Paper 156; Institute of Social Ecology, Alpen-Adria-Universitaet: Vienna, Austria, 2015.
10. Tello, E.; Galán, E.; Sacristán, V.; Cunfer, G.; Guzmán, G.I.; de Molina, M.G.; Krausmann, F.; Gingrich, S.; Padró, R.; Marco, I.; et al. Opening the black box of energy throughputs in farm systems: A decomposition analysis between the energy returns to external inputs, internal biomass reuses and total inputs consumed (the Vallès County, Catalonia, c. 1860 and 1999). *Ecol. Econ.* **2016**, *121*, 160–174.
11. Gingrich, S.; Marco, I.; Aguilera, E.; Padro, R.; Cattaneo, C.; Cunfer, G.; Guzmán Casado, G.; MacFayden, J.; Watson, A. Agroecosystem energy transitions in the old and new worlds: Trajectories and determinants at the regional scale. *Reg. Environ. Chang.* **2017**, in press.
12. Tello, E.; Valldeperas, N.; Ollés, A.; Marull, J.; Coll, F.; Warde, P.; Wilcox, P.T. Looking Backwards into a Mediterranean Edge Environment: Landscape Changes in El Congost Valley (Catalonia), 1850–2005. *Environ. Hist. Camb.* **2014**, *20*, 347–384. [CrossRef]
13. Marull, J.; Otero, I.; Stefanescu, C.; Tello, E.; Miralles, M.; Coll, F.; Pons, M.; Diana, G.L. Exploring the links between forest transition and landscape changes in the Mediterranean. Does forest recovery really lead to better landscape quality? *Agrofor. Syst.* **2015**, *89*, 705–719. [CrossRef]
14. Marull, J.; Tello, E.; Fullana, N.; Murray, I.; Jover, G.; Font, C.; Coll, F.; Domene, E.; Leoni, V.; Decolli, T. Long-term bio-cultural heritage: Exploring the intermediate disturbance hypothesis in agro-ecological landscapes (Mallorca, c. 1850–2012). *Biodivers. Conserv.* **2015**, *24*, 3217–3251.
15. Garcia-Ruiz, R.; de Molina, M.G.; Guzmán, G.; Soto, D.; Infante-Amate, J. Guidelines for Constructing Nitrogen, Phosphorus, and Potassium Balances in Historical Agricultural Systems. *J. Sustain. Agric.* **2012**, *36*, 650–682. [CrossRef]
16. De Molina, M.G.; Garcia-Ruiz, R.; Soto Fernandez, D.; Guzmán Casado, G.; Cid, A.; Infante-Amate, J. Nutrient Balances and Management of Soil Fertility Prior to the Arrival of Chemical Fertilizers in Andalusia, Southern Spain. *Hum. Ecol. Rev.* **2015**, *21*, 23–48.
17. Güldner, D.; Krausmann, F.; Winiwarter, V. From farm to gun and no way back: Habsburg gunpowder production in the eighteenth century and its impact on agriculture and soil fertility. *Reg. Environ. Chang.* **2016**, *16*, 151–162. [CrossRef]
18. Güldner, D.; Krausmann, F. Nutrient Recycling and Soil Fertility Management in the Course of the Industrial Transition of Traditional, Organic Agriculture: The Case of Bruck Estate, 1787–1906. *Agric. Ecosyst. Environ.* **2017**, in press.

19. Martinez-Alier, J. Ecological Distribution Conflicts and Indicators of Sustainability. *Int. J. Political Econ.* **2004**, *34*, 13–30.

20. Martinez-Alier, J. Social Metabolism, Ecological Distribution Conflicts, and Languages of Valuation. *Cap. Nat. Soc.* **2009**, *20*, 58–87. [CrossRef]

21. Badia-Miro, M.; Tello, E. Vine-growing in Catalonia: The main agricultural change underlying the earliest industrialization in Mediterranean Europe (1720–1939). *Eur. Rev. Econ. Hist.* **2014**, *18*, 203–226. [CrossRef]

22. Parcerisas, L. Landownership Distribution, Socio-Economic Precariousness and Empowerment: The Role of Small Peasants in Maresme County (Catalonia, Spain) from 1850 to the 1950s. *J. Agrar. Chang.* **2014**, *15*, 261–285. [CrossRef]

23. Gizicki-Neundlinger, M.; Gingrich, S.; Güldner, D.; Krausmann, F.; Tello, E. Land, Food and Labour in Pre-Industrial Agro-Ecosystems: A Socio-Ecological Perspective on Early 19th Century Seigneurial Systems. *Historia Agrar.* **2017**, in press.

24. De Molina, M.G.; Amate, J.I.; Molina, A.H.G. Cuestionando los relatos tradicionales: desigualdad, cambio liberal y crecimiento agrario en el Sur peninsular (1752–1901). *Historia Agrar.* **2014**, *63*, 55–88.

25. Garrabou, R.; Planas Maresma, J.; Enric, S.; Vicedo, E. Propiedad de la Tierra y Desigualdad Social en el Mundo Rural Catalán de Mediados del Siglo XIX 2014. Available online: http://repositori.uji.es/xmlui/handle/10234/150308 (accessed on 09 February 2017).

26. Kautsky, K. *Die Agrarfrage: Eine Übersicht über die Tendenzen der Modernen Landwirthschaft und Die Agrarpolitik der Sozialdemokratie*; Dietz: Stuttgart, Germany, 1899.

27. Wolf, E.R. *Peasants*, 14th ed.; Prentice-Hall: Englewood Cliffs, NJ, USA, 1966.

28. Geertz, C. *Agricultural Involution: The Process of Ecological Change in Indonesia*, 5th ed.; University of California Press: Berkeley, CA, USA, 1971.

29. Čajanov, A.V.; Thorner, D. (Eds.) *Chayanov on the Theory of Peasant Economy*, 1st ed.; University of Wisconsin Press: Madison, WI, USA, 1986.

30. Hussain, A.; Tribe, K.; Tribe, K.P. *Marxism and the Agrarian Question*, 2nd ed.; Macmillan: Basingstoke, UK, 1983.

31. Aston, T.H.; Philpin, C.H.E. (Eds.) *The Brenner Debate: Agrarian Class Structure and Economic Development in Pre-Industrial Europe*, 1st ed.; Cambridge University Press: Cambridge, UK, 1987.

32. Brenner, R. Agrarian Class Structure and Economic Development in Pre-Industrial Europe. In *The Brenner Debate: Agrarian Class Structure and Economic Development in Pre-Industrial Europe*, 1st ed.; Aston, T.H., Philpin, C.H.E., Eds.; Cambridge University Press: Cambridge, UK, 1987.

33. Allen, R.C. *Enclosure and the Yeoman*; Clarendon Press: England, UK; Oxford University Press: Oxford, UK; New York, NY, USA, 1992.

34. Byres, T.J. *Capitalism from above and Capitalism from Below: An Essay in Comparative Political Economy*; Macmillan: Basingstoke, UK, 1996.

35. Foster, J.B. Marx's Theory of Metabolic Rift: Classical Foundations for Environmental Sociology 1. *Am. J. Soc.* **1999**, *105*, 366–405. [CrossRef]

36. Moore, J.W. Environmental Crises and the Metabolic Rift in World-Historical Perspective. *Organ. Environ.* **2000**, *13*, 123–157. [CrossRef]

37. Wittman, H. Reworking the metabolic rift: La Vía Campesina, agrarian citizenship, and food sovereignty. *J. Peasant Stud.* **2009**, *36*, 805–826. [CrossRef]

38. Schneider, M.; McMichael, P. Deepening, and repairing, the metabolic rift. *J. Peasant Stud.* **2010**, *37*, 461–484. [CrossRef] [PubMed]

39. Friedmann, H. From Colonialism to Green Capitalism: Social Movements and Emergence of Food Regimes. In *New Directions in the Sociology of Global Development*; Emerald (MCB UP): Bingley, UK, 2006.

40. Patel, R. Food sovereignty. *J. Peasant Stud.* **2009**, *36*, 663–706. [CrossRef]

41. Martinez-Alier, J. The EROI of agriculture and its use by the Via Campesina. *J. Peasant Stud.* **2011**, *38*, 145–160. [CrossRef]

42. Altieri, M.A.; Funes-Monzote, F.R.; Petersen, P. Agroecologically efficient agricultural systems for smallholder farmers: Contributions to food sovereignty. *Agron. Sustain. Dev.* **2012**, *32*, 1–13. [CrossRef]

43. Altieri, M.A. Agroecology, Small Farms, and Food Sovereignty. *Monthly Review*, 2009, 61. Available online: http://monthlyreview.org/2009/07/01/agroecology-small-farms-and-food-sovereignty (accessed on 9 February 2017).

44. Gliessman, S.R. *Agroecology: The Ecology of Sustainable Food Systems*, 3rd ed.; CRC Press: Hoboken, NJ, USA, 2015.
45. Borras, S.M.; Hall, R.; Scoones, I.; White, B.; Wolford, W. Towards a better understanding of global land grabbing: An editorial introduction. *J. Peasant Stud.* **2011**, *38*, 209–216. [CrossRef]
46. McMichael, P. The land grab and corporate food regime restructuring. *J. Peasant Stud.* **2012**, *39*, 681–701. [CrossRef]
47. Margulis, M.E.; McKeon, N.; Borras, S.M. Land Grabbing and Global Governance: Critical Perspectives. *Globalizations* **2013**, *10*, 1–23. [CrossRef]
48. Borras, S.M.; Franco, J.C. Global Land Grabbing and Trajectories of Agrarian Change: A Preliminary Analysis. *J. Agrar. Chang.* **2012**, *12*, 34–59. [CrossRef]
49. Hornborg, A. Towards an ecological theory of unequal exchange: Articulating world system theory and ecological economics. *Ecol. Econ.* **1998**, *25*, 127–136. [CrossRef]
50. Hornborg, A. Zero-Sum World: Challenges in Conceptualizing Environmental Load Displacement and Ecologically Unequal Exchange in the World-System. *Int. J. Comp. Soc.* **2009**, *50*, 237–262. [CrossRef]
51. Muradian, R.; Martinez-Alier, J. Trade and the environment: From a 'Southern' perspective. *Ecol. Econ.* **2001**, *36*, 281–297. [CrossRef]
52. Singh, S.J.; Eisenmenger, N. How Unequal is International Trade? An Ecological Perspective Using Material Flow Accounting (MFA). *J. Entwickl.* **2010**, *26*, 57–88.
53. Bernstein, H. *Class Dynamics of Agrarian Change*; Fernwood Publication: Halifax, UK; Kumarian Press: Sterling, VA, USA, 2010.
54. McNeill, J.R. *Soils and Societies: Perspectives from Environmental History*, 2nd ed.; White Horse Press: Knapwell, UK, 2010.
55. Davis, J.; Abbott, L. Soil fertility in organic farming systems. In *Organic Agriculture: A Global Perspective*; Reganold, J.P., Taji, A.M., Kristiansen, P., Eds.; CSIRO Publishing: Clayton, Austraila, 2006.
56. Hester, R.E.; Harrison, R.M. (Eds.) *Soils and Food Security*; The Royal Society for Chemistry Publishing: Cambridge, UK, 2012.
57. Gingrich, S.; Haidvogl, G.; Krausmann, F.; Preis, S.; Garcia-Ruiz, R. Providing Food While Sustaining Soil Fertility in Two Pre-industrial Alpine Agroecosystems. *Hum. Ecol.* **2015**, *43*, 395–410. [CrossRef]
58. Mazoyer, M.; Roudart, L. *A History of World Agriculture: From the Neolithic Age to the Current Crisis*; Monthly Review Press: New York, NY, USA, 2006.
59. Krausmann, F. Milk, Manure, and Muscle Power. Livestock and the Transformation of Preindustrial Agriculture in Central Europe. *Hum. Ecol.* **2004**, *32*, 735–772.
60. Olarieta, J.R.; Padrò, R.; Masip, G.; Rodríguez-Ochoa, R.; Tello, E. 'Formiguers', a historical system of soil fertilization (and biochar production?). *Agric. Ecosyst. Environ.* **2011**, *140*, 27–33. [CrossRef]
61. Guzmán Casado, G.; de Molina, M.G. Preindustrial agriculture versus organic agriculture. *Land Use Policy* **2009**, *26*, 502–510. [CrossRef]
62. Guzmán, G.I.; de Molina, M.G.; Alonso, A.M. The land cost of agrarian sustainability. An assessment. *Land Use Policy* **2011**, *28*, 825–835. [CrossRef]
63. Cerman, M. Agrardualismus in Europa? Geschichtsschreibung über Gutsherrschaft und ländliche Gesellschaft in Mittel- und Osteuropa. *Jahrb. Geschichte Ländlichen Raumes* **2004**, *1*, 12–29.
64. Congost, R. Property Rights and Historical Analysis: What Rights? What History? *Past Present* **2003**, *181*, 70–106. [CrossRef]
65. Béaur, G. (Ed.) *Property Rights, Land Markets and Economic Growth in the European Countryside: (Thirteenth-Twentieth Centuries)*; Brepols: Turnhout, Belgium, 2013.
66. Soto Fernández, D. Community, institutions and environment in conflicts over commons in Galicia, Northwest Spain (18th–20th centuries). *Workers World* **2014**, *1*, 58–74.
67. Gutkas, K. Die Probleme der Landwirtschaft zur Zeit Maria Theresias. In *Die Auswirkungen der Theresianisch-Josephinischen Reformen auf die Landwirtschaft und Die Ländliche Sozialstruktur Niederösterreichs: Vorträge und Diskussionen des ersten Symposiums des Niederösterreichischen Institutes für Landeskunde, Geras, 9–11 Oktober 1980*; Feigl, H., Ed.; Selbstverl. d. NÖ Inst. für Landeskunde: Wien, Austria, 1982.
68. Berthold, W. Die Einkommensstruktur der adeligen Herrschaften um die Mitte des 18. Jahrhunderts. In *Nutzen, Renten, Erträge: Struktur und Entwicklung Frühneuzeitlicher Feudaleinkommen in Niederösterreich*; Knittler, H., Ed.; Verlag für Geschichte und Politik: Wien, Austria, 1989.

69. Bruckmüller, E. Die Anfänge der Landwirtschaftsgesellschaften und die Wirkung ihrer Tätigkeiten. In *Die Auswirkungen der Theresianisch-Josephinischen Reformen auf die Landwirtschaft und die Ländliche Sozialstruktur Niederösterreichs: Vorträge und Diskussionen des Ersten Symposiums des Niederösterreichischen Institutes für Landeskunde, Geras, 9–11 Oktober 1980*; Feigl, H., Ed.; Selbstverl. d. NÖ Inst. für Landeskunde: Wien, Austria, 1982.

70. Planas, J.; Saguer, E. Accounting records of large rural estates and the dynamics of agriculture in Catalonia (Spain), 1850–1950. *Account. Bus. Financ. Hist.* **2005**, *15*, 171–185. [CrossRef]

71. Gutsverwaltung Grafenegg. *Ausweis Uiber die zur Herrschaft Grafengg Gehörigen Aecker, Gaerten, Weingaerten und Hutwaiden mit Bemerkung der Zehent-Verhältnisse*; Austrian State Archives: Grafenegg, Austria, 1846.

72. Guzmán Casado, G.; Aguilera, E.; Soto Fernández, D.; Cid, A.; Infante Amate, J.; García Ruiz, R.; Herrera, A.; Villa, I.; de Molina, M.G. Methodology and Conversion Factors to Estimate the Net Primary Productivity of Historical and Contemporary Agroecosystems; Sociedad Española de Historia Agraria. 2014. Available online: http://repositori.uji.es/xmlui/bitstream/10234/91670/3/DT-SEHA%201407.pdf (accessed on 9 February 2017).

73. Sandgruber, R. Der Franziszeische Kataster als Quelle für die Wirtschaftsgeschichte und historische Volkskunde. *Mitteilungen des Niederösterreichischen Landesarchivs* **1979**, *3*, 16–28.

74. Marquart, E. *Grundlagen für eine umwelthistorische Bearbeitung des Franziszeischen Katasters*; Diplomarbeit: Wien, Austria, 2006.

75. Bauer, M. *Agrarsysteme in Niederösterreich im frühen 19. Jahrhundert: Eine Analyse auf Basis der Schätzungsoperate des Franziszeischen Katasters*; Rural History Working Papers 20; Institut für Geschichte des ländlichen Raumes (IGLR): St. Pölten, Austria, 2014.

76. Freudenberger, H. Human Energy and Work in a European Village. *Anthropol. Anz.* **1998**, *56*, 239–249. [PubMed]

77. Smil, V. *Feeding the World: A Challenge for the Twenty-First Century*; MIT Press: Cambridge, UK, 2001.

78. Mühlpeck, V.; Sandgruber, R.; Woitek, H. *Index der Verbraucherpreise 1800–1914: Eine Rückberechnung für Wien und den Gebietsstand des Heutigen Österreichs*; Österreichische Staatsdruckerei: Vienna, Austria, 1979.

79. Gutsverwaltung Grafenegg. *Verzeichnis Sämmtlicher Grundstücke Welche der Obbenannten Zehent-Herrschaft Zum Grund—Feld Zehent Verpflichtet Sind*; und Nachweisung des Umfanges und Geldwerthes dieser Verpflichtungen; Austrian State Archives: Grafenegg, Austria, 1844.

80. Gutsverwaltung Grafenegg. *Roboth-Register*; Austrian State Archives: Grafenegg, Austria, 1845.

81. Oenema, O.; Kros, H.; Vries, W. Approaches and uncertainties in nutrient budgets: Implications for nutrient management and environmental policies. *Eur. J. Agron.* **2003**, *20*, 3–16. [CrossRef]

82. Gootas, H.B. *Composting: Sanitary Disposal and Reclamation of Organic Wastes*; World Health Organization (WHO): Geneva, Switzerland, 1956.

83. Håkan Jönsson, A. *Guidelines on the Use of Urine and Faeces in Crop Production*; Stockholm Environment Institute: Stockholm, Sweden, 2004.

84. Gierlinger, S.; Haidvogl, G.; Gingrich, S.; Krausmann, F. Feeding and cleaning the city: The role of the urban waterscape in provision and disposal in Vienna during the industrial transformation. *Water Hist.* **2013**, *5*, 219–239. [CrossRef]

85. Barles, S. Feeding the city: Food consumption and flow of nitrogen, Paris, 1801–1914. *Sci. Total Environ.* **2007**, *375*, 48–58. [CrossRef] [PubMed]

86. Billen, G.; Barles, S.; Garnier, J.; Rouillard, J.; Benoit, P. The food-print of Paris: Long-term reconstruction of the nitrogen flows imported into the city from its rural hinterland. *Reg. Environ. Chang.* **2009**, *9*, 13–24. [CrossRef]

87. Krausmann, F. A city and its hinterland: Vienna's energy metabolism 1800–2006. In *Long Term Socio-Ecological Research: Studies in Society: Nature Interactions across Spatial and Temporal Scales*; Singh, S.J., Haberl, H., Chertow, M., Mirtl, M., Schmid, M., Eds.; Springer: Dordrecht, The Netherlands, 2013.

88. Sandgruber, R. *Die Anfänge der Konsumgesellschaft: Konsumgüterverbrauch, Lebensstandard und Alltagskultur in Österreich im 18 und 19 Jahrhundert*; Verlag für Geschichte und Politik: Vienna, Austria, 1982.

89. Feigl, H. *Die Niederösterreichische Grundherrschaft vom Ausgehenden Mittelalter bis zu den Theresianisch-Josephinischen Reformen*, 2nd ed.; Verein für Landeskunde von Niederösterreich: St. Pölten, Austria, 1998.

90. Blaikie, P.M.; Brookfield, H.C. *Land Degradation and Society*; Methuen: London, UK; New York, NY, USA, 1987.

91.  Wrigley, E.A. *Continuity, Chance and Change: The Character of the Industrial Revolution in England*; Cambridge University Press: Cambridge, UK, 1993.

92.  Boserup, E.; Chambers, R. *The Conditions of Agricultural Growth: The Economics of Agrarian Change under Population Pressure*, 1st ed.; Earthscan Publications: London, UK, 1965.

93.  Kander, A.; Malanima, P.; Warde, P. *Power to the People: Energy in Europe over the Last Five Centuries*; Princeton University Press: Princeton, NJ, USA, 2013.

94.  Franziszeischer Kataster. *Schätzungsoperat der Gemeinde Kamp*; Niederösterreichisches Landesarchiv: St. Pölten, Austria, 1830.

*sustainability*

MDPI

*Commentary*

# The Material Stock–Flow–Service Nexus: A New Approach for Tackling the Decoupling Conundrum

Helmut Haberl *, Dominik Wiedenhofer, Karl-Heinz Erb, Christoph Görg and Fridolin Krausmann

Institute of Social Ecology, Alpen-Adria Universitaet Klagenfurt, Wien, Graz, Schottenfeldgasse 29, Vienna 1070, Austria; dominik.wiedenhofer@aau.at (D.W.); karlheinz.erb@aau.at (K.-H.E.); christoph.goerg@aau.at (C.G.); fridolin.krausmann@aau.at (F.K.)
* Correspondence: helmut.haberl@aau.at; Tel.: +43-1-5224000-406

Received: 6 February 2017; Accepted: 10 May 2017; Published: 26 June 2017

**Abstract:** Fundamental changes in the societal use of biophysical resources are required for a sustainability transformation. Current socioeconomic metabolism research traces flows of energy, materials or substances to capture resource use: input of raw materials or energy, their fate in production and consumption, and the discharge of wastes and emissions. This approach has yielded important insights into eco-efficiency and long-term drivers of resource use. But socio-metabolic research has not yet fully incorporated material stocks or their services, hence not completely exploiting the analytic power of the metabolism concept. This commentary argues for a material stock–flow–service nexus approach focused on the analysis of interrelations between material and energy flows, socioeconomic material stocks ("in-use stocks of materials") and the services provided by specific stock/flow combinations. Analyzing the interrelations between stocks, flows and services will allow researchers to develop highly innovative indicators of eco-efficiency and open new research directions that will help to better understand biophysical foundations of transformations towards sustainability.

**Keywords:** socioeconomic metabolism; economy-wide material flow analysis; in-use stocks of materials; stock-flow relations; dynamic stock model; decoupling; socioecological transformation

---

## 1. Introduction

Human impacts on the earth are escalating, as evidenced by the ongoing debates on global environmental change [1], proposals to introduce the Anthropocene as a new geological epoch [2], calls for sustainability science [3] and planetary stewardship [4,5]. Key environmental impacts have accelerated significantly in the last century [6]. Seventeen Sustainable Development Goals (SDGs) have been agreed at the UN General Assembly in September 2015, including goals such as ending poverty and hunger by 2030 and combating climate change. The UNFCCC COP21 agreement adopted in Paris in December 2015 aims at limiting global warming to 2 °C or less. The challenge is to achieve key social and economic goals such as the provision of high-quality education and good sanitation for all, the global reduction of infant mortality or the eradication of poverty and hunger, while keeping humanity's use of natural resources and wastes/emissions within earth's safe operating space [7,8].

The concept of socioeconomic metabolism has become a cornerstone of sustainability science [9,10]. It has been used in international assessment reports [11] and can help improving integrated assessment models [12]. Many sustainability problems are related to the mass of resources extracted (e.g., depletion of non-renewable resources, effects of renewable resources) or wastes/emissions discharged per unit of time, thus motivating analyses of yearly flows of materials or energy. This focus allowed researchers to analyze links between human agency, institutions, policy, economic prosperity, trade, development or social conflicts on the one hand with environmental issues like climate change, wastes and emissions, pressures on biodiversity or deterioration of ecosystems on the other hand [13]. It also underpinned analyses of relations between economic activities and resource use. One example are analyses of environmental impacts as a function of population, affluence and technology [14,15], where affluence is often measured as Gross Domestic Product (GDP) and impacts are assessed using a wealth of different environmental indicators, including material and energy flows.

The concept of eco-efficiency, i.e., the amount of resources used or pollutants respectively greenhouse gases (GHG) emitted per unit of GDP, became central to strategies aimed at progressing toward sustainability, from the micro-level (e.g., enterprises, households) up to the macro-level (national economies) as well across spatial scales, from local to global [16]. However, improvements of eco-efficiency have so far not reduced global resource use, as they were overcompensated by economic growth and rebound effects [17]. Growing skepticism regarding the possibility of reducing resource use adequately through eco-efficiency [18] motivates discussions on transformations towards sustainability [19–21], sufficiency [22] and de-growth [23]. Although comprehensive analyses of transformation processes at the global level are missing [24], many experts agree that moving towards sustainability will require fundamental changes in resource use patterns [25].

In recent years, the crucial role of in-use stocks of infrastructures and buildings for resource use patterns and as the biophysical spatial structures of society have increasingly come into focus [26–28]. Early pioneering work already highlighted the importance of the specific services demanded by society, for example m$^2$ of living space, as the driver of building stock accumulation and subsequent material flows [29,30]. The majority of that research was focused on specific substances or materials, especially metals [31], and increasingly construction minerals [32]. Economy-wide material stock accumulation across all resource flows and consistent with harmonized methods of economy-wide material flow accounting have recently been published [26,33,34]. This line of research is a prerequisite for a more systematic and comprehensive approach to investigating stock-flow relationships because it aims to cover all resource flows and subsequent material stock dynamics. Recent research then started to investigate the socio-economic drivers [35] and comparatively the patterns of stock accumulation [34], the state of global economy-wide circularity [36], and more comprehensive perspectives on the resource efficiency of economy-wide stocks and flows [26].

We here argue that major progress could be achieved by systematically complementing the current flow-centered approach of economy-wide socio-metabolic research with two key elements, namely (a) in-use stocks of materials (short: material stocks) and (b) services provided by these stocks and flows. Similar concepts have been introduced in the "dynamic material flow analysis concept" proposed by Müller et al. [29], but not based on the economy-wide material flow analysis concept in focus in this article. Systematic investigation of these two elements will allow analyzing critical interrelations between the flows used to build-up, maintain and use these stocks and the services they provide (Figure 1).

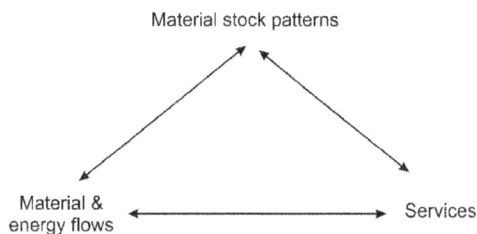

**Figure 1.** The material stock–flow–service nexus approach could provide new orientations for Social Ecology, Industrial Ecology and Ecological Economics. Source: own figure, see text for explanation.

Analyzing the stock–flow–service nexus shown in Figure 1 will provide key pieces of information missing in current definitions of eco-efficiency. Such analyses will enable researchers to undertake comprehensive and powerful analyses of options to decouple societal wellbeing from resource requirements:

1.  Resource flows do not suffice to provide services (e.g., shelter, mobility, communication), i.e., creating benefits for societal wellbeing—they can only do so in combination with material stocks such as machinery, buildings or infrastructures [28]. The location as well as functional types of material stocks and their specific qualities determine both the resource requirements related to the provision of key services, e.g., transport, and spatio-temporal characteristics of their provision, as well as disposal in terms of end-of-life wastes, or their availability for recycling; the latter being crucial knowledge for closing material loops [27,36].

2.  Approximating "affluence" or even societal wellbeing with measures of economic activity has been criticized. Many scholars believe that indicators such as GDP may be a part of the problem; see the "beyond GDP" [37,38] and "degrowth" [23] debates. Service indicators can provide complementary insights that will help forging alternative or at least complementary concepts of socio-economic wellbeing whose relations to material stocks and flows may be at least as important as efficiency measures such as materials or energy used per unit of GDP. We think that a shift from mainstream economics towards a service-based approach can help determining the societal needs underlying certain flows and the stock–flow–service linkages due to its ability to compare different options to address the same services and the societal needs behind them.

3.  Many individual and collective decisions concern purchases of long-lived goods and investments into buildings, infrastructures and machinery, i.e., into building up material stocks. However, this also results in legacies that narrow future option spaces.

This commentary outlines the stock–flow–service nexus approach and discusses how it could help advancing scientific understanding of social metabolism. We first clarify the difference of stocks and flows in economy-wide socio-metabolic research and review the current state-of-the art in that area (Section 2). Then we discuss how the stock–flow–service nexus approach could enrich studies of eco-efficiency and outline its potential contributions to sustainability transformation research (Section 3), followed by short conclusions.

## 2. Stock-Flow Relations in Socioeconomic Metabolism Research

### 2.1. The Concept of Socioeconomic Metabolism

The notion of socioeconomic metabolism transfers the biological concept of metabolism, i.e., the material and energy in-and outflows of organisms and the biochemical processes providing them with energy, maintaining their biophysical structures, reproduction and functioning, to human society [39–41]. One major strand of socioeconomic metabolism research is focused on analyzing flows of materials [42] and energy [43] following a top-down research strategy ("economy-wide material and energy flow analysis", abbreviated EW-MEFA) using data from statistical offices. EW-MEFA has started with nation states (national economies) as basic unit, hence its name, but has been extended to lower levels such as provinces, cities and villages or households, as well as higher levels such as groups of countries or world regions [44,45]. It complements other approaches by comprehensively depicting the material and energy throughput of a defined socioeconomic system. By contrast, substance flow analysis zooms in on specific substances (e.g., metals or chemical elements [46]), whereas life cycle assessment (LCA) is focused on the specific supply chains and uses of products [47].

The study of material and energy flows associated with socioeconomic activities can be traced back into the 19th century [40,41]. Concepts of material flow analysis were pioneered in the 1970s [48,49]. Meanwhile, material flows are reported by statistical offices and international bodies like UNEP, using standardized EW-MEFA methods [42]. Systematically appraising the uncertainties of MEFA data is still a major research topic, however [42,50,51]. Energy is a key resource of all biophysical systems [52–54]. It drives all biophysical processes, including the biogeochemical cycles of water or carbon as well as socioeconomic material flows. The study of socioeconomic energy flows has a venerable tradition [55–58] and plays a key role in the current sustainability discourse [59]. Studying the energetic metabolism of society in a manner that is compatible with material flow analysis is key for linking social and natural sciences in interdisciplinary studies [43,60,61]. Many aspects of socioeconomic metabolism have recently received much attention, as summarized in Table 1.

**Table 1.** Examples for socio-metabolic research.

| Empirical Content Generated | Questions Addressed | References |
|---|---|---|
| Analyses of long-term changes of material and energy flows on national and other levels | Changes in supply and demand of biophysical resources and related sustainability problems | [62–67] |
| Assessment of material and energy flows in different regions | Cross-country comparison of resource demand related to production and consumption | [17,64,68–70] |
| Establishment of physical trade data bases, including upstream requirements | Analysis of unequal exchange and problem shifting between regions | [71–73] |
| Creation of databases of flows of specific chemical elements such as metals, carbon or plant nutrients | Analysis of specific problems such as eutrophication, toxic emissions, scarcity or climate change | [46,74–78] |
| Indicators relating material or energy flows and land use, e.g., the ecological footprint or the human appropriation of net primary production | Evaluation of the role of land as a resource; Analysis of land-use competition; Data basis for analysis of land grabbing; Clarification of the role of land as limited resource | [79–82] |
| Investigation of specific material cycles through socio-ecological systems, mainly for stocks and flows of metals | Materials management potentials in material cycles and potential future secondary resources and wastes | [31,83–87] |
| Linking cycles of material stocks and flows through society to energy use and GHG implications | Investigation of stock accumulation patterns during economic development and subsequent GHG implications due to life cycle emissions | [26,28,88–90] |
| Quantification of stocks and flows related with building dynamics and construction minerals at various scales | Recycling potentials in the building sector, resource demands for expansion | [32,91–94] |
| Dynamic modeling of stocks for scenarios of stocks and flows | Forecasting of potentials for improved loop closing due to growing secondary resource supply from obsolete stocks | [29,75,86,89,91,95] |
| Spatially explicit databases of in-use stocks for various scales | Spatial optimization of waste management strategies, investigations of urbanization dynamics | [27,96–99] |

## 2.2. Stock-Flow Relations in Socioeconomic Metabolism Research

A stock is a variable measured at a specific point in time, whereas a flow is a variable measured over a period, i.e., per unit of time. Stocks and flows are incommensurable, but their relations (stock/flow and its inverse, flow/stock) are meaningful and can be interpreted as mean residence time [year] respectively turnover rate [1/year] of resources. In material flow analysis, a stock is measured as mass [kg], while a flow is mass per unit of time [kg/year].

Early conceptual statements already stressed the importance of material stocks in defining and analyzing society's metabolism [100–102], Figure 2. However, the systematic analysis of material stocks and stock-flow relations has only recently begun (see review below). Usually denoted as "in-use stocks of materials" (abbreviated here as "material stocks"), material stocks may be defined as ordered and interrelated biophysical entities created and reproduced by the continuous flows of energy and materials that constitute society's metabolism. According to conventions of socioeconomic metabolism studies, material stocks comprise humans, livestock and artefacts (i.e., manufactured capital *sensu* [103]).

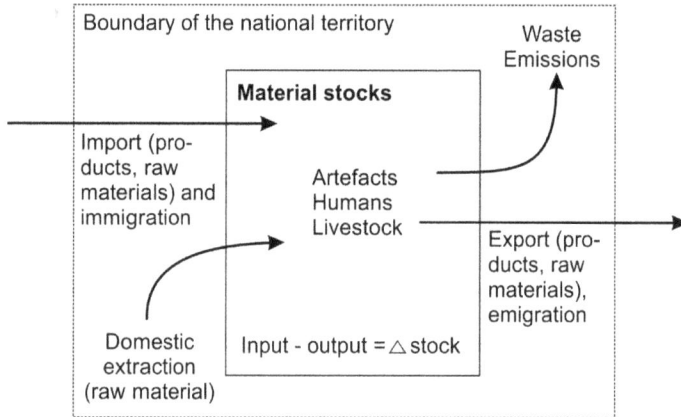

**Figure 2.** Stock-flow relations in socioeconomic metabolism. Source: own graph, modified after [104].

Material stocks represent the physical infrastructure for production and consumption, and hence the material basis of societal wellbeing [103]. They play a central role in deriving services required for the functioning of society such as shelter, mobility and communication from material and energy flows [28,105]. Socioeconomic material and energy flows are required to create stocks such as buildings, infrastructures or machinery in the first place, to maintain or improve them to keep them in a usable state and to serve dissipative uses required to provide services from stocks [28,76,88,91,105]. Material stocks influence flows, not only because resources are required for building, maintaining and removing them, but also due to the path dependencies for the future resource use patterns they create (e.g., transport infrastructures or heating and cooling requirements of buildings).

Some studies underline the importance of stocks for determining future resource demands, availability of secondary resources for recycling and emissions. GHG emissions from fossil fuels required to use existing energy infrastructures until the end of their lifetime in the period 2010–2060 amount to approximately one-half of the remaining emission budget consistent with a 50% chance of reaching the 2 °C target [106,107]. GHG emissions from the production of three major material resources (steel, cement and aluminum) required for a globalization of current Western levels of infrastructure stocks amount to approximately one-third of that GHG emission budget [88].

*2.3. Recent Progress in Quantifying Socioeconomic Material Stocks*

In Industrial Ecology, empirical studies of material stocks are rapidly advancing in the last few years, although with varying focus and different concepts according to which the metabolic framework is implemented [27,28,32,76,88,103]. Much research on material stocks has focused on specific substances or materials at various scales. Metals have received most attention [31], from global reconstructions of specific in-use metals stocks and flows [28,85,105] to national level long-term studies on specific metal uses and accumulated stocks [105]. The study of stocks and flows of construction minerals has also increased in the last years [32].

In EW-MEFA, the study of material stocks has long been almost absent mainly due to the lack of robust data. Recently much progress has been made by employing dynamic stock-flow models to estimate stock accumulation from material use and lifetime dynamics [26,33,35,108]. Dynamic stock-flow models distinguish cohorts of stocks, e.g., age of passenger cars or efficiency of industrial plants depending on the time of construction. They typically allow characterizing stocks in terms of the resource requirements involved in building, renewing and using them, as well as the related emissions [29,31,91]. Quantification of uncertainty remains underdeveloped in EW-MEFA [42,50,51], and even more so in regard to material stocks. Few authors have provided sensitivity analyses [30,85,108], multi-method approaches [95] or Monte-Carlo simulations [26,86]. Clearly, uncertainty is a topic of growing importance for MFA research and applications of the socioeconomic metabolism concept in general [31,32,50,51,109].

The role of in-use stocks for energy use and climate-change mitigation in important sectors such as housing, the steel industry or passenger cars has been analyzed using global models [88,89,110]. Aggregate material stocks have been modelled from long-term historical material flow data [26,33,34]. Combinations of bottom-up and top-down approaches have recently been published, showing the complexity of adequately capturing all in-use stock dynamics and linking them to economy-wide material flows [91,94]. The distinction between material flows for dissipative use from stock-building materials allowed assessing potentials for a circular economy [36,94]. Recently the drivers of economy-wide stock accumulation were investigated for Japan's prefectures [35] and a typology of the dynamics of stock accumulation was developed [34]. The global dynamics of socio-economic metabolism, including all material flows and stocks have recently been estimated for the 20th century [26], see Figure 3.

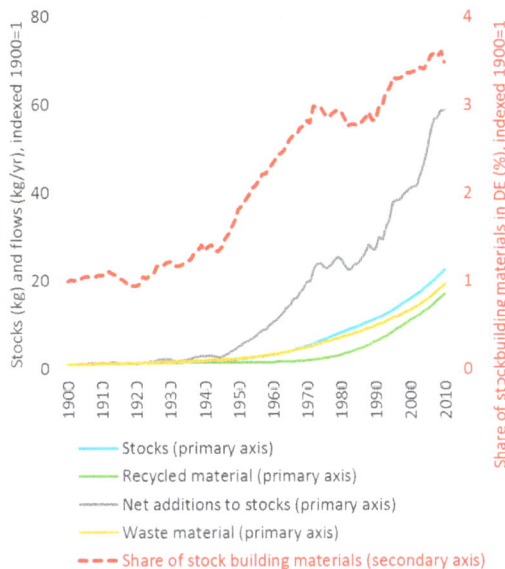

**Figure 3.** Long-term development of global material stocks and related flows. Material stocks (blue line, primary axis) grow >20 fold over the 20th century, net additions to stocks even almost 60 fold. The share of stock-building materials (red line, secondary axis) rises 3-fold over this time period to approximately 50% (not shown). The amount of end of life waste that is recycled grew 17 fold, but accounts for only a small share of inputs to stock. Source: own figure drawn using results from [26].

Material stocks depicted in Figure 3 were measured as kg of materials of certain aggregated categories (e.g., concrete) but not classified according to their respective functions (e.g., concrete in residential buildings versus concrete in roads, railroads or power plants). Stock estimates were modelled from material flow data using assumptions on the residence time of resources in the form of lifetime distributions. Uncertainty estimations were based on error propagation via Monte-Carlo Simulations, driven by assumptions on uncertainty for material inputs and all other model parameters derived from the literature [26]. Results show that global material stocks grow exponentially over the entire period. The study finds large differences between per capita stocks in industrial countries and the rest of the world, and high growth rates of stocks in emerging economies, which is likely to drive global stock growth in the future. While many commentators have designated the 20th century as the epoch marking the advent of throwaway societies, these numbers rather suggest that in fact we move toward stockpiling societies: The fraction of resources extracted allocated to the build-up of material stocks has tripled in the last century and amounted to 55% in 2010, the share of recycled materials in inputs to stocks remains low at 12% [26]. Reductions in resource extraction seem difficult or impossible to achieve as long as that stockpiling continues.

### 2.4. Resource Flows, Material Stocks, and Services

A substantial part of the sustainability debate has focused on options to raise eco-efficiency; i.e., to "decouple" economic growth and societal wellbeing from material and energy flows [16–18,111]. However, these eco-efficiency concepts have been criticized as too narrow [16,111–113], among others because societal wellbeing depends only partly on the products provided each year, but also on the quantity and quality of existing stocks of buildings and infrastructure. Raising efficiency may be important, but it is most likely not sufficient to move toward sustainability [19,114]. A focus on services from stocks such as infrastructures, buildings or machinery [28,88] can provide a new angle for that debate, as indicators for services have so far been almost entirely absent in eco-efficiency research, as was the importance of stocks for the emissions or resource flows that were sought to be decoupled from GDP.

Two recent strands of research where service indicators have been intensively discussed are those concerned with energy services [115,116] and ecosystem services [117,118]. In both debates, the aim is to understand how human societies gain benefits, either from using energy (energy services) or from ecosystems and natural capital (ecosystem services). In both contexts, services are diverse and recalcitrant against attempts to be measured in one single unit. They include items as different as provision of comfortable living space or adequate lighting, recreation, adequate nutrition, shaping of workpieces, computational services, cleaning of water, air, homes or human bodies, maintenance of biological diversity, cultural and religious services and many more. Both debates have developed concepts to analyze the multiple steps involved in the provision of services from resources.

Energy services are derived in several steps involving energy conversions from raw or primary energy extracted from deposits or diverted from natural flows to final energy sold to consumers, to useful energy delivered by machinery operated by consumers and eventually to energy services, a sequence that has been denoted as "energy chain" [115]. In a somehow similar manner, a "cascade" concept is used to understand the multi-step process through which benefits and finally their values are derived from biophysical structures and processes, which in turn influences the functioning of ecosystems and their services [119]. The importance of human agency for the delivery of ecosystem services is increasingly recognized [118]. Even more explicitly than ecosystem services, however, services from stocks are not naturally given, but a product of past work and reoccurring material and energy flows, while at the same time delimiting future development options.

Material stocks such as machinery, buildings or infrastructures are important because they are required to transform resources which are in themselves barely useful (e.g., crude oil, mixtures of minerals in a deposit, or even trees in a forest) into services such as comfortable living space, the movement of people or goods across space, or the delivery and storage of information through electronic devices or paper. Infrastructures determine which services are available where (e.g., shelter, transport or production capacity). Spatial patterns of settlements and production sites as well as available transport infrastructures co-determine the distances and energy demand required to provide transport-related services [120,121].

*2.5. The Importance of Spatial Patterns and Urbanization for Resource Demand*

Lack of comprehensive data on the spatial patterns of material stock data so far impedes systematic integration of spatial analysis with socioeconomic metabolism research, except for human population and livestock, which are also considered as parts of societies' material stocks (Figure 2). Population maps [122,123] are widely used to downscale national or regional data to maps, following the assumption that human drivers or impacts scale with population [124,125]. Such maps are constructed by combining census statistics with proxies from remote sensing, e.g., nighttime light [126,127]. Similar approaches allowed mapping the global distribution and environmental impacts of livestock [128–130].

One key spatial aspect related to material stocks is urbanization. Settlement patterns, urban form and population density influence transportation energy use and emissions [114,120,131–134]. It is expected that urban population will reach 5.6–7.1 billion in the mid of the century [121]. Relationships between urban material and energy flows, city size and city structure were analyzed for 27 global megacities [135]. Consumption patterns are embedded in specific urban forms. In densely settled areas, use of transportation energy use is usually lower, but higher incomes in cities lead to more consumption and larger energy and carbon footprints of households [136–139]. An integrated perspective hence needs to consider the systemic interactions between production and consumption in urban areas and rural hinterlands [140–142].

Studies of urban metabolism have quantified urban resource use, and some even stocks [27,141,143,144]. European material stocks and recycling potentials in buildings and infrastructures were assessed in a bottom-up study [91]. The interrelations between energy transitions and urbanization processes have recently been explored [145]. Spatially explicit material stock data have been explored for substances like metals [97,146], or more comprehensively for Japan [27]. Overall, spatially explicit mapping of stocks has made much progress in the last decade, although so far mostly on regional levels [27,96–98,144].

A comprehensive perspective consistent with EW-MEFA is important because services are delivered by material stocks consisting of mixtures of materials. For example, a building can be made of steel-reinforced concrete, bricks, mortar, timber and many other materials. Hence only a comprehensive perspective can address substitution. In a study of stocks and flows of iron, a building made only of wood and bricks would become invisible, but it might still fulfill functions and deliver services, and it would still require material inputs during construction. The environmental impacts of the related resource flows would be different from those of a building made primarily of steel, aluminum and glass, but they would not be irrelevant. A comprehensive, systemic approach is important due to its ability to address problem-shifting and substitution, whereas substance-specific studies are required to address specific impacts [41,44].

## 3. Contributions of the Stock-Flow-Service Nexus Approach to Socioeconomic Metabolism Research

*3.1. New Conceptualizations of Eco-Efficiency*

Analyses of eco-efficiency so far primarily used GDP as a measure of affluence and related GDP to either resource requirements (e.g., energy or material inputs) or emissions (e.g., GHG or $CO_2$ emissions). Service indicators can complement GDP in such analyses, as services highlight the relevance of specific combinations of material stocks and resource flows for contributions to societal wellbeing. The availability of indicators for stocks and services would allow defining innovative measures of resource intensity and emission intensity, such as the measurement of resources required or emissions resulting from achieving certain levels of material stocks or services in different world regions or changes of resource requirements per unit of service delivered along temporal trajectories. For example, one could study the resource requirements of different options to achieve sustainable development goals such as the universal provision of clean water and sanitation or clean and affordable energy. The stock–flow–service nexus approach will hence allow researchers to go significantly beyond relating resource use or emissions to economic indicators such as GDP [72], the Social Progress Index [147], or the Human Development Index (HDI) [17,90,148] by providing indicators for the services derived from specific stock-flow constellations. Novel concepts of eco-efficiency that could be derived from the analysis of the stock–flow–service nexus approach (Figure 1) will enable researchers to estimate the amount of resources sufficient for essential societal wellbeing at different spatial levels in a scientifically sound and comprehensive way.

Efficiency is usually defined as the relation between two flows (e.g., materials/GDP), hence stocks played no important role in the eco-efficiency debate, except where stocks such as population or area are used to compare flows in different countries by calculating flows per capita or per unit area [14,70,149]. Such concepts of eco-efficiency mask the importance of stocks in determining resource requirements and service provision. Indicators of material stocks as well as material stock maps could help elucidating the importance of specific qualities of material stocks respectively their spatial patterns for resource demand, waste production and closing material loops. Mapping of human activities is so far largely restricted to population density maps and other datasets mainly derived from proxies such as nighttime lights [97,127,150]. Maps of material stocks would be useful to characterize spatial patterns of material stocks and thereby analyze stock–flow–service relations and the spatial distribution of human activities on earth much more consistently than is possible today. Insights from such empirical efforts could also help improve integrated assessment models [12] that currently play a key role in evaluating transformation pathways, e.g., in analyses of options for climate-change mitigation [151].

*3.2. Food for Thought for Socioecological Transformation Research*

Following the recognition that a continuation of current trajectories, even when combined with efforts to raise eco-efficiency, will likely be insufficient to reduce climate change to 2 °C or less [152] and tackle other sustainability challenges [25], much research is currently focused on wide-ranging changes in society-nature relations denoted as "transformations towards sustainability". For such a transformation, material flows need to be reduced to an ecologically sustainable level, which requires at least a stabilization of stocks (Figure 4)—quite contrary to the global trends depicted in Figure 3. Sustainability transformation research builds on approaches such as the "sociotechnical transition" concept [153–155] as well as analyses of technological transitions related to energy [151,156–159]. The importance of infrastructures for providing services and stabilizing practices of production and consumption [160], for future energy system transformations [59] or for reaching climate-change mitigation targets [151,152] has been acknowledged. Lock-ins into emission-intensive pathways resulting from existing stocks have recently be analyzed [90]. The material and energy repercussions of a future global low-emission electricity system were found to be significant, but manageable [161].

However, so far transition management approaches mainly focus on developed countries and lack a global perspective [162]. Their focus on technology and innovation underpins important leverage points for targeted interventions, but will require complementary approaches to address rebound effects and other transition barriers and to provide a more realistic understanding of the biophysical and societal constraints involved. A "Great Transformation" [20,163] involves not only technological innovations, but also major shifts in resource production and consumption [164] and socioeconomic metabolism [9,19] at various spatial and temporal scales.

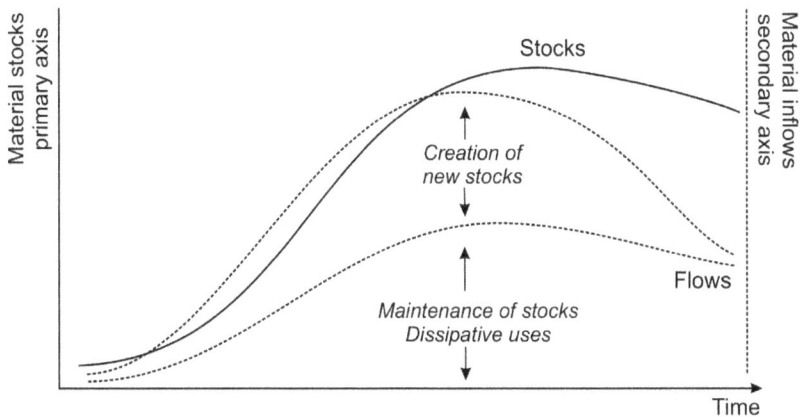

**Figure 4.** Hypothetical stylized representation of stock-flow relations over time in a national economy. So far, stocks do not seem to be stabilizing even in industrialized countries, which could explain lacking success in reducing flows. Source: own graph.

The study of shifts between socioecological regimes such as hunter-gatherers, agrarian, industrial, or a yet unknown future sustainable society [19,165] has so far been focused on changes in resource flows [64] and land use [166] at larger scales. Links to changes in population trends related to so-called demographic transitions [167] have been analyzed in these approaches [165] but could profit from being better linked to the analysis of socioeconomic interests and the power relations behind them that may hinder or block sustainability transformations. An important aspect are transformations in energy systems, e.g., from biomass to fossil fuels during the agrarian-industrial transition [53,168,169] which today, due to climate change and the quest for alternatives to fossil fuels, stimulates far reaching conflicts at and between several spatial scales, including increasing land use conflicts.

Until now transformation research is marked by an analytical gap to address ongoing tendencies towards unsustainability and challenges for shaping societal relations with nature. What is required is a more comprehensive analysis of the crisis-driven and contested character of the appropriation of nature and the power relations involved as discussed under the concept of social-ecological transformations (see [170], in this issue). Neither material stocks nor services have so far featured prominently in these analyses. However, the arguments laid out above suggest that the inclusion of material stocks and services can improve the analysis of both past and possible future transformations and thereby contribute to transformation research.

Linking stocks, flows and services can elucidate constraints to as well as leverage points for sustainability transformations that are so far poorly understood. Material stocks are products of past actions, i.e., past interventions in natural systems or transformations of socioecological systems, which constrain as well as enable further human activities and societal development by providing services for societal wellbeing. They have high value and often persist for a long time, thereby creating legacies, lock-in effects and path-dependencies. They result from complex interactions of societal and natural processes under certain institutional arrangements [171]. Thus, an analysis of these historically specific conditions can provide starting points for the identification of alternative resource use pattern. This becomes visible if compared with existing mainstream economic approaches, which are either supply-based or demand-side driven. Supply-based and demand-based approaches both lack criteria for fulfillment of societal needs and are therefore implicitly directed towards unlimited growth. In contrast, a service-based approach is able to compare different options to address a sufficient level of services and the societal needs behind them, thereby helping to determine alternative options to fulfill these needs through specific stock-flow configurations. Material stocks hence depict important conditions for transformations toward sustainability. Research can focus on questions like their malleability through targeted societal interventions, which is yet poorly understood, or the relevance of alternative stock/flow/service configurations for sustainability transformations, which currently have not been addressed systematically. Quantifications of material stocks and investigations into their relations with flows and provision of services are hence a key component of attempts to scientifically underpin sustainability transformations.

## 4. Conclusions

Moving toward sustainability will require far-reaching changes in socioeconomic metabolism, i.e., in the stocks and flows of materials and energy related with societal activities. The consistent integration of material stock data and maps one the one hand, and of the services provided through specific constellations of stocks and flows, can inspire new empirical research (e.g., quantification and mapping of material stocks), modelling and analysis to better understand past trajectories as well as future patterns of society-nature interaction. One critical research gap that well have to be addressed is the need for material stock accounts that do not only distinguish between different kinds of materials [26] but also their different functions. When such accounts become available, the stock–flow–service nexus approach will be able to underpin new concepts for analyzing eco-efficiency in the sense of a decoupling between societal well-being and resource demand, the contingencies and lock-ins resulting from past build-up of material stocks, as well as possible leverage points to foster sustainability transformations. Comparing alternative options for providing the same service with lower resource needs, i.e., through different stock-flow configurations than those in place today, will open up space for determining leverage points for sustainability transformations and addressing the issue of sufficiency beyond a purely normative claim.

**Acknowledgments:** We thank two anonymous reviewers for providing hugely valuable constructively-critical comments. Any errors that might have survived the revision process are of course ours. Funding by the Austrian Science Funds (FWF), projects no. P27890 and P29130-G27, is gratefully acknowledged. Conceptual work behind this article contributed to the formulation of ERC-AdG-proposal MAT_STOCKS, No. 741950 (grant agreement currently under negotiation). This research contributes to the Global Land Programme (http://www.futureearth.org/projects/glp-global-land-programme).

**Author Contributions:** The authors jointly developed the ideas presented and wrote the article. Empirical work shown in Figure 3 was designed and carried out primarily by D.W. and F.K.

**Conflicts of Interest:** The authors declare no conflict of interest.

## References

1. Turner, B.L.I.; Clark, W.C.; Kates, R.W.; Richards, J.F.; Mathews, J.T.; Meyer, W.B. *The Earth as Transformed by Human Action: Global and Regional Changes in the Biosphere over the Past 300 Years*; Cambridge University Press: Cambridge, UK, 1990.
2. Crutzen, P.J. Geology of mankind. *Nature* **2002**, *415*, 23. [CrossRef] [PubMed]
3. Kates, R.W.; Clark, W.C.; Corell, R.; Hall, J.M.; Jaeger, C.C.; Lowe, I.; McCarthy, J.J.; Schellnhuber, H.J.; Bolin, B.; Dickson, N.M.; et al. Sustainability science. *Science* **2001**, *292*, 641–642. [CrossRef] [PubMed]
4. Seitzinger, S.P.; Svedin, U.; Crumley, C.L.; Steffen, W.; Abdullah, S.A.; Alfsen, C.; Broadgate, W.J.; Biermann, F.; Bondre, N.R.; Dearing, J.A.; et al. Planetary Stewardship in an Urbanizing World: Beyond City Limits. *Ambio* **2012**, *41*, 787–794. [CrossRef] [PubMed]
5. Steffen, W.; Persson, Å.; Deutsch, L.; Zalasiewicz, J.; Williams, M.; Richardson, K.; Crumley, C.; Crutzen, P.; Folke, C.; Gordon, L.; et al. The Anthropocene: From Global Change to Planetary Stewardship. *Ambio* **2011**, *40*, 739–761. [CrossRef] [PubMed]
6. Steffen, W.; Broadgate, W.; Deutsch, L.; Gaffney, O.; Ludwig, C. The trajectory of the Anthropocene: The Great Acceleration. *Anthr. Rev.* **2015**, *2*, 81–98. [CrossRef]
7. Rockström, J.; Steffen, W.; Noone, K.; Persson, Å.; Chapin, F.S., III; Lambin, E.; Lenton, T.M.; Scheffer, M.; Folke, C.; Schellnhuber, H.; et al. Planetary boundaries: Exploring the safe operating space for humanity. *Ecol. Soc.* **2009**, *14*, 32. [CrossRef]

8.  Steffen, W.; Richardson, K.; Rockstrom, J.; Cornell, S.E.; Fetzer, I.; Bennett, E.M.; Biggs, R.; Carpenter, S.R.; de Vries, W.; de Wit, C.A.; et al. Planetary boundaries: Guiding human development on a changing planet. *Science* **2015**, *347*. [CrossRef] [PubMed]

9.  Haberl, H.; Fischer-Kowalski, M.; Krausmann, F.; Winiwarter, V. *Social Ecology, Society-Nature Relations across Time and Space*; Human-Environment Interactions, no. 5; Springer: Dordrecht, The Netherlands, 2016.

10. Pauliuk, S.; Hertwich, E.G. Socioeconomic metabolism as paradigm for studying the biophysical basis of human societies. *Ecol. Econ.* **2015**, *119*, 83–93. [CrossRef]

11. Krey, V.; Masera, O.; Hanemann, M.; Blanford, G.; Bruckner, T.; Haberl, H.; Hertwich, E.; Müller, D. Annex II: Methods and metrics. In *Climate Change 2014: Mitigation of Climate Change, Working Group III Contribution to the IPCC Fifth Assessment Report (AR5) of the Intergovernmental Panel for Climate Change*; Edenhofer, O., Pichs-Madruga, R., Sokona, Y., Minx, J.C., Farahani, E., Kadner, S., Seyboth, K., Adler, A., Baum, I., Brunner, S., et al., Eds.; Cambride University Press: Cambridge, UK; New York, NY, USA, 2014; pp. 1281–1328.

12. Pauliuk, S.; Arvesen, A.; Stadler, K.; Hertwich, E.G. Industrial ecology in integrated assessment models. *Nat. Clim. Chang.* **2017**, *7*, 13–20. [CrossRef]

13. Fischer-Kowalski, M.; Haberl, H. Social metabolism—A metric for biophysical growth and degrowth. In *Handbook of Ecological Economics*; Martinez-Alier, J., Muradian, R., Eds.; Edward Elgar: Cheltenham, UK; Northampton, MA, USA, 2015; pp. 100–138.

14. Dietz, T.; Frank, K.A.; Whitley, C.T.; Kelly, J.; Kelly, R. Political influences on greenhouse gas emissions from US states. *Proc. Natl. Acad. Sci. USA* **2015**, *112*, 8254–8259. [CrossRef] [PubMed]

15. York, R.; Rosa, E.A.; Dietz, T. STIRPAT, IPAT and ImPACT: Analytic tools for unpacking the driving forces of environmental impacts. *Ecol. Econ.* **2003**, *46*, 351–365. [CrossRef]

16. Steinberger, J.K.; Krausmann, F. Material and Energy Productivity. *Environ. Sci. Technol.* **2011**, *45*, 1169–1176. [CrossRef] [PubMed]

17. Steinberger, J.K.; Krausmann, F.; Getzner, M.; Schandl, H.; West, J. Development and Dematerialization: An International Study. *PLoS ONE* **2013**, *8*, e70385. [CrossRef] [PubMed]

18. UNEP. *Decoupling Natural Resource Use and Environmental Impacts from Economic Growth*; A Report of the Working Group on Decoupling to the International Resource Panel; United Nations Environment Programme: Nairobi, Kenya, 2011.

19. Haberl, H.; Fischer-Kowalski, M.; Krausmann, F.; Martinez-Alier, J.; Winiwarter, V. A socio-metabolic transition towards sustainability? Challenges for another Great Transformation. *Sustain. Dev.* **2011**, *19*, 1–14. [CrossRef]

20. WBGU. *Welt im Wandel. Gesellschaftsvertrag für eine Große Transformation*; Wissenschaftlicher Beirat Globale Umweltveränderungen (WBGU): Berlin, Germany, 2011.

21. Korten, D.C. *Change the Story, Change the Future: A Living Economy for A Living Earth*, 1st ed.; Berrett-Koehler Publishers, Inc.: Oakland, CA, USA, 2015.

22. Sachs, W. Fair Wealth. Pathways into Post Development. In *Rethinking Development in a Carbon-Constrained World. Development Cooperation and Climate Change*; Palosuo, E., Ed.; Ministry of Foreign Affairs: Helsinki, Finland, 2009; pp. 196–206.

23. Kallis, G.; Kerschner, C.; Martinez-Alier, J. The economics of degrowth. *Ecol. Econ.* **2012**, *84*, 172–180. [CrossRef]

24. Brand, U. "Transformation" as a New Critical Orthodoxy: The Strategic Use of the Term "Transformation" Does Not Prevent Multiple Crises. *GAIA—Ecol. Perspect. Sci. Soc.* **2016**, *25*, 23–27. [CrossRef]

25. UNEP. *Policy Coherence of the Sustainable Development Goals, a Natural Resources Perspective*; International Resource Panel Report; United Nations Environment Programme (UNEP): Nairobi, Kenya, 2015.

26. Krausmann, F.; Wiedenhofer, D.; Lauk, C.; Haas, W.; Tanikawa, H.; Fishman, T.; Miatto, A.; Schandl, H.; Haberl, H. Global socioeconomic material stocks rise 23-fold over the 20th century and require half of annual resource use. *Proc. Natl. Acad. Sci. USA* **2017**, *114*, 1880–1885. [CrossRef] [PubMed]

27. Tanikawa, H.; Fishman, T.; Okuoka, K.; Sugimoto, K. The Weight of Society over Time and Space: A Comprehensive Account of the Construction Material Stock of Japan, 1945–2010: The Construction Material Stock of Japan. *J. Ind. Ecol.* **2015**, *19*, 778–791. [CrossRef]

28. Pauliuk, S.; Müller, D.B. The role of in-use stocks in the social metabolism and in climate change mitigation. *Glob. Environ. Chang.* **2014**, *24*, 132–142. [CrossRef]

29. Müller, D.B.; Wang, T.; Duval, B.; Graedel, T.E. Exploring the engine of anthropogenic iron cycles. *Proc. Natl. Acad. Sci. USA* **2006**, *103*, 16111–16116. [CrossRef] [PubMed]

30. Müller, D. Stock dynamics for forecasting material flows—Case study for housing in The Netherlands. *Ecol. Econ.* **2006**, *59*, 142–156. [CrossRef]

31. Müller, E.; Hilty, L.M.; Widmer, R.; Schluep, M.; Faulstich, M. Modeling Metal Stocks and Flows: A Review of Dynamic Material Flow Analysis Methods. *Environ. Sci. Technol.* **2014**, *48*, 2102–2113. [CrossRef] [PubMed]

32. Augiseau, V.; Barles, S. Studying construction materials flows and stock: A review. *Resour. Conserv. Recycl.* **2016**, in press. [CrossRef]

33. Fishman, T.; Schandl, H.; Tanikawa, H.; Walker, P.; Krausmann, F. Accounting for the Material Stock of Nations. *J. Ind. Ecol.* **2014**, *18*, 407–420. [CrossRef] [PubMed]

34. Fishman, T.; Schandl, H.; Tanikawa, H. Stochastic Analysis and Forecasts of the Patterns of Speed, Acceleration, and Levels of Material Stock Accumulation in Society. *Environ. Sci. Technol.* **2016**, *50*, 3729–3737. [CrossRef] [PubMed]

35. Fishman, T.; Schandl, H.; Tanikawa, H. The socio-economic drivers of material stock accumulation in Japan's prefectures. *Ecol. Econ.* **2015**, *113*, 76–84. [CrossRef]

36. Haas, W.; Krausmann, F.; Wiedenhofer, D.; Heinz, M. How Circular is the Global Economy? An Assessment of Material Flows, Waste Production, and Recycling in the European Union and the World in 2005. *J. Ind. Ecol.* **2015**, *19*, 765–777. [CrossRef]

37. Fleurbaey, M. Beyond GDP: The Quest for a Measure of Social Welfare. *J. Econ. Lit.* **2009**, *47*, 1029–1075. [CrossRef]

38. Kubiszewski, I.; Costanza, R.; Franco, C.; Lawn, P.; Talberth, J.; Jackson, T.; Aylmer, C. Beyond GDP: Measuring and achieving global genuine progress. *Ecol. Econ.* **2013**, *93*, 57–68. [CrossRef]

39. Ayres, R.U.; Simonis, U.E. *Industrial Metabolism: Restructuring for Sustainable Development*; United Nations University Press: Tokyo, Japan; New York, NY, USA; Paris, France, 1994.

40. Martinez-Alier, J. *Ecological Economics. Energy, Environment and Society*; Blackwell: Oxford, UK, 1987.

41. Fischer-Kowalski, M. Society's Metabolism: The Intellectual History of Materials Flow Analysis, Part I, 1860–1970. *J. Ind. Ecol.* **1998**, *2*, 107–136. [CrossRef]

42. Fischer-Kowalski, M.; Krausmann, F.; Giljum, S.; Lutter, S.; Mayer, A.; Bringezu, S.; Moriguchi, Y.; Schütz, H.; Schandl, H.; Weisz, H. Methodology and Indicators of Economy-wide Material Flow Accounting—State of the Art and Reliability Across Sources. *J. Ind. Ecol.* **2011**, *15*, 855–876. [CrossRef]

43. Haberl, H. The Energetic Metabolism of Societies Part I: Accounting Concepts. *J. Ind. Ecol.* **2001**, *5*, 11–33. [CrossRef]

44. Bringezu, S.; Moriguchi, Y. Material flow analysis. In *A Handbook of Industrial Ecology*; Ayres, R.U., Ayres, L.W., Eds.; Edward Elgar: Cheltenham, UK, 2002; pp. 79–91.

45. Bringezu, S.; van de Sand, I.; Schütz, H.; Moll, S. Analysing global resource use of national and regional economies across various levels. In *Sustainable Resource Management. Global Trends, Visions and Policies*; Bringezu, S., Bleischwitz, R., Eds.; Greenleaf Publishing: Sheffield, UK, 2009; pp. 10–52.

46. Graedel, T.E.; Harper, E.M.; Nassar, N.T.; Reck, B.K. On the materials basis of modern society. *Proc. Natl. Acad. Sci. USA* **2015**, *112*, 6295–6300. [CrossRef] [PubMed]

47. Finnveden, G.; Hauschild, M.Z.; Ekvall, T.; Guinée, J.; Heijungs, R.; Hellweg, S.; Koehler, A.; Pennington, D.; Suh, S. Recent developments in Life Cycle Assessment. *J. Environ. Manag.* **2009**, *91*, 1–21. [CrossRef] [PubMed]

48. Ayres, R.U.; Kneese, A.V. Production, Consumption, and Externalities. *Am. Econ. Rev.* **1969**, *59*, 282–297.

49. Boulding, K. The Economics of the Coming Spaceship Earth. In *Steady State Economics*; Daly, H.E., Ed.; W.H. Freeman: San Francisco, CA, USA, 1972; pp. 121–132.

50. Laner, D.; Rechberger, H.; Astrup, T. Systematic Evaluation of Uncertainty in Material Flow Analysis. *J. Ind. Ecol.* **2014**, *18*, 859–870. [CrossRef]

51. Patrício, J.; Kalmykova, Y.; Rosado, L.; Lisovskaja, V. Uncertainty in Material Flow Analysis Indicators at Different Spatial Levels. *J. Ind. Ecol.* **2015**, *19*, 837–852. [CrossRef]

52. Lotka, A.J. *Elements of Physical Biology*; Williams & Wilkins Company: Baltimore, MA, USA, 1925.

53. Smil, V. *General Energetics: Energy in the Biosphere and Civilization*; Wiley Interscience: New York, NY, USA, 1991.

54. Lindeman, R.L. The tropic-dynamic aspect of ecology. *Ecology* **1942**, *23*, 399–418. [CrossRef]

55.  Cottrell, F. *Energy and Society: The Relation between Energy, Social Change, and Economic Development*; McGraw-Hill: New York, NY, USA; Toronto, ON, Canada; London, UK, 1955.

56.  Odum, H.T. *Environment, Power and Society*; Wiley-Interscience: New York, NY, USA, 1971.

57.  Hall, C.A.S.; Cleveland, C.J.; Kaufmann, R. *Energy and Resource Quality: The Ecology of the Economic Process*; University Press of Colorado: Niwot, CO, USA, 1992.

58.  Malcolm, S. International Federation of Institutes for Advanced Study. In *Energy Analysis Workshop on Methodology and Conventions: 25th–30th August, 1974, Guldsmedshyttan, Sweden*; IFIAS: Stockhom, Sweden, 1974.

59.  GEA. *Global Energy Assessment—Toward a Sustainable Future*; IIASA: Laxenburg, Austria; Cambridge University Press: Cambridge, UK, 2012.

60.  Giampietro, M. Comments on "The Energetic Metabolism of the European Union and the United States" by Haberl and Colleagues: Theoretical and Practical Considerations on the Meaning and Usefulness of Traditional Energy Analysis. *J. Ind. Ecol.* **2006**, *10*, 173–185. [CrossRef]

61.  Weisz, H. The Probability of the Improbable: Society–Nature Coevolution. *Geogr. Ann. Ser. B Hum. Geogr.* **2011**, *93*, 325–336. [CrossRef]

62.  Infante-Amate, J.; Soto Fernandez, D.; Aguilera, E.; Garcia-Ruiz, R.; Guzmán, G.I. The Spanish Transition to Industrial Metabolism: Long-Term Material Flow Analysis (1860–2010). *J. Ind. Ecol.* **2015**, in press. [CrossRef]

63.  Sieferle, R.-P.; Krausmann, F.; Schandl, H.; Winiwarter, V. *Das Ende der Fläche: Zum Gesellschaftlichen Stoffwechsel der Industrialisierung*; Böhlau: Köln, Germay, 2006.

64.  Schaffartzik, A.; Mayer, A.; Gingrich, S.; Eisenmenger, N.; Loy, C.; Krausmann, F. The global metabolic transition: Regional patterns and trends of global material flows, 1950–2010. *Glob. Environ. Chang.* **2014**, *26*, 87–97. [CrossRef] [PubMed]

65.  Krausmann, F.; Gingrich, S.; Eisenmenger, N.; Erb, K.-H.; Haberl, H.; Fischer-Kowalski, M. Growth in global materials use, GDP and population during the 20th century. *Ecol. Econ.* **2009**, *68*, 2696–2705. [CrossRef]

66.  Gierlinger, S.; Krausmann, F. The physical economy of the United States of America: Extraction, trade and consumption of materials from 1870 to 2005. In *The Socio-Metabolic Transition. Long Term Historical Trends and Patterns in Global Material and Energy Use*; Krausmann, F., Ed.; Social Ecology Working Papers; IFF Social Ecology: Vienna, Austria, 2011; pp. 24–49.

67.  Gingrich, S.; Schmid, M.; Gradwohl, M.; Krausmann, F. How Material and Energy Flows Change Socio-Natural Arrangements: The Transformation of Agriculture in the Eisenwurzen Region, 1860–2000. In *Long Term Socio-Ecological Research. Studies in Society-Nature Interactions across Spatial and Temporal Scales*; Springer Science + Business Media B.V.: Dordrecht, The Netherlands, 2013; Volume 2, pp. 297–313.

68.  Pothen, F.; Schymura, M. Bigger cakes with fewer ingredients? A comparison of material use of the world economy. *Ecol. Econ.* **2015**, *109*, 109–121. [CrossRef]

69.  Weisz, H.; Krausmann, F.; Amann, C.; Eisenmenger, N.; Erb, K.-H.; Hubacek, K.; Fischer-Kowalski, M. The physical economy of the European Union: Cross-country comparison and determinants of material consumption *Ecol. Econ.* **2006**, *58*, 676–698. [CrossRef]

70.  Schandl, H.; West, J. Resource use and resource efficiency in the Asia–Pacific region. *Glob. Environ. Chang.* **2010**, *20*, 636–647. [CrossRef]

71.  Dittrich, M.; Bringezu, S. The physical dimension of international trade: Part 1: Direct global flows between 1962 and 2005. *Ecol. Econ.* **2010**, *69*, 1838–1847. [CrossRef]

72.  Wiedmann, T.O.; Schandl, H.; Lenzen, M.; Moran, D.; Suh, S.; West, J.; Kanemoto, K. The material footprint of nations. *Proc. Natl. Acad. Sci. USA* **2015**, *112*, 6271–6276. [CrossRef] [PubMed]

73.  Dorninger, C.; Hornborg, A. Can EEMRIO analyses establish the occurrence of ecologically unequal exchange? *Ecol. Econ.* **2015**, *119*, 414–418. [CrossRef]

74.  Erb, K.; Gingrich, S.; Krausmann, F.; Haberl, H. Industrialization, Fossil Fuels, and the Transformation of Land Use. *J. Ind. Ecol.* **2008**, *12*, 686–703. [CrossRef]

75.  Chen, W.-Q.; Graedel, T.E. Dynamic analysis of aluminum stocks and flows in the United States: 1900–2009. *Ecol. Econ.* **2012**, *81*, 92–102. [CrossRef]

76.  Chen, M.; Graedel, T.E. A half-century of global phosphorus flows, stocks, production, consumption, recycling, and environmental impacts. *Glob. Environ. Chang.* **2016**, *36*, 139–152. [CrossRef]

77.  Peiró, L.T.; Méndez, G.V.; Ayres, R.U. Material Flow Analysis of Scarce Metals: Sources, Functions, End-Uses and Aspects for Future Supply. *Environ. Sci. Technol.* **2013**, *47*, 2939–2947. [CrossRef] [PubMed]

78. Rauch, J.N.; Pacyna, J.M. Earth's global Ag, Al, Cr, Cu, Fe, Ni, Pb, and Zn cycles: GLOBAL METAL CYCLES. *Glob. Biogeochem. Cycles* **2009**, *23*. [CrossRef]
79. Wackernagel, M.; Schulz, N.B.; Deumling, D.; Linares, A.C.; Jenkins, M.; Kapos, V.; Monfreda, C.; Loh, J.; Myers, N.; et al. Tracking the ecological overshoot of the human economy. *Proc. Natl. Acad. Sci. USA* **2002**, *99*, 9266–9271. [CrossRef] [PubMed]
80. Haberl, H.; Erb, K.H.; Krausmann, F.; Gaube, V.; Bondeau, A.; Plutzar, C.; Gingrich, S.; Lucht, W.; Fischer-Kowalski, M. Quantifying and mapping the human appropriation of net primary production in earth's terrestrial ecosystems. *Proc. Natl. Acad. Sci. USA* **2007**, *104*, 12942–12947. [CrossRef] [PubMed]
81. Schaffartzik, A.; Haberl, H.; Kastner, T.; Wiedenhofer, D.; Eisenmenger, N.; Erb, K.-H. Trading Land: A Review of Approaches to Accounting for Upstream Land Requirements of Traded Products: A Review of Upstream Land Accounts. *J. Ind. Ecol.* **2015**, *19*, 703–714. [CrossRef] [PubMed]
82. Kastner, T.; Kastner, M.; Nonhebel, S. Tracing distant environmental impacts of agricultural products from a consumer perspective. *Ecol. Econ.* **2011**, *70*, 1032–1040. [CrossRef]
83. Gerst, M.D.; Graedel, T.E. In-Use Stocks of Metals: Status and Implications. *Environ. Sci. Technol.* **2008**, *42*, 7038–7045. [CrossRef] [PubMed]
84. Pauliuk, S.; Wang, T.; Müller, D.B. Steel all over the world: Estimating in-use stocks of iron for 200 countries. *Resour. Conserv. Recycl.* **2013**, *71*, 22–30. [CrossRef]
85. Liu, G.; Müller, D.B. Centennial Evolution of Aluminum In-Use Stocks on Our Aluminized Planet. *Environ. Sci. Technol.* **2013**, *47*, 4882–4888. [CrossRef] [PubMed]
86. Glöser, S.; Soulier, M.; Tercero Espinoza, L.A. Dynamic Analysis of Global Copper Flows. Global Stocks, Postconsumer Material Flows, Recycling Indicators, and Uncertainty Evaluation. *Environ. Sci. Technol.* **2015**, *47*, 6564–6572. [CrossRef] [PubMed]
87. Wang, T.; Müller, D.B.; Hashimoto, S. The Ferrous Find: Counting Iron and Steel Stocks in China's Economy: Counting Iron and Steel Stocks in China's Economy. *J. Ind. Ecol.* **2015**, *19*, 877–889. [CrossRef]
88. Müller, D.B.; Liu, G.; Løvik, A.N.; Modaresi, R.; Pauliuk, S.; Steinhoff, F.S.; Brattebø, H. Carbon Emissions of Infrastructure Development. *Environ. Sci. Technol.* **2013**, *47*, 11739–11746. [CrossRef] [PubMed]
89. Liu, G.; Bangs, C.E.; Müller, D.B. Stock dynamics and emission pathways of the global aluminium cycle. *Nat. Clim. Chang.* **2012**, *3*, 338–342. [CrossRef]
90. Lin, C.; Liu, G.; Müller, D.B. Characterizing the role of built environment stocks in human development and emission growth. *Resour. Conserv. Recycl.* **2017**, *123*, 67–72. [CrossRef]
91. Wiedenhofer, D.; Steinberger, J.K.; Eisenmenger, N.; Haas, W. Maintenance and Expansion: Modeling Material Stocks and Flows for Residential Buildings and Transportation Networks in the EU25. *J. Ind. Ecol.* **2015**, *19*, 538–551. [CrossRef] [PubMed]
92. Brattebø, H.; Bergsdal, H.; Sandberg, N.H.; Hammervold, J.; Müller, D.B. Exploring built environment stock metabolism and sustainability by systems analysis approaches. *Build. Res. Inf.* **2009**, *37*, 569–582. [CrossRef]
93. Pauliuk, S.; Sjöstrand, K.; Müller, D.B. Transforming the Norwegian Dwelling Stock to Reach the 2 Degrees Celsius Climate Target: Combining Material Flow Analysis and Life Cycle Assessment Techniques. *J. Ind. Ecol.* **2013**, *17*, 542–554. [CrossRef]
94. Schiller, G.; Müller, F.; Ortlepp, R. Mapping the anthropogenic stock in Germany: Metabolic evidence for a circular economy. *Resour. Conserv. Recycl.* **2017**, *123*, 93–107. [CrossRef]
95. Pauliuk, S.; Milford, R.L.; Müller, D.B.; Allwood, J.M. The Steel Scrap Age. *Environ. Sci. Technol.* **2013**, *47*, 3448–3454. [CrossRef] [PubMed]
96. Tanikawa, H.; Hashimoto, S. Urban stock over time: Spatial material stock analysis using 4d-GIS. *Build. Res. Inf.* **2009**, *37*, 483–502. [CrossRef]
97. Rauch, J.M. Global mapping of Al, Cu, Fe, and Zn in-use stocks and in-ground resources. *Proc. Natl. Acad. Sci. USA* **2009**, *106*, 18920–18925. [CrossRef] [PubMed]
98. Kleemann, F.; Lederer, J.; Rechberger, H.; Fellner, J. GIS-based Analysis of Vienna's Material Stock in Buildings: GIS-based Analysis of Material Stock in Buildings. *J. Ind. Ecol.* **2016**. [CrossRef]
99. Wallsten, B.; Magnusson, D.; Andersson, S.; Krook, J. The economic conditions for urban infrastructure mining: Using GIS to prospect hibernating copper stocks. *Resour. Conserv. Recycl.* **2015**, *103*, 85–97. [CrossRef]
100. Baccini, P.; Brunner, P.H. *Metabolism of the Anthroposphere*; Springer-Verlag: Berlin, Germany; New York, NY, USA, 1991.

101. Fischer-Kowalski, M.; Weisz, H. Society as hybrid between material and symbolic realms. *Adv. Hum. Ecol.* **1999**, *8*, 215–251.

102. Giampietro, M.; Mayumi, K.; Ramos-Martin, J. Multi-scale integrated analysis of societal and ecosystem metabolism (MuSIASEM): Theoretical concepts and basic rationale. *Energy* **2009**, *34*, 313–322. [CrossRef]

103. Weisz, H.; Suh, S.; Graedel, T.E. Industrial Ecology: The role of manufactured capital in sustainability. *Proc. Natl. Acad. Sci. USA* **2015**, *112*, 6260–6264. [CrossRef] [PubMed]

104. Haberl, H.; Fischer-Kowalski, M.; Krausmann, F.; Weisz, H.; Winiwarter, V. Progress towards sustainability? What the conceptual framework of material and energy flow accounting (MEFA) can offer. *Land Use Policy* **2004**, *21*, 199–213. [CrossRef]

105. Chen, W.-Q.; Graedel, T.E. In-use product stocks link manufactured capital to natural capital. *Proc. Natl. Acad. Sci. USA* **2015**, *112*, 6265–6270. [CrossRef] [PubMed]

106. Davis, S.J.; Caldeira, K.; Matthews, H.D. Future $CO_2$ Emissions and Climate Change from Existing Energy Infrastructure. *Science* **2010**, *329*, 1330–1333. [CrossRef] [PubMed]

107. Raupach, M.R.; Davis, S.J.; Peters, G.P.; Andrew, R.M.; Canadell, J.G.; Ciais, P.; Friedlingstein, P.; Jotzo, F.; van Vuuren, D.P.; Le Quéré, C. Sharing a quota on cumulative carbon emissions. *Nat. Clim. Chang.* **2014**, *4*, 873–879. [CrossRef]

108. Lauk, C.; Haberl, H.; Erb, K.H.; Gingrich, S.; Krausmann, F. Global socioeconomic carbon stocks and carbon sequestration in long-lived products 1900–2008. *Environ. Res. Lett.* **2012**, *7*, 034023. [CrossRef]

109. Pauliuk, S.; Majeau-Bettez, G.; Mutel, C.L.; Steubing, B.; Stadler, K. Lifting Industrial Ecology Modeling to a New Level of Quality and Transparency: A Call for More Transparent Publications and a Collaborative Open Source Software Framework: Open Source Software for Industrial Ecology. *J. Ind. Ecol.* **2015**, *19*, 937–949.

110. Modaresi, R.; Pauliuk, S.; Løvik, A.N.; Müller, D.B. Global Carbon Benefits of Material Substitution in Passenger Cars until 2050 and the Impact on the Steel and Aluminum Industries. *Environ. Sci. Technol.* **2014**, *48*, 10776–10784. [CrossRef] [PubMed]

111. Allwood, J.M.; Ashby, M.F.; Gutowski, T.G.; Worrell, E. Material efficiency: providing material services with less material production. *Philos. Trans. R. Soc. Math. Phys. Eng. Sci.* **2013**, *371*. [CrossRef] [PubMed]

112. Herring, H. Energy efficiency—A critical view. *Energy* **2006**, *31*, 10–20. [CrossRef]

113. Schipper, L.; Grubb, M. On the rebound? Feedback between energy intensities and energy uses in IEA countries. *Energy Policy* **2000**, *28*, 367–388. [CrossRef]

114. Creutzig, F.; Fernandez, B.; Haberl, H.; Khosla, R.; Mulugetta, Y.; Seto, K.C. Beyond Technology: Demand-Side Solutions for Climate Change Mitigation. *Annu. Rev. Environ. Resour.* **2016**, *41*, 173–198. [CrossRef]

115. Haas, R.; Nakicenovic, N.; Ajanovic, A.; Faber, T.; Kranzl, L.; Müller, A.; Resch, G. Towards sustainability of energy systems: A primer on how to apply the concept of energy services to identify necessary trends and policies. *Energy Policy* **2008**, *36*, 4012–4021. [CrossRef]

116. Sovacool, B.K. Conceptualizing urban household energy use: Climbing the "Energy Services Ladder". *Energy Policy* **2011**, *39*, 1659–1668. [CrossRef]

117. Daily, G.C.; Alexander, S.; Ehrlich, P.R.; Goulder, L.; Lubchenco, J.; Matson, P.A.; Mooney, H.A.; Postel, S.L.; Schneider, S.; Tilman, D.R.; et al. *Ecosystem Services: Benefits Supplied to Human Societies by Natural Ecosystems*; Issues in Ecology, No. 2; Ecological Society of America: Washington, DC, USA, 1997.

118. Spangenberg, J.H.; Görg, C.; Truong, D.T.; Tekken, V.; Bustamante, J.V.; Settele, J. Provision of ecosystem services is determined by human agency, not ecosystem functions. Four case studies. *Int. J. Biodivers. Sci. Ecosyst. Serv. Manag.* **2014**, *10*, 40–53. [CrossRef]

119. Potschin, M.B.; Haines-Young, R.H. Ecosystem services exploring a geographical perspective. *Prog. Phys. Geogr.* **2011**, *35*, 575–594. [CrossRef]

120. Creutzig, F.; Baiocchi, G.; Bierkandt, R.; Pichler, P.-P.; Seto, K.C. Global typology of urban energy use and potentials for an urbanization mitigation wedge. *Proc. Natl. Acad. Sci. USA* **2015**, *112*, 6283–6288. [CrossRef] [PubMed]

121. Seto, K.C.; Dhakal, S.; Bigio, A.; Blanco, H.; Delgado, G.C.; Dewar, D.; Huang, L.; Inaba, A.; Kansal, A.; Lwasa, S.; et al. Human Settlements, Infrastructure and Spatial Planning. In *Climate Change 2014: Mitigation of Climate Change. Working Group III Contribution to the IPCC Fifth Assessment Report (AR5) of the Intergovernmental Panel for Climate Change*; Edenhofer, O., Pichs-Madruga, R., Sokona, Y., Farahani, E., Kadner, S., Seyboth, K., Adler, A., Baum, I., Brunner, S., Eickemeier, P., et al., Eds.; Cambride University Press: Cambridge, UK; New York, NY, USA, 2014; pp. 923–1000.

122. CIESIN. *Gridded Population of the World (GPW), v3. SEDAC*; Center for International Earth Science Information Network: New York, NY, USA, 2016.

123. Goldewijk, K.K. Estimating global land use change over the past 300 years: The HYDE Database. *Glob. Biogeochem. Cycles* **2001**, *15*, 417–433. [CrossRef]

124. Imhoff, M.L.; Bounoua, L.; Ricketts, T.; Loucks, C.; Harriss, R.; Lawrence, W.T. Global patterns in human consumption of net primary production. *Nature* **2004**, *429*, 870–873. [CrossRef] [PubMed]

125. Sanderson, E.W.; Jaiteh, M.; Levy, M.A.; Redford, K.H.; Wannebo, A.V.; Woolmer, G. The Human Footprint and the Last of the Wild. *BioScience* **2002**, *52*, 891–904. [CrossRef]

126. Chen, X.; Nordhaus, W.D. Using luminosity data as a proxy for economic statistics. *Proc. Natl. Acad. Sci. USA* **2011**, *108*, 8589–8594. [CrossRef] [PubMed]

127. Jean, N.; Burke, M.; Xie, M.; Davis, W.M.; Lobell, D.B.; Ermon, S. Combining satellite imagery and machine learning to predict poverty. *Science* **2016**, *353*, 790–794. [CrossRef] [PubMed]

128. Wint, W.; Robinson, T. *Gridded Livestock of the World 2007*; Food and Agriculture Organisation of the United Nations: Rome, Italy, 2007.

129. Steinfeld, H.; Gerber, P.; Wassenaar, T.; Castel, V.; Rosales, M.; de Haan, C. *Livestock's Long Shadow, Environmental Issues and Options*; Food and Agriculture Organization of the United Nations (FAO): Rome, Italy, 2006.

130. Herrero, M.; Havlik, P.; Valin, H.; Notenbaert, A.; Rufino, M.C.; Thornton, P.K.; Blummel, M.; Weiss, F.; Grace, D.; Obersteiner, M. Biomass use, production, feed efficiencies, and greenhouse gas emissions from global livestock systems. *Proc. Natl. Acad. Sci. USA* **2013**, *110*, 20888–20893. [CrossRef] [PubMed]

131. Baiocchi, G.; Creutzig, F.; Minx, J.; Pichler, P.-P. A spatial typology of human settlements and their $CO_2$ emissions in England. *Glob. Environ. Chang.* **2015**, *34*, 13–21. [CrossRef]

132. Weisz, H.; Steinberger, J.K. Reducing energy and material flows in cities. *Curr. Opin. Environ. Sustain.* **2010**, *2*, 185–192. [CrossRef]

133. Gately, C.K.; Hutyra, L.R.; Sue Wing, I. Cities, traffic, and $CO_2$: A multidecadal assessment of trends, drivers, and scaling relationships. *Proc. Natl. Acad. Sci. USA* **2015**, *112*, 4999–5004. [CrossRef] [PubMed]

134. Gardner, G.T.; Prugh, T.; Renner, M.; Mastny, L. *State of the World: Can a City Be Sustainable?* Worldwatch Institute: Washington, DC, USA, 2016.

135. Kennedy, C.A.; Stewart, I.; Facchini, A.; Cersosimo, I.; Mele, R.; Chen, B.; Uda, M.; Kansal, A.; Chiu, A.; Kim, K.; et al. Energy and material flows of megacities. *Proc. Natl. Acad. Sci. USA* **2015**, *112*, 5985–5990. [CrossRef] [PubMed]

136. Jones, C.; Kammen, D.M. Spatial Distribution of U.S. Household Carbon Footprints Reveals Suburbanization Undermines Greenhouse Gas Benefits of Urban Population Density. *Environ. Sci. Technol.* **2014**, *48*, 895–902. [CrossRef] [PubMed]

137. Wiedenhofer, D.; Lenzen, M.; Steinberger, J.K. Energy requirements of consumption: Urban form, climatic and socio-economic factors, rebounds and their policy implications. *Energy Policy* **2013**, *63*, 696–707. [CrossRef]

138. Heinonen, J.; Jalas, M.; Juntunen, J.K.; Ala-Mantila, S.; Junnila, S. Situated lifestyles: I. How lifestyles change along with the level of urbanization and what the greenhouse gas implications are—A study of Finland. *Environ. Res. Lett.* **2013**, *8*, 025003. [CrossRef]

139. Minx, J.; Baiocchi, G.; Wiedmann, T.; Barrett, J.; Creutzig, F.; Feng, K.; Förster, M.; Pichler, P.-P.; Weisz, H.; Hubacek, K. Carbon footprints of cities and other human settlements in the UK. *Environ. Res. Lett.* **2013**, *8*, 035039. [CrossRef]

140. Anderson, J.E.; Wulfhorst, G.; Lang, W. Energy analysis of the built environment—A review and outlook. *Renew. Sustain. Energy Rev.* **2015**, *44*, 149–158. [CrossRef]

141. Yetano Roche, M.; Lechtenböhmer, S.; Fischedick, M.; Gröne, M.-C.; Xia, C.; Dienst, C. Concepts and Methodologies for Measuring the Sustainability of Cities. *Annu. Rev. Environ. Resour.* **2014**, *39*, 519–547. [CrossRef]

142. Ottelin, J.; Heinonen, J.; Junnila, S. New Energy Efficient Housing Has Reduced Carbon Footprints in Outer but Not in Inner Urban Areas. *Environ. Sci. Technol.* **2015**, *49*, 9574–9583. [CrossRef] [PubMed]

143. Athanassiadis, A.; Bouillard, P.; Crawford, R.H.; Khan, A.Z. Towards a Dynamic Approach to Urban Metabolism: Tracing the Temporal Evolution of Brussels' Urban Metabolism from 1970 to 2010. *J. Ind. Ecol.* **2016**. [CrossRef]

144. Reyna, J.L.; Chester, M.V. The Growth of Urban Building Stock: Unintended Lock-in and Embedded Environmental Effects: Urban Buildings, Lock-in, and Environmental Effects. *J. Ind. Ecol.* **2015**, *19*, 524–537. [CrossRef]

145. Chabrol, M. Re-examining historical energy transitions and urban systems in Europe. *Energy Res. Soc. Sci.* **2016**, *13*, 194–201. [CrossRef]

146. Liang, H.; Dong, L.; Tanikawa, H.; Zhang, N.; Gao, Z.; Luo, X. Feasibility of a new-generation nighttime light data for estimating in-use steel stock of buildings and civil engineering infrastructures. *Resour. Conserv. Recycl.* **2016**, *123*, 11–23. [CrossRef]

147. Mayer, A.; Haas, W.; Wiedenhofer, D. How Countries' Resource Use History Matters for Human Well-being—An Investigation of Global Patterns in Cumulative Material Flows from 1950 to 2010. *Ecol. Econ.* **2017**, *134*, 1–10. [CrossRef]

148. UNEP; Schandl, H.; Fischer-Kowalski, M.; West, J.; Giljum, S.; Dittrich, M.; Eisenmenger, N.; Geschke, A.; Lieber, M.; Wieland, H.; et al. *Global Material Flows and Resource Productivity. An Assessment Study of the UNEP International Resource Panel*; United Nations Environment Programme: Paris, France, 2016.

149. Haberl, H.; Krausmann, F. Changes in Population, Affluence, and Environmental Pressures during Industrialization: The Case of Austria 1830–1995. *Popul. Environ.* **2001**, *23*, 49–70. [CrossRef]

150. Zhang, Q.; Seto, K.C. Mapping urbanization dynamics at regional and global scales using multi-temporal DMSP/OLS nighttime light data. *Remote Sens. Environ.* **2011**, *115*, 2320–2329. [CrossRef]

151. Clarke, L.; Jiang, K.; Akimoto, K.; Babiker, M.; Blanford, G.; Fisher-Vanden, K.; Hourcade, J.-C.; Krey, V.; Kriegler, E.; Löschel, A.; et al. Assessing Transformation Pathways. In *Climate Change 2014: Mitigation of Climate Change. Contribution of Working Group III to the Fifth Assessment Report of the Intergovernmental Panel on Climate Change*; Edenhofer, O., Pichs-Madruga, R., Sokona, Y., Farahani, E., Kadner, S., Seyboth, K., Adler, A., Baum, I., Brunner, S., Eickemeier, P., Eds.; Cambridge University Press: Cambridge, UK; New York, NY, USA, 2014; pp. 413–511.

152. IPCC. *Climate Change 2014: Mitigation of Climate Change, Working Group III contribution to the IPCC Fifth Assessment Report (AR5), Summary for Policy Makers; Intergovernmental Panel for Climate Change*; Cambridge University Press: Geneva, Switzerland; Cambridge, UK, 2014.

153. Geels, F.W. Technological transitions as evolutionary reconfiguration processes: A multi-level perspective and a case-study. *Res. Policy* **2002**, *31*, 1257–1274. [CrossRef]

154. Grin, J.; Rotmans, J.; Schot, J. *Transitions to Sustainable Development: New Directions in the Study of Long Term transformative Change*; Routledge: New York, NY, USA, 2010.

155. Geels, F.W. *Technological Transitions and System Innovation: A Co-Evolutionary and Socio-Technical Analysis*; Edward Elgar: Cheltenham, UK; Northampton, MA, USA, 2005.

156. Grübler, A.; Nakićenović, N.; Victor, D.G. Dynamics of energy technologies and global change. *Energy Policy* **1999**, *27*, 247–280. [CrossRef]

157. Stirling, A. Transforming power: Social science and the politics of energy choices. *Energy Res. Soc. Sci.* **2014**, *1*, 83–95. [CrossRef]

158. Grübler, A. *Technology and Global Change*; Cambridge University Press: Cambridge, UK, 2009.

159. Nakicenovic, N.; Grübler, A.; McDonald, A. *Global Energy: Perspectives*; Cambridge University Press: New York, NY, USA, 1998.

160. Niewöhner, J. Infrastructures of Society (Anthropology of). In *International Encyclopedia of the Social & Behavioral Sciences*; Elsevier: Amsterdam, The Netherlands, 2015; pp. 119–125.

161. Hertwich, E.G.; Gibon, T.; Bouman, E.A.; Arvesen, A.; Suh, S.; Heath, G.A.; Bergesen, J.D.; Ramirez, A.; Vega, M.I.; Shi, L. Integrated life-cycle assessment of electricity-supply scenarios confirms global environmental benefit of low-carbon technologies. *Proc. Natl. Acad. Sci. USA* **2015**, *112*, 6277–6282. [CrossRef] [PubMed]

162. Falcone, P.M. Sustainability Transitions: A Survey of an Emerging Field of Research. *Environ. Manag. Sustain. Dev.* **2014**, *3*, 61–83. [CrossRef]

163. Polanyi, K. *The Great Transformation*; Farrar and Reinhart: New York, NY, USA, 1944.

164. Sovacool, B.K. How long will it take? Conceptualizing the temporal dynamics of energy transitions. *Energy Res. Soc. Sci.* **2016**, *13*, 202–215. [CrossRef]

165. Fischer-Kowalski, M.; Haberl, H. *Socioecological Transitions and Global Change. Trajectories of Social Metabolism and Land Use*; Advances in Ecological Economics (series editor: Jeroen van den Bergh); Edward Elgar: Cheltenham, UK; Northampton, UK, 2007.

166. Haberl, H.; Erb, K.-H.; Krausmann, F. Human Appropriation of Net Primary Production: Patterns, Trends, and Planetary Boundaries. *Annu. Rev. Environ. Resour.* **2014**, *39*, 363–391. [CrossRef]

167. Lutz, W.; Sanderson, W.C.; Scherbov, S. *The End of World Population Growth in the 21st Century: New Challenges for Human Capital Formation and Sustainable Development*; Routledge: Oxfordshire, UK, 2013.

168. Krausmann, F.; Weisz, H.; Eisenmenger, N. Transitions in Socio-Metabolic Regimes through Human History. In *Social Ecology: Society-Nature Relations Across Time And Space*; Haberl, H., Fischer-Kowalski, M., Krausmann, F., Winiwarter, V., Eds.; Springer: Cham, Switzerland, 2016; pp. 63–92.

169. Sieferle, R.-P. *The Subterranean Forest: Energy Systems and the Industrial Revolution*; The White Horse Press: Cambridge, UK, 2001.

170. Görg, C.; Brand, U.; Haberl, H.; Hummel, D.; Jahn, T.; Liehr, S. Challenges for Social-Ecological Transformations, Contributions from Social and Political Ecology. *Sustainability* **2017**, in press.

171. Winiwarter, V.; Schmid, M.; Haberl, H.; Singh, S.J. Why Legacies Matter: Merits of a Long-Term Perspective. In *Social Ecology, Society-Nature Interaction across Space and Time*; Haberl, H., Fischer-Kowalski, M., Krausmann, F., Winiwarter, V., Eds.; Springer International: Cham, Switzerland, 2016; pp. 149–168.

MDPI AG
St. Alban-Anlage 66
4052 Basel, Switzerland
Tel. +41 61 683 77 34
Fax +41 61 302 89 18
http://www.mdpi.com

*Sustainability* Editorial Office
E-mail: sustainability@mdpi.com
http://www.mdpi.com/journal/sustainability

www.ingramcontent.com/pod-product-compliance
Lightning Source LLC
Chambersburg PA
CBHW051851210326
41597CB00033B/5862